T0235133

COHESION
A Scientific History of Intermolecular Forces

Why does matter stick together? Why do gases condense to liquids, and liquids freeze to solids? This book provides a detailed historical account of how some of the leading scientists of the past three centuries have tried to answer these questions.

The topic of cohesion and the study of intermolecular forces has been an important component of physical science research for hundreds of years. This book is organised into four broad periods of advance in our understanding. The first three are associated with Newton, Laplace and van der Waals. The final section gives an account of the successful use in the 20th century of quantum mechanics and statistical mechanics to resolve most of the remaining problems.

Throughout the last 300 years there have been periods of tremendous growth in our understanding of intermolecular forces but such interest proved to be unsustainable, and long periods of stagnation usually followed. The causes of these fluctuations are also discussed.

The book will be of primary interest to historians of science as well as physicists and physical chemists interested in the historical origins of our modern-day understanding of cohesion.

JOHN SHIPLEY ROWLINSON is Dr Lee's Professor of Chemistry Emeritus in the Physical and Theoretical Chemistry Laboratory at the University of Oxford.

John Rowlinson obtained his MA and D. Phil. from Oxford in 1950, after which he took up a position in the Chemistry Department at the University of Manchester. In 1961 he was appointed Professor of Chemical Technology at the Imperial College of Science and Technology. After 13 years in London Professor Rowlinson returned to Oxford to become the Dr Lee's Professor of Chemistry, a position he held for 19 years. In 1970 he was made a Fellow of the Royal Society. During his distinguished career Professor Rowlinson was awarded a number of prizes including the Leverhulme medal from the Royal Society and the Meldola and Marlow medals from the Royal Society of Chemistry. He was the Andrew D. White Professor-at-large at Cornell University for 6 years and in the year 2000 he was knighted.

COHESION

A Scientific History of Intermolecular Forces

J. S. ROWLINSON

CAMBRIDGE
UNIVERSITY PRESS

CAMBRIDGE UNIVERSITY PRESS
Cambridge, New York, Melbourne, Madrid, Cape Town, Singapore, São Paulo

Cambridge University Press
The Edinburgh Building, Cambridge CB2 2RU, UK

Published in the United States of America by Cambridge University Press, New York

www.cambridge.org
Information on this title: www.cambridge.org/9780521810081

First published 2002
This digitally printed first paperback version 2005

A catalogue record for this publication is available from the British Library

ISBN-13 978-0-521-81008-1 hardback
ISBN-10 0-521-81008-6 hardback

ISBN-13 978-0-521-67355-6 paperback
ISBN-10 0-521-67355-0 paperback

Contents

Preface

The aim and scope of this work are set out in the first chapter. Here I explain the conventions that I have used and thank those who have been kind enough to criticize my efforts.

The work is based on primary printed sources. A few letters and other informal documents have been used but only if they have already been printed. Secondary sources are given when they refer directly to the matter in hand or when they seem to be particularly useful. No attempt has been made, however, to cite everything that is relevant to the background of the subject since this would have led to the inflation of an already long bibliography. This policy has led to a fuller coverage of the 18th century than of the 19th where the secondary literature is potentially vast. In contrast, there are almost no directly useful secondary sources for the 20th century, but here the number of primary sources is impossibly large. It would have been easy to have given ten or more times the number listed. The choice is inevitably biased by the recent aspects of the subject upon which I have chosen to concentrate; others might have made other choices, but no one could give a comprehensive coverage of the last century.

The references are listed in four main groups, one at the end of each of the Chapters 2 to 5. There is so little overlap between those in each chapter that this method seemed less clumsy than a consolidated list for the whole book and leaves each chapter almost self-contained. The form in which the titles of journals is abbreviated follows the usual conventions. A few journals that are often known by their editor's name are shown by inserting this name in brackets before the title, e.g. (*Silliman's*) *Amer. Jour. Sci. Arts*. The journal that is now called the *Annalen der Physik* was often abbreviated, after its editors, *Pogg. Ann.* or *Wied. Ann.* etc. during the 19th century, and was formally the *Annalen der Physik und Chemie* until 1899, when Paul Drude became the editor; the simple form *Ann. Physik* is used here throughout. The dates at which some journals appeared differ from the nominal date on the volume. This problem is particularly acute for the publications of the French Academy. Here the nominal date is used and the actual date of the appearance of the paper is noted if this relevant. The *Annual Reports of the British Association* are dated by the year in which the meeting was held; they were usually published a year later. The place of publication of books is given but not the name of the publisher. Cross-references to 'Collected Works' are given for some foreign authors but not for most British ones such as Maxwell or William Thomson.

Experimental work is described in the units of the time when it was made but a translation into the current units of the Système International is added. The ångström has, however, been retained to describe intermolecular separations. This unit is more convenient than the correct SI unit, the nanometre ($10 \text{ Å} = 1$ nm), since almost all the distances quoted are in the range of 1 to 10 Å.

The index of names covers only those whose scientific work is being discussed; authors of secondary sources are not indexed, although I admit that the distinction between primary and secondary is not easily defined. Biographical references are given for the major workers in the field who had died by the end of the 20th century in December 2000, but not for those believed to be still alive. These references are given at the point in the text where the scientist's work first becomes important to this narrative, and so not necessarily at the first citation. If he or she is one of those in *The Dictionary of Scientific Biography*, ed. C.C. Gillispie, 18 vols., New York, 1970–1981, then a reference to that work is generally thought to be sufficient; it is abbreviated DSB. Additional sources are given only if they are particularly important for the subject of this book, or have been published later than the DSB article. If the scientist is not in this work then the next source is the volumes of J.C. Poggendorff, *Biographisch-Literarisches Handwörterbuch zur Geschichte der exacten Wissenschaften*, Leipzig, now Berlin, 1863 onwards. This is abbreviated Pogg. References to the British *Dictionary of National Biography* are abbreviated DNB but details are omitted since the work is ordered alphabetically and since a new edition is now being prepared.

I thank those who have been good enough to read parts of the book and give me advice on how they might be improved: Robert Fox, Ivor Grattan-Guinness, Rupert and Marie Hall, Peter Harman, John Heilbron, John Lekner, Anneke Levelt Sengers and Brian Smith. Others are thanked in the references for more specific information.

Oxford   *J.S.R.*
October, 2001

1

Introduction and summary

Some problems have always been with us. No one knows when man first asked 'What is the origin of our world?' or 'What is life?', and progress towards satisfactory answers has been slow and exceedingly difficult. One aim of this study is to take such a perennial theme, although one narrower than either of these two problems, and see how it has been tackled in the Western world in the last three hundred years. The topic is that of cohesion – why does matter stick together? Why do gases condense to liquids, liquids freeze to solids or, as it has been put more vividly, why, when we lift one end of a stick, does the other end come up too? Such questions make sense at all times and the attempts to answer them have an intrinsic interest, for the subject of cohesion has at many times in the last three centuries been an important component of the physical science of the day. It has attracted the attention of some of the leading scientists of each era, as well as a wide range of the less well known. It is a part of our history that is worth setting out in some detail, a task that I think has not yet been attempted.

This study has, however, a wider aim also. Historians have rightly given much attention to the great turning-points of science – Newton's mechanics, Lavoisier's chemistry, Dalton's atomic theory, Maxwell's electrodynamics, Planck's quantum theory, and Einstein's theories of relativity, to name but half a dozen in the physical sciences. These are the points that Thomas Kuhn described as revolutions [1]. The study of cohesion shows no such dramatic moments, the closest being, perhaps, the discovery of the quantal origin of the universal force of attraction between molecules in 1927–1930. This is, therefore, an account of a branch of 'normal' science that exemplifies how such work is done.

Science is not a logical and magisterial progress in which experimental discoveries lead directly to new theories and in which these theories then guide new experimental work. The practitioners know this on a small scale. Research workers can see how their progress is helped or hindered by chance discoveries, misleading experiments, half-remembered lectures, chance finds in the 'literature', unexpected

1

discussions at a conference, and all the other perturbations of laboratory life. More-over science can be fun. Investigations can be made just out of curiosity even when it is clear that the answer, when found, will solve no particular experimental or theoretical problem. We shall see that similar disorderliness marks progress on a larger scale. Matters move forward rapidly for a decade or so, and then stagnate for many decades. Here three broad periods of advance have been identified and named after Newton, Laplace and van der Waals. They were, of course, not the only gen-erators of the advances but their contributions were decisive and, perhaps, stretch the concept of normality to its limits. Their names may, however, conveniently be used to identify their periods.

It is of interest to seek for the causes of this punctuated advance. Some of the periods of stagnation are related to weaknesses in the contemporary infra-structure, either experimental or, more usually, theoretical. Thus we shall see that many of the natural philosophers of the 18th century were hampered by their inadequate knowledge of mechanics and of the calculus. What Newton and Leibniz had cre-ated needed to be completed by the Bernoullis, Euler and others before it passed into general scientific circulation. This passage occurred notably in the institutions established in 'revolutionary' France at the end of the century, and it is not surpris-ing that a second period of advance in understanding came with Laplace. There were also other less direct reasons for the relative stagnation of the 18th century. Some were cultural. One cannot imagine a present-day undergraduate or research student being told by his or her teacher that there was a worrying metaphysical problem with forces between molecules acting at a distance, or with a model sys-tem of hard spheres undergoing elastic collisions, but these were very real concerns in the 18th century. By the 19th they were not so much banished as ignored. An indifference to metaphysical problems seems to be one of the features of normal science. We shall see that scientists have a well-developed defensive mechanism when faced with theoretical obstacles. They ignore them, hope that what they are doing will turn out to be justified, and leave it to their deeper brethren or to their successors to resolve the difficulty. In the 18th century and beyond, this proved to be the right way forward both for gravity and for interparticle forces; they functioned for all practical purposes as if they acted at a distance. It was not until the 1940s that the problem of how this intermolecular action was transmitted had to be faced. This defensive mechanism can go wrong; we shall see that in the early years of the 20th century there were repeated attempts to seek a classical electrostatic origin for the intermolecular forces, in spite of what is to us, and perhaps should have been to them, clear evidence that these were bound to fail.

Another problem in the 18th century that we can broadly call cultural was what we now see as an inadequate way of assessing new theories. The same metaphysical bias that objected to action at a distance without a discernible mechanism to effect it, led to theories that laid too much emphasis on plausible mechanisms, and not enough on means of testing the theories or of seeing if they had any predictive power. By the end of the century (again judging by our notions) matters had improved, and this change, coupled with the 'revolutionary' mathematics of the French, meant that by the early 19th century theoretical physics had taken a form in which we can recognise many of the ways of working that we still use.

But beyond these internal weaknesses and metaphysical doubts there remains an unexplained cause of the flow and stagnation of progress that we can only call fashion. It was obvious to Réaumur as early as 1749 that science was as prone to fashion as any other human activity [2], and these swings may be strongest when there are few in the field. The spectacular experiments that could be made in the 18th century in electricity, and the solid advances in the study of chemistry and of heat, attracted the best men, and left only a few, mainly of the second-rank, to study capillarity and other manifestations of cohesion. To call this fashion is perhaps to go too far in imputing irrationality. Research programmes do degenerate and are justifiably overtaken by rising fields in which progress is easier. Science is like a rising tide; if certain areas are perceived to be open to flooding then the practitioners rush in, leaving other research programmes as unconquered and ignored islands of resistance. But once this is said there remains an element, if not of irrationality, than of adventitiousness about scientific advance.

There are also in the background those changes in the sociological, political, religious and economic aspects of each era whose influence on the science of the day is now the main concern of many historians. If I have not pursued these with the rigour that current practice seems to demand it is not because I doubt their importance but because it becomes hard to discern their effects in a specialised and 'philosophic' subject such as cohesion. In the 18th and 19th centuries religious convictions certainly influenced philosophical thought but I have not seen a direct or strong enough link to the problem of cohesion to follow the subject beyond an occasional remark. No doubt others would tackle the subject differently.

The 19th century is more complex than the 18th but analysis is helped by the greater attention paid to it by historians. Laplace and his colleagues had much success in the first twenty years of the century, in which his solving of the problems of capillarity is the one that is the most central to our story. Then came about what has been called 'the fall of Laplacian physics' [3]. His belief in a corpuscular theory of light, in matter as a static array of interacting particles, and of heat as a caloric fluid that was responsible for the repulsive component of the force between the particles, all told against him and his followers when physics advanced beyond these ideas. But it was again the competition of the rising fields of electricity, magnetism, optics, and later, thermodynamics that attracted the attention; the one field where Laplace's ideas were still important was that of the elasticity of solids, a subject in which the imperfections of his physics were of little consequence.

The big struggle of the 19th century was that between the picture of interacting particles of matter, each surrounded by a vacuum, that had been held by Newton and Laplace, and the continuum picture of matter and space that came to be embodied in field theories. This was not a competition between different scientists, for many adopted both views at different times, or even apparently at the same time, but it was a competition between methods of interpretation. For example the classical thermodynamics of the 1850s and 1860s, a subject apparently independent of any view of the structure of matter, grew up alongside the developing kinetic theory of gases which required a corpuscular theory. The continuum mechanics that proved most successful in describing the elastic properties of solids lived in uneasy conjunction with the Laplacian attempts to interpret these properties in terms of interparticle forces. Cauchy could switch from one view to the other within a few months.

The struggle between field theories and particulate theories is only one example of the great debates that are relevant to the subject of cohesion but whose full discussion would take us too far from the main line. Here we can only follow what was found at the time to be successful in practice. Not until 1954 did a field theory of cohesion appear, and even now it is only of specialist interest. This account is therefore weighted towards those who believed in interparticle forces and so drove the subject forward. Other cognate topics that might have been explored but are not, are 18th century chemistry, which overlaps with what we now call physics, the theory of the optical aether which inspired much of the 19th century work on elasticity, and the final resolution of the atomic debates in the early years of the 20th century.

By the early 19th century chemistry and physics were regarded as distinct subjects. The physical aspects of chemistry had a brief Laplacian flourish at the hands of Berthollet, Gay-Lussac and Dumas but then fell out of fashion under the competition from the electrochemistry of Davy and Berzelius, and the successes of organic chemistry and the problems of atomic weight and molecular structure. Physical chemistry revived towards the end of the century, first as the chemistry of solutions, ions and electrolytes, and then more widely under the impact of quantum theory in the first half of the 20th century. Most of those working on intermolecular forces in the second half of the century would describe themselves as physical or theoretical chemists, not as physicists.

The 20th century brought new dangers. The number of scientists grew rapidly and with this growth came the problems of specialisation. When a field fell out of fashion, as did that of cohesion in the early part of the century, then important work could be forgotten when the next generation returned to the field. The achievements of van der Waals and his school were ignored from about 1910 onwards; work on cohesion and the properties of liquids could not compete with the great developments of the day in quantum theory on the one hand and the experimental work on radioactivity and fundamental particles on the other. The work of many of the leading physicists of the passing generation, published in hundreds of papers in the leading journals of the day, became almost overnight a forgotten backwater of physics. This was not the field where great discoveries were to be made, reputations to be gained, and honours to be won. The same thing still happens, if not so dramatically. The topic of intermolecular forces, a matter of great debate in the 1950s, 1960s, and early 1970s, has now dropped from the front rank. This exit followed one important success, the accurate determination of the force between a pair of argon atoms, but that achievement left plenty of work still to be done. Nevertheless the subject was thought to have gone off the boil, and in the 1980s and 1990s few of those earning the star salaries in American universities were to be found in this field.

With increased specialisation came also a certain arrogance. One can sense in the writing of some of those active in the 1930s and later, a reluctance to believe that anything of importance could have happened before the great days of quantum theory in the 1920s. Spectroscopy is a field that generated many interesting numerical results in the 19th century but which owes its quantitative theory to quantum mechanics. Its practitioners made some late but valuable contributions to the determination of simple intermolecular forces, but they did not bother with the older field of statistical mechanics, and their interpretation of their results was often flawed. These had to be analysed by others before their value could be appreciated. At the very end of the century, however, the spectroscopists made one spectacular advance with the determination of the forces between two water molecules, a system so complicated that it had defied the efforts of those who had been trying to find these forces from the macroscopic properties of water. Little is said here, however, about experimental advances or problems since throughout its history cohesion has been a subject where the experiments have usually been simple but their interpretation difficult. There are exceptions, of which the most obvious is, perhaps, the absence of direct evidence of the particulate structure of crystals which hampered 19th century attempts at a theory of elasticity. But, as so often, this difficulty was resolved by a totally unrelated discovery – that of x-rays and the realisation that they were electromagnetic waves.

Making generalisations about how science is done from the example of one rather narrow field is hazardous. Many may dispute those drawn here, even on the evidence provided, but they are put forward as an attempt to show how this field has advanced over three hundred years. I would not wish to be dogmatic; others should try to draw their own conclusions from this field, and other fields may lead to different conclusions. One can read Popper, Kuhn, Lakatos and other philosophers of science and recognise there many truths that call to mind instances of how it is done, but it is difficult to fit even one physical science into their moulds. Science does in practice seem to move in less logical ways than philosophers would wish. Feyerabend would surely find here examples with which to justify his claim that "Science is an essentially anarchic enterprise" [4].

It is, of course, the common-sense view of practising scientists that the movement of science is an advance, and that, although the advance itself may be irregular, the result is a coherent structure. This narrative would not make sense without that belief. That the advance is not always logical, rarely neat, and occasionally repetitious, is not a theme that can be summarised in the trite phrase 'history repeats itself'. That does happen; a curious example is the repetition in the second half of the 20th century of arguments about the representation of the pressure tensor that duplicate, in ignorance, and almost word for word, some of those of a hundred years earlier. But such repetitions are, I think, curiosities of little consequence. I end, however, with some quotations that show that a certain simile came to mind repeatedly for 150 years, and then apparently disappeared for the next 130. Why, I cannot say, unless it be that astronomy has lost something of its former prestige, so these quotations are offered for their interest only.

We behold indeed, in the motions of the celestial bodies, some effects of it [the attraction] that may be call'd more august or pompous. But methinks these little *hyperbola's*, form'd by a fluid between two glass planes, are not a-whit less fine and curious, than the spacious ellipses describ'd by the planets, in the bright expanse of Heaven.
                  (Humphry Ditton, mathematics master at Christ's Hospital, 1714) [5]

Peut-être un jour la précision des données sera-t-elle amenée au point que le Géomètre pourra calculer, dans son cabinet, les phénomènes d'une combinaison chimique quelconque, pour ainsi dire de la même manière qu'il calcule le mouvement des corps célestes. Les vues que M. de la Place a sur cet objet, & les expériences que nous avons projétées, d'après ses idées, pour exprimer par des nombres la force des affinités des différens corps, permettent déjà de ne pas regarder cette espérance absolument comme une chimère.
                                                                    (A.L. Lavoisier, 1785) [6]

Quelques expériences déjà faites par ce moyen, donnent lieu d'espérer qu'un jour, ces lois seront parfaitement connues; alors, en y appliquant le calcul, on pourra élever la physique des corps terrestres, au degré de perfection, que la découverte de la pesanteur universelle a donné à la physique céleste.
                                                                    (P.-S. Laplace, 1796) [7]

We are not wholly without hope that the real weight of each such atom may some day be known . . . ; that the form and motion of the parts of each atom, and the distance by which they are separated, may be calculated; that the motions by which they produce heat, electricity, and light may be illustrated by exact geometrical diagrams. . . . Then the motion of the planets and music of the spheres will be neglected for a while in admiration of the maze in which the tiny atoms turn.
(H.C. Fleeming Jenkin, Professor of Engineering at Edinburgh in a review of a book on Lucretius, 1868, repeated by William Thomson in his Presidential Address to the British Association, 1871, and quoted from there, in Dutch, by J.D. van der Waals as the closing words of his doctoral thesis at Leiden in 1873) [8]

Notes and references

1 T.S. Kuhn, *The structure of scientific revolutions*, Chicago, 1962.
2 See Section 2.5. For modern instances of the same view, see F. Hoyle, *Home is where the wind blows*, Oxford, 1994, pp. 279–80, on recent fashions in astronomy; F. Franks, *Polywater*, Cambridge, MA, 1981, for the frantic pursuit of a non-existent anomaly in the 1960s; and P. Laszlo, *La découverte scientifique*, Paris, 1999, chap. 8, for a vivid account of a 1969 fad in research on nuclear magnetic resonance. The rapid dissemination of some papers on the Internet and the ease with which the number of times that they have been 'read' can be recorded, has made worse the irrational pursuit of current fashions, according to the report on a discussion at a recent Seven Pines Symposium, by J. Glanz in the *International Herald Tribune* of 20 June, 2001.
3 R. Fox, 'The rise and fall of Laplacian physics', *Hist. Stud. Phys. Sci.* **4** (1975) 89–136.
4 P. Feyerabend, *Against method*, 3rd edn, London, 1993, p. 9. The first edition was published in 1975. J.D. Watson made the same point for the biological sciences in the opening words of the Preface to *The double helix*, New York, 1968.
5 H. Ditton, *The new law of fluids or, a discourse concerning the ascent of liquors, in exact geometrical figures, between two nearly contiguous surfaces*; . . . , London, 1714, p. 41.
6 A.L. Lavoisier, 'Sur l'affinité du principe oxygène avec les différentes substances auxquelles il est susceptible d'unir', *Mém. Acad. Roy. Sci.* (1782) 530–40, published 1785, see pp. 534–5.
7 P.-S. Laplace, *Exposition du système du monde*, Paris, 1796, v. 2, p. 198.
8 [Anon.], 'Lucretius and the atomic theory', *North British Review* **6** (1868) 227–42, see pp. 241–2, and in *Papers, literary, scientific, etc. by the late Fleeming Jenkin*, ed. S. Colvin and J.A. Ewing, London, 1887, v. 1, pp. 177–214, see pp. 213–14; W. Thomson, Presidential address, *Rep. Brit. Assoc.* **41** (1871) lxxxiv–cv, see p. xciv; J.D. van der Waals, *Over de continuiteit van den gas- en vloeistoftoestand*, Thesis, Leiden, 1873, p. 128. Thomson wrote the last word as 'run', not 'turn': either a slip or a reference to the prevailing view of the 1870s that molecular motions were primarily translational, not rotational, as had sometimes been supposed in the early 19th century.

2

Newton

2.1 Newton's legacy

The natural philosophers of the eighteenth century knew Newton's work [1] through his two books, the *Principia mathematica* of 1687 [2] and the *Opticks* of 1704 [3]. His belief in a corpuscular philosophy is clear in both, and is particularly prominent in the later editions of the *Opticks*, but the cohesive forces between the particles of matter are not the prime subject of either book. Together, however, they contain enough for his views on cohesion to be made clear. We, who are now privy to many of his unpublished writings, know how much more he might have said, or said earlier in his life, had he not been so fearful of committing himself in public on so controversial a topic. He was not the first to speculate in this field but his views were better articulated than those of his predecessors [4] and, what is perhaps more important, they carried in the 18th century the force of his ever-increasing authority. It was his vision that was transmitted to the physicists of the early 19th century, and we examine first the legacy that he left to his philosophical heirs. The account is restricted to the subject in hand; that is, how does matter stick together, and wider aspects of Newton's thought remain untouched.

In the Preface to the *Principia* he describes the success of his treatment of mechanics and gravitation, and then continues:

I wish we could derive the rest of the phaenomena of Nature by the same kind of reasoning from mechanical principles. For I am induced by many reasons to suspect that they may all depend upon certain forces by which the particles of bodies, by some causes hitherto unknown, are either mutually impelled towards each other and cohere in regular figures, or are repelled and recede from each other; which forces being unknown, philosophers have hitherto attempted the search of Nature in vain. But I hope the principles here laid down will afford some light either to that, or some truer, method of philosophy. [5]

Here he alludes not only to the short-ranged forces of attraction that he held to be responsible for the cohesion of liquids and solids but also to those other forces that he was to propose later in the book as a possible explanation of the pressure of a gas as a repulsion between stationary particles [6]. Readers of the *Principia* were to learn little more about the cohesive forces although he had at one time intended to take the subject further. In a draft version of the Preface, he had described the cohesion between its parts as being responsible for mercury being able to stand in a Torricellian vacuum at a height greatly in excess of the atmospheric pressure of thirty inches, and he had intended to enquire further into these forces. Then, in a phrase he was to use more than once, he wrote:

For if Nature be simple and pretty conformable to herself, causes will operate in the same kind of way in all phenomena, so that the motions of smaller bodies depend upon certain smaller forces just as the motions of larger bodies are ruled by the greater force of gravity. [7]

His comment on the relative sizes of the forces betrays a looseness of thought that he was to correct before he published anything in this field.

He made a second attempt to say more about cohesive forces and the forces that lead to solution, to chemical action, to fermentation and similar processes, in a draft Conclusion that was also intended for the first edition of the *Principia*. In this he expressed the same thoughts but now couched more as hopes than intentions. "If any one shall have the good fortune to discover all these [causes of local motion], I might almost say that he will have laid bare the whole nature of bodies so far as the mechanical causes of things are concerned." [8] He discussed the rise of liquids in small tubes, a phenomenon that was later to play an important role in the study of cohesion since it was such an obvious departure from the known laws of hydrostatics. He (like Robert Hooke [9]) thought then that the rise was caused by a repulsion of air by glass, a consequent rarefaction of the air in the tube, and the rise of liquid to replace it.

Newton was holding back twenty-five years later when Roger Cotes [10] was preparing the second edition of the *Principia*. He wrote to Cotes on 2 March 1712/13: "I intended to have said much more about the attraction of small particles of bodies, but upon second thoughts I have chose rather to add but one short paragraph about that part of philosophy. This Scholium finishes the book." [11, 12] Again there are draft versions of this Scholium that go beyond what was printed [13].

In spite of these hesitations and withdrawals the *Principia* of 1687 contains much that hints at the tenor of his thoughts. This material is often in the form of mathematical theorems that could have been used to discuss cohesion, but the application is never made. Thus Section 13 of Book 1 contains in Proposition 86 the statement that for forces that "decrease, in the recess of the attracted body, in a triplicate or more than triplicate ratio of the distance from the particles; the attraction will be vastly stronger in the point of contact than when the attracting and attracted bodies are separated from each other though by never so small an interval." [14] In Proposition 91 the discussion is extended to "forces decreasing in any ratio of the distances whatsoever", and in Proposition 93 he shows that if the particles attract as r^{-m}, where r is the separation, then a particle is attracted by a slab composed of such particles by a force proportional to R^{-m+3}, where R is the distance of the particle from the planar surface of the body. Similarly his discussion of the repulsive forces between contiguous particles in a gas [6] is generalised to forces proportional to r^{-m} which, he shows, lead to a pressure proportional to the density to a power of $(m+2)/3$, so that what we now call Boyle's law requires that m is 1. Propositions 94–96 of Section 14 of Book 1 are "Of the motion of very small bodies when agitated by centripetal [i.e. attractive] forces tending to the several parts of any very great body", but it is soon clear that the application he has in mind is to optics; the "very small bodies" are his particles of light.

John Harris [15], in the first volume of his *Lexicon technicum* of 1704, commented accurately that the word 'attraction' is "retained by good naturalists and, in particular, by the excellent Mr. Isaac Newton in his *Principia*; but without there determining any thing of the *quale* of it, for he doth not consider things so much physically as mathematically." [16] This was true in 1704 but six years later, in his second volume, when he had read the Latin edition of the *Opticks*, he changed his mind and accepted the physical reality of these forces. He was briefly a Secretary at the Royal Society and had seen the experiments performed there, often under Newton's direction as President.

When, in the *Principia*, Newton does discuss the physical consequences of forces steeper than inverse square then his thoughts turn more naturally to magnetism than to cohesion. In Book 3, Proposition 6, Theorem 6, Cor. 4 of the 1687 edition he says of magnetism that "it surely decreases in a ratio of distance greater than the duplicate." [17] By the time of time of the second edition of 1713 he is more precise, and in what is re-numbered Cor. 5, he writes that the force "decreases not in the duplicate, but almost in the triplicate proportion of the distance, as nearly as I could judge from some rude observations." [18] His early remarks may have been based on some observations of Hooke [19] but his later ones stemmed from the experiments made at the Royal Society by Brook Taylor [20] and Francis Hauksbee [21] that started in June 1712 [22]. Taylor deduced that "at the distance of nine feet, the power alters faster, than as the cubes of the distances, whereas at the distances of one and two feet, the power alters nearly as their squares". The interpretation of these results is not simple. Newton speaks of "magnetic attraction", which might imply the force of attraction between two magnets, but Taylor and Hauksbee measured the field of the magnet (in modern terms) by observing the deflection of a small test or compass magnet at different distances from the lodestone. The distances were measured both from the centre of the lodestone or, more usually, from its "extremity", and it is not clear what function of the angle of deflection is taken as a measure of the "power", presumably the angle itself. Such far from simple results did not hold out much hope that the less easily studied forces of cohesion would prove to have a simple algebraic form.

Within a few years of the publication of the *Principia* Newton was collecting his papers for the book that was to become the *Opticks* of 1704. In some frank but unpublished notes, which were probably written about 1692, he says why he could afford now to be more open about the cohesive forces:

And if Nature be most simple and fully consonant to herself she observes the same method in regulating the motions of smaller bodies which she doth in regulating those of the greater. The principle of nature being very remote from the conceptions of philosophers I forbore to describe in that book [i.e. the *Principia*] lest I [it?] should be accounted an extravagant freak and so prejudice my readers against all those things which were the main designe of the book: but and yet I hinted at it both in the Preface and in the book it self where I speak of the inflection of light and of the elastick power of the air but the design of that book being secured by the approbation of mathematicians, I have not scrupled to propose this principle in plane words. The truth of this hypothesis I assert not, because I cannot prove it, but I think it very probable because a great part of the phaenomena of nature do easily flow from it which seems otherwise inexplicable. [23]

This passage is from the second of five hypotheses that were intended to provide the conclusion of a fourth book of the *Opticks* "concerning the nature of light and the power of bodies to refract and reflect it". Nothing of this appeared, however, in the first edition of 1704, in which the Queries in Book 3 are strictly 'optical', but we see in these hypotheses the germs of those Queries that appeared first in the Latin edition of 1706. The best known of these, Query 23, dealt with cohesive and chemical forces and it was the last form of this, Query 31 of the later English editions [24], that became, in the eyes of Newton's followers, the final distillation of his views on cohesion. It opens:

Quest. 31. Have not the small particles of bodies certain powers, virtues, or forces, by which they act at a distance, not only upon the rays of light for reflecting, refracting, and inflecting them, but also upon one another for producing a great part of the phaenomena of Nature? For it's well known, that bodies act one upon another by the attractions of gravity, magnetism, and electricity; and these instances shew the tenor and course of Nature, and make it not improbable but that there may be more attractive powers than these. For Nature is very consonant and conformable to her self. How these attractions may be perform'd, I do not here consider. What I call attraction may be perform'd by impulse, or by some other means unknown to me. I use that word here to signify only in general any force by which bodies tend towards one another, whatsoever be the cause. For we must learn from the phaenomena of Nature what bodies attract one another, and what are the laws and properties of the attraction, before we enquire the cause by which the attraction is perform'd. The attractions of gravity, magnetism, and electricity, reach to very sensible distances, and so have been observed by vulgar eyes, and there may be others which reach to so small distances as hitherto escape observation; and perhaps electrical attraction may reach to such small distances, even without being excited by friction? [25]

This introduction is followed by a long section of some 3000 words in which a substantial part of chemistry is dissected by means of questions that are phrased in a way that almost compels the reader's assent. Some forces are stronger than others and so apparent repulsions, such as that between oil and water, can be explained in terms of different strengths of attraction, a simple form of what later came to be formalised by chemists as the doctrine of 'elective affinities'. The heat that accompanies many chemical changes is ascribed to rapid movement, ". . . does not this heat argue a great motion in the parts of the liquors?" His enthusiasm for chemistry is clear on every page, and it was in this subject that he foresaw many applications of his doctrine of corpuscular attractions, once his successors had worked out the quantitative details [26]. He returns eventually to problems that we should call physical rather than chemical, noting, by way of transition, that the diffusion of a solute through a solution argues that there is an effective repulsion between its particles, "or at least, that they attract the water more strongly than they do one another." The crystallisation of a salt from a liquor suggests a regularity in the forces between the particles of the salt, so that in the crystal "the particles not only ranged themselves in rank and file for concreting in regular figures, but also by some kind of polar virtue turned their homogeneal sides the same way." We are now back to cohesion which, he says some (he means Descartes and his followers) have explained by

. . . hooked atoms, which is begging the question; and others tell us that bodies are glued together by rest, that is, by an occult quality, or rather by nothing; and others, that they stick together by conspiring motions, that is by relative rest amongst themselves. I had rather infer from their cohesion, that their particles attract one another by some force, which in immediate contact is exceeding strong, at small distances performs the chymical operations above-mention'd, and reaches not far from the particles with any sensible effect.

He believed matter to be porous; its basic units all identical:

. . . it seems probable to me, that God in the beginning form'd matter in solid, massy, hard, impenetrable, moveable particles, of such sizes and figures, and with such other properties, and in such proportion to space, as most conduced to the end for which he form'd them; and that these primitive particles being solids, are incomparably harder than any porous bodies compounded of them; even so very hard, as never to wear or to break in pieces; no ordinary power being able to divide what God himself made one in the first Creation.

The compound particles of, say, water or gold are formed from arrays of these primitive particles with greater or less proportions of empty space to matter. He did not at this point explain how these compound particles might be constructed but he had, earlier in the book, considered a possible ramified structure [27], one that he had discussed in December 1705 with David Gregory [28, 29].

In Query 31 he infers a force of attraction also from the "cohering of two polish'd marbles *in vacuo*", and, as earlier, from the fact that mercury when "well-purged of air" can stand at a height of 70 inches or more in a barometer tube. This was an observation based originally on work by Christiaan Huygens but, more directly, on a demonstration by Hooke before the Royal Society in 1663 [30]. He then moves naturally into the field of capillary rise, such a baffling but striking manifestation of the cohesive tendency of matter (to use a neutral term) that it became throughout the 18th century the testing ground for theories of corpuscular attraction for those that believed in such theories, and a means of refuting them for those that did not. In this field Newton draws heavily on the experiments carried out under his supervision and often at his suggestion by Francis Hauksbee [21], who was the demonstrator at the Royal Society from 1704, shortly after Newton's election to the Presidency, until Hauksbee's death in 1713. Most of his first experiments were electrical and have been credited with reviving Newton's belief in an aether in his later years [31], but in 1709 and 1712 he carried out important experiments on capillarity. In an early paper [32] he had corrected Newton's opinion that capillary rise was due to a lowering of air pressure in a narrow tube; he did this by showing that the same rise is found *in vacuo* as in air, a result that had been found as early as 1667 at the Accademia del Cimento, and later by others [33]. He then established, apparently for the first time, that water also rises between parallel vertical plates of glass and that the rise was proportional to the separation of the plates [34]. In the same series he made a simple but potentially decisive experiment which showed that it was only the forces emanating from the innermost layer of glass in the tube that attracted the water:

I found, that neither the figure of vessel, nor the presence of the air did in any ways assist in the production of the forementioned appearance [i.e. the rise]. To try therefore whether a quantity of matter would help unriddle the mistery; I produc'd two tubes of an equal bore, as near as I could, but of very unequal substances, one of them being at least ten times the thickness of the other; yet when I came to plunge them into the premention'd liquid the ascent of it seem'd to be alike in both. [34]

He is intrigued by the analogy with magnets, which also retain their potency when broken into smaller pieces, and we shall see that only hesitantly does he draw the natural implication of the short range of the forces.

In his Query 31 Newton states clearly that the rise between parallel plates is inversely proportional to their separation. His obviously rough figure of a rise of water of about one inch for plates separated by one-hundredth of an inch is only about half what is expected for clean plates that are perfectly wetted by the water. He says that the rise between plates is equal to that in a tube "if the semi-diameter of the cavity of the pipe be equal to the distance between the planes, or thereabouts." This important result is not to be found in Hauksbee's papers nor in the later experiments before the Royal Society made by James Jurin [35, 36], which are discussed below. Hauksbee showed that the rise between plates is inversely proportional to their separation and Jurin states the same result for tubes in such a way as to make it seem then to be an accepted truth. Neither claims the relation between the two configurations but it is hard to imagine that Newton's source was any but the experiments carried out before the Society.

Perhaps surprisingly, Newton ignores Hauksbee's experiments with tubes of different wall thicknesses, but he does devote some space to one experiment that we know he had proposed himself [37]. In this a drop of 'oil of oranges' or other liquid is placed between two large plates of glass that touch along a horizontal edge and make a small angle with each other. If the lower plate is horizontal, and the upper therefore nearly so, the drop of liquid moves rapidly towards the line where the two plates touch. The experiment consists in finding how much the pair of plates must be tilted, keeping a fixed angle between them, for the force of gravity to balance the force of attraction and the drop to be maintained at a fixed distance from the line of contact of the plates [38]. Hauksbee makes no calculation of the strength of the forces, perhaps because such a calculation was not his province, or perhaps because of the onset of his final illness, but in Query 31, by an argument that is not there made clear, Newton says that the attraction between the oil and the glass "seems to be so strong, as within a circle of an inch in diameter, to suffice to hold up a weight equal to that of a cylinder of water of an inch in diameter, and two or three furlongs in length." He follows this estimate with the exhortation that:

There are therefore agents in Nature able to make the particles of bodies stick together by very strong attractions. And it is the business of experimental philosophy to find them out.

The basis of his estimate is to be found in an unpublished manuscript of 1713, *De vi electrica* [39]. His measure of the adhesion of liquid to glass as a pressure was paralleled a hundred years later by Young and Laplace, but his estimate of the magnitude, about 40 to 60 bar in modern units, is nearly a thousand times smaller than was thought reasonable early in the 19th century. He does not commit himself explicitly to an estimate of the range of the forces, except to say that it is exceedingly small. He discusses the adhesion of a liquid layer whose thickness is that of the innermost black zone of the light fringes between two curved glass surfaces, namely "three eighths of the ten hundred thousandth part of an inch", so this may be his best guess at the range of the forces; it is about 100 Å in modern measure.

In his book Hauksbee attempts to explain capillary rise by the horizontal force of attraction between the glass wall and the contiguous particles of water (*aa* and *bb* in Fig. 2.1), but without saying how this horizontal force is converted to a vertical force that lifts the liquid (particles *ee* and *gg*) [40]. He is also uncertain about the range of the forces, saying first that his experiments with tubes of different wall

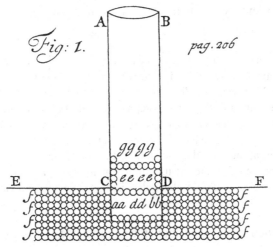

Fig. 2.1 Hauksbee's picture of the rise of particles of a liquid in a capillary tube; from his *Physico-mechanical experiments* [38].

thicknesses show that "the attractive power of small particles of matter acts only on such corpuscules as are in contact with them, or remov'd at infinitely little distance from them." On the next page, however, he supposes that the particles of water at the centre of the tube (*dd* in Fig. 2.1) "are near enough to be within the reach of the powerful attraction of the surface." He does not tell us whether the bore of his tubes was greater or less than the thickness of the wall of his thinner-walled tube so his experiment was not entirely conclusive, but his intention was clearly to show that it was only the innermost layer of glass that acted on the water and the converse is that only the outermost layer of water is affected by the glass. No doubt the assumption that all the water was attracted was needed to save his theory, but it was a confusion of thought that was to persist; better mathematicians than he such as Clairaut and Lalande were later led into the same apparent contradiction which they were to justify by saying that it needed experiments with tubes whose wall thickness is less than their internal radius to be quite certain that the forces could not reach the liquid at the centre of the tube.

Hauksbee's last experiment, of which Newton makes no mention in Query 31 although it is described in *De vi electrica*, is to confirm a rough result of Brook Taylor [41] by showing that the rise of water between two glass plates that meet along a vertical (or even tilted) edge leads to a bounding liquid surface or meniscus that is part of a hyperbola [42]. Humphry Ditton, the mathematics master at Christ's Hospital [43], tried to explain the form of this curve by treating the wedge-shaped space between the plates as a set of ever narrower capillary tubes [44].

A further experiment of Taylor's was attached, almost as an afterthought, to a short paper on magnetism [45]. It aroused little interest at the time but was to be revived later in the century as 'Dr Taylor's experiment' and was then much repeated and extended. He wrote:

I took several very thin pieces of fir-board, and having hung them sucessively in a convenient manner to a nice pair of scales, I tried what weight was necessary, (over and above their own, after they had been well soak'd in water) to separate them at once from the surface of stagnating water. I found 50 grains to separate a surface of one inch square; and the weight at every trial being exactly proportional to the surface, I was encourag'd to think the experiment well made. The distance of the under surface of the board from the surface of the stagnating water, at the time they separated, I found to be 16/100 of an inch; though I believe it would be found greater, if it could be measured at a greater distance from the edge of the board, than I could do it, the water rising a little before it came quite under the edge of the board.

There was to be much speculation about the significance of this force of detachment.

Repulsive forces feature less in Newton's exposition; many apparent effects of repulsion were, as we have seen, attributed to the effects of unequal attractive forces [46]. The 'elastic' properties of air called for a repulsive force which he assumed to be general:

And as in algebra, where affirmative quantities vanish and cease, there negative ones begin; so in mechanicks, where attraction ceases, there a repulsive virtue ought to succeed. . . . The particles when they are shaken off from bodies by heat or fermentation, so soon as they are beyond the reach of the attraction of the body, receding from it, and also from one another with great strength, and keeping at a distance, so as sometimes to take up a million of times more space than they did before in the form of a dense body. . . . From the same repelling power it seems that flies walk upon water without wetting their feet; and that object glasses of long telescopes lie upon another without touching; and that dry powders are difficultly made to touch one another so as to stick together, ... [47]

Newton does not say so but presumably this moderately long-ranged repulsion changes again and becomes a gravitational attraction at even larger distances. There is here the germ of an idea that was to be expressed more explicitly later in the century by Rowning and Boscovich.

A repulsive force at short distances might seem to be necessary to account for the space-filling properties of solid and liquid matter, but as long as his particles had volume and were held to be almost incompressible, and as long as he did not enquire into the elasticity of solids or into the small and then unknown compressibilities of liquids, he could ignore this refinement. It was a point of view that could still be held well into the 19th century.

We need not enter deeply into Newton's private speculations on the cause of gravity and, by implication, on the cause of cohesion. He was not prepared to accept that gravity was an inherent property of matter, and attraction at a distance, without a mediating cause, was as absurd a notion to him as it was to his Continental critics.

In 1692 Richard Bentley, then Chaplain to the Bishop of Worcester, was preparing the first set of Boyle Lectures, and wanted advice. Newton wrote to him: "You sometimes speak of gravity as essential and inherent to matter: pray do not ascribe that notion to me, for the cause of gravity is what I do not pretend to know. . . ." He believed that it is "unconceivable that inanimate brute matter should (without the mediation of something else which is not material) operate upon, and affect other matter without mutual contact, . . .", and that: "Gravity must be caused by an agent acting constantly according to certain laws; but whether this agent be material or immaterial, I have left to the consideration of my readers." [48] John Locke echoed the same sentiments and, in a parenthetical phrase (originally medieval [49]) that was to be repeated throughout the 18th century, laid down that it was "impossible to conceive that a body should operate on what it does not touch (which is all one to imagine it can operate where it is not)." [50]

In his early years, influenced by his reading of chemical, theological, and magical authors, Newton believed that an aether was the effective cause of gravity [26]. In his middle years he was more inclined to put his faith in the literal omnipresence of God, whose actions filled all space and so effected the attraction [28]. A memorandum of David Gregory of 20 February, 1697/8 records that: "Mr C. Wren says that he is in possession of a method of explaining gravity mechanically. He smiles at Mr Newton's belief that it does not occur by mechanical means, but was introduced originally by the Creator." [51] Alas, we hear no more of Wren's mechanical theory. In his later years, influenced by Hauksbee's spectacular electrical experiments, Newton returned to an aether, or to

. . . a certain most subtle spirit, which pervades and lies hid in all gross bodies; by the force and action of which spirit, the particles of bodies mutually attract one another at near distances, and cohere, if contiguous. . . . But these are things that cannot be explain'd in a few words, nor are we furnish'd with that sufficiency of experiments which is required to an accurate determination and demonstration of the laws by which this [electric and elastic][†] spirit operates.

These are the closing words of the last edition of the *Principia*.

These twists and turns of Newton's thoughts [52] make it hard to summarise his views but it was as an exponent of attractive forces between independent particles that he was to be remembered in later times. His changes of emphasis arose in part from his sensitivity to the views of his critics, particularly Leibniz [53] and his followers who thought that Newton's gravitational force, without a mechanical explanation, was a resurrection of those 'occult qualities' that they believed had been banished from natural philosophy in the 17th century. Newton had demolished

---

[†] The words 'electric and elastic' are not in the Latin text of the third edition but were added by the translator, Andrew Motte, from a hand-written addition by Newton in his own copy of the second edition.

the best-known mechanical explanation, the great vortices [*tourbillons*] of invisible material that Descartes had supposed carried the planets round the Sun [54], but the demand for a mechanical cause did not go away and was to plague Newton's followers for many years after his death.

2.2 Newton's heirs

First in Edinburgh and then in Oxford and Cambridge, Newtonian philosophy made its way into the universities. In 1683 David Gregory [55, 56] succeeded his uncle, James, as professor of mathematics at Edinburgh and at once started teaching the mathematics and astronomy he had learned from the works of Descartes and Wallis. After the publication of the *Principia* in 1687 he became "the first who introduced the Newtonian philosophy into the schools" [56]. With the support of Newton and Flamsteed he was appointed to the Savilian chair of astronomy at Oxford in 1691, where he was joined three years later by his pupil, John Keill (or Keil) [56, 57], who in 1699 became the deputy to Thomas Millington, the Sedleian professor of natural philosophy [58].

Both Gregory and Keill were soon familiar with Newton's as yet unpublished thoughts on matter and its cohesion [28]. Gregory's discussions with him took place in London; Newton's only visit to Oxford was not until 1720, in the company of Keill [59]. Gregory is known to have had a copy of Newton's unpublished manuscript *De natura acidorum* [29, 60]. In his lectures as Millington's deputy or in his rooms in Balliol College, Keill introduced experiments into his teaching, using equipment that he had paid for himself. He was, wrote Desaguliers [61], the "first who publickly taught natural philosophy by experiments in a mathematical manner . . . instructing his auditors in the laws of motion, the principles of hydrostaticks and opticks, and some of the chief propositions of Sir Isaac Newton concerning light and colours" [62], to which Keill's biographer adds that this "yet had not 'till then been attempted in either university" [56]. (Burchard de Volder had introduced experiments into the course at Leiden as early as 1675, on his return from London where he had seen them performed before the Royal Society [63].) Keill's lectures were first published in Latin in 1702, and in English in 1720, with many later editions in both languages [64]. In his published lectures he confined himself to Newton's mechanics and its applications; astronomy he left, at that stage of his career, to Gregory, and cohesion he omitted. This omission was soon repaired in two ways; first, through a paper that he published in the *Philosophical Transactions* of 1708 (issued in 1710) which contained thirty theorems on matter and its cohesion [65, 66], and secondly, through some lectures, soon to be followed by a book, by his colleague John Freind, the reader or professor of chemistry [56, 67].

In his paper of 1708 Keill laid down three principles, two of which, the existence of a vacuum and the mutual attractions of the particles of matter, followed Newton's views, and a third which did not: a belief in the infinite divisibility of matter [68]; it seems, however, to play no part in his theorems. The first three of these repeated Newton's arguments for a porous structure of matter, and the fourth asserted that:

Besides that attractive force [i.e. gravity], . . . there is also another power in matter, by which all its particles mutually attract; and are mutually attracted, by each other, which power decreases in a greater ratio, than the duplicate ratio of the increase of the distances. This theorem may be proved by several experiments: but it does not yet so well appear by experiments, whether the ratio, by which this power decreases, as the particles recede from each other, be in a triplicate, quadruplicate, or any other ratio of the increase of the distances.

Theorems 5 to 11 point out, as Newton had done, that the attractive forces dominate the gravitational force at short distances, and that it is only the forces between the immediate points of contact that contribute to the cohesion of two bodies. These clear arguments then pass, in the remaining theorems, into less precise but still essentially Newtonian explanations of how fluidity, elasticity, diffusion, solution, precipitation, etc., can be explained in terms of these forces. He is clear, however, on the distinction between what we now call elastic and plastic bodies. In the first, an applied force moves the particles a little, without destroying their configuration and leaving them subject to the restraining force of their mutual attractions. Plastic, or 'soft' bodies, as he calls them, have the configuration of their particles destroyed by weak applied forces. A more fully developed version of this idea was put forward by Coulomb [69] in 1784, in a paper that can now be seen as the link between the simple ideas on elasticity of the early 18th century and the more detailed corpuscular theories of Navier, Poisson and Cauchy in the early 19th [70].

One of the last phenomena that Keill sought to reduce to a mechanical explanation was the rising of sap in trees, thus foreshadowing the later attempts to extend Newton's philosophy into biological and botanical fields made by his younger brother James [56, 71] and by Stephen Hales [72, 73].

In 1704 John Freind gave nine lectures in the Museum at Oxford which, when he published them five years later (probably in revised form), he acknowledged were based on Keill's ideas [74]. His aim was to derive chemistry from Newtonian principles. He reduces Keill's thirty theorems to eight and, like his mentor, is clear that the attractive force responsible for cohesion falls off "in a ratio of increasing distances, which is more than duplicate." [75] Melting is caused by particles of fire insinuating themselves into matter and so weakening the attraction. Since lead melts at a lower temperature than many less dense metals it follows that the attractive forces are not proportional to mass, and so they are not gravitational. Later in the century more subtle French minds were to find such arguments unconvincing. Solution and precipitation are reduced to the effects of differential attractions, and distillation is assisted by a rarefaction of the liquid by air. Perhaps his most ambitious attempt to reduce chemistry to mathematical laws is his explanation of why *aqua fortis* dissolves silver but not gold, while *aqua regia* dissolves gold but not silver, a paradox that had engaged Newton's attention in Query 31, and also that of others [76]. Freind's explanation is in terms of differences in the sizes of the particles and of the strengths of the attractions, all expressed in algebraic symbols [77]. Crystallisation is a result of the forces being stronger on one side of the particles than the other. The geometric shape of crystals was therefore, he thought, a consequence of the different shapes of the particles [78]. He closes on a cautionary note:

There remain indeed many other things, which cannot be accounted for, without great difficulty; but we hope the difficulty, sometime or another, may be surmounted, when people take the pains to pursue these inquiries in a right method. . . . but if these can't be reduc'd to the laws of mechanism, we had better confess, that they are out of our reach, than advance notions and speculations about 'em, which no ways agree with sound philosophy. [79]

Freind's lectures of 1704 were too early to have been influenced by Newton's published words, and he was abroad from 1705 to1707, but there is no doubt that he benefited indirectly from Newton's contacts with Gregory and Keill, and the publication of the lectures came after the Latin edition of the *Opticks*. Thomas Hearne of the Bodleian Library went so far as to accuse "some Scotch men, (who would make a great figure in mathematical learning)" of stealing Newton's results [80], and it is now known that Gregory used Newton's manuscripts, presumably with permission, in preparing his own book on astronomy [81]. Certainly the whole of Freind's book is imbued with the spirit of Query 31 and, as his translator in 1712 ('J.M.', not identified) puts it in his Preface, by "the principle of attraction, which so happily accounts for the phaenomena of Nature". Freind's lectures were the most ambitious attempt yet to reduce the operations of chemistry to mechanics, but this was not to be the way forward; the world was not yet ready for quantitative physical chemistry.

Most of Newton's followers in Cambridge were less ambitious than Keill and Freind; they were in the main translators, editors, and textbook writers [82]. Samuel Clarke, a Fellow of Gonville and Caius College [83], translated the Cartesian textbook of Jacques Rohault into Latin in 1697 and embellished it with Newtonian comments that often contradicted the sense of the original text. Roger Cotes, who became the first Plumian professor of astronomy in 1706 was an original mathematician but his main contribution to physics was as editor, and writer of a Preface, for the second edition of the *Principia* in 1713. Cotes died three years later, with little in the way of thanks from Newton for his considerable labours, and was succeeded by his cousin, Robert Smith [84], who wrote a thoroughly Newtonian account of geometrical optics in which he adduced arguments to show that the force of attraction of matter for the particles of light was "infinitely stronger than the power of gravity" [85]. William Whiston [86] succeeded Newton as Lucasian professor in 1701 but was ejected from the chair for heresy in 1710; his interests were more in theology and popular astronomy than in mathematics and physics. He was involved with Humphry Ditton in a hare-brained scheme for determining longitude at sea by discharging cannon from lines of ships moored in mid-ocean [87].

Little more was done experimentally at the Royal Society in the field of cohesion after the death of Hauksbee in 1713. He was succeeded as demonstrator by J.T. Desaguliers [61, 88], the son of a Huguenot refugee. He had been educated at Oxford and had succeeded Keill at Hart Hall when Keill had gone abroad in 1710; there he learnt to lecture and demonstrate. His experiments before the Royal Society were many and ingenious but were mainly optical, electrical and mechanical; his *Course of experimental philosophy* [62] became an important Newtonian textbook. He was one of the first to appreciate that Newton's 'force' (generally our momentum) and Leibniz's 'force' (the *vis viva*, or twice our kinetic energy) were different constructs, and that many of the arguments about the much-used word were misconceived.

James Jurin [35], a physician educated at Cambridge and Leiden, was a Secretary of the Royal Society during the last six years of Newton's Presidency. In 1718 he made an important experiment that added a new fact to those discovered by Hauksbee; the height to which water rose in a tube depended only on the diameter at the position of the meniscus. A tube that was wide at the bottom but narrow at the top could therefore hold in suspension a greater volume of water than one of uniform bore. This fact undermined Hauksbee's not very coherent explanation that the rise was due to a diminution of the "gravitating force" by a horizontal attraction of the whole of the glass wall, in essence the same view that Newton had expressed in *De vi electrica* [39]. Jurin claimed to have found "the real cause of that phaenomenon, which is the attraction of the periphery, or section of the surface of the tube, to which the upper surface of the water is contiguous and coheres" [36]. He expounded six propositions: such as, for example, that water particles attract water but not as strongly as they are attracted to glass, whereas mercury attracts mercury more strongly than mercury is attracted to glass. He established that the depression of mercury in a capillary tube, like the rise of water, is as the reciprocal of the bore.

Ephraim Chambers published his *Cyclopaedia* in the year after Newton's death [89]. In opening his article on 'Attraction' he seems to subscribe to the view that attractive forces are innate; he writes, "Attractive force, in physicks, is a natural power inherent in certain bodies, whereby they act on other distant bodies, and draw them towards themselves." In the fifth edition of 1741 he (or his editor, he died in 1740) showed that this view was not the one then held by changing the word 'physicks' to 'ancient physics'. He then outlines the opinions of Newton, Keill and Freind, surmises that the last two may have gone too far ("but this seems a little too precipitate") and then sets out 25 theorems. These derive from the 30 in Keill's paper of 1708, either directly, or from the 19 in Harris's *Lexicon* of 1710. His article on cohesion opens: "The cause of this cohesion, or the *nexus materiae*, has extremely perplex'd the philosophers of all ages. In all the systems of physicks, matter is suppos'd originally to be in minute, indivisible atoms." The rest of the article consists of long quotations from Newton's writings. On 'Capillary tubes' he writes: "The ascent of water etc. in capillary tubes is a famous phaenomenon which has long embarrass'd the philosophers." These phrases were to be repeated throughout the century, and the opening of the article on cohesion was, as we shall see, to be distorted by d'Alembert for the French *Encyclopédie* of 1751.

Repulsive forces played even less part in the expositions of Keill and Freind than they did in that of Newton, but they were given a more prominent role by Stephen Hales and Desaguliers. The former, perhaps the most original of the Cambridge Newtonians, took seriously the 'fixation' of air in solid bodies, from which it could be expelled again by heat or fermentation. It was a thesis of his *Vegetable staticks* [73] that such fixation was not merely the accommodation of 'airs' within the the solid but that it required the annulment of the repulsive forces. Later his work on airs was an important influence on Continental 'pneumatic chemistry', particularly on Lavoisier, via Buffon's translation of *Vegetable staticks* in 1735 [90].

Desaguliers, who wrote a long abstract of *Vegetable staticks* for the *Philosophical Transactions* [91], took the matter further by considering the relevance of repulsive forces to the apparently unrelated phenomena of the evaporation of liquids [92] and the elasticity of solids [62, 93]. He notes first that Newton "has demonstrated" that the elasticity of air arises from the repulsion of contiguous particles, claims that he and Henry Beighton [94] had shown that water increases in volume by a factor of "about 14000" on boiling, and then tries to marry these ideas to a repulsive force at short distances. He says that such a force is needed because water is known to be incompressible. He writes that this property of resisting compression

... must be intirely owing to a centrifugal [i.e. repulsive] force of its parts, and not its want of vacuity; since salts may be imbib'd by water without increasing its bulk, as appears by the encrease of its specifick gravity.... The attraction and repulsion exert their forces differently: The attraction only acts upon the particles, which are in contact, or very near it; in which it overcomes the repulsion so far, as to render the fluid unelastick, which otherwise would be so; but it does not wholly destroy the repulsion of the parts of the fluid, because it is on account of that repulsion that the fluid is then incompressible. [92]

His facts are not quite correct; his estimate of the increase in volume of water on boiling is too large by a factor of about 8, but the same erroneous figure was still being quoted twenty years later in the widely used textbook of the Abbé Nollet [95]. Salts do not (in modern language) have zero partial volumes in solution, although these volumes are often much smaller than the volumes of the solid salt. The problem of how the supposed pores in water could take up solutes was one that received spasmodic attention throughout the century. Daniel Bernoulli claimed that the dissolution of sugar in water also led to no increase in volume [96]. Richard Watson of Cambridge established the facts most clearly in 1770; the solution occupies more space than pure water but less than the sum of the volumes of the water and the solid solute [97]. He attempted no explanation of this result.

Evaporation of a liquid into air continued to be a puzzle for some time after its discussion by Hales and Desaguliers. Hugh Hamilton [98], in Dublin, ascribed it to an attraction between the particles of air and those of water, and added that he had been told the Abbé Nollet held the same view. When his paper was sent to the Royal Society in 1765 it was remembered that Benjamin Franklin had placed similar views before the Society nine years earlier, and so his paper was appended to Hamilton's. Franklin had added a Newtonian repulsion of the air particles to the air–water attraction [99].

Ten years after his paper on evaporation and solution Desaguliers extended his ideas on repulsive forces to the field of the elasticity of solids [93]. He believed that attractive forces alone between spherical particles would result in the material forming an easily deformed spherical body. He went beyond Keill's ideas in thinking that something more than attraction was needed to explain, for example, the elasticity of a blade of steel. He opened his paper with the ringing Newtonian declaration: "Attraction and repulsion seem to be settled by the Great Creator as first principles in Nature; that is, as the first of second causes; so that we are not solicitous about their causes, and think it enough to deduce other things from them." [93] He then mentions Hales's experiments on the release of fixed air by distillation, a reference that suggests that he was not entirely clear on how repulsive forces could act at both large and small distances, with attraction in between, and (presumably) again at very large distances as gravity takes over. The repulsive forces he introduces are polar, and probably magnetic; only such different-sided forces could account for the preference of an array of particles to adopt a linear configuration, and for that line to resist bending.

Desaguliers was, perhaps, the first to suppose that the concept of impenetrability could be replaced by the potentially more quantifiable concept of a short-ranged repulsive force, and his later work may have owed something to a clearer expression of this proposal in a recently published popular account of Newtonian philosophy. In his *Compendious system of natural philosophy*, the Revd John Rowning [100] of Anderby in Lincolnshire, and sometime Fellow of Magdalene College, Cambridge, had written that "matter . . . has also certain powers or active principles, known by the names of *attraction* and *repulsion*, probably not essential or necessary to its existence, but impressed upon it by the Author of its being, for the better performance of the offices for which it was designed." [101] His words are similar to those used later by Desaguliers. Two facts, Rowning says, show the existence of "the attraction of cohesion"; the rise of a liquid in a capillary tube and the joining of two small spheres of mercury to form one. He sets out the rules of attraction as, first, that it acts only on contact or at very small distances, second, that it is proportional to the "breadth of the surfaces of the attracting bodies, not according to their quantities of matter", and, third, that "'tis observ'd to decrease much more than as the squares of the attracting bodies from each other increase". [102] All this follows, he says, from Keill's work. Later, when writing on hydrostatics, he goes further and says that

. . . since it has been proved that if the parts of fluids are placed just beyond their natural distances from each other, they will approach and run together; and if placed further asunder still, will repel each other; it follows, upon the foregoing supposition that each particle of a fluid must be surrounded with three spheres of attraction and repulsion one within another: the innermost of which is a sphere of repulsion, which keeps them from approaching into contact; the next a sphere of attraction diffused around this of repulsion, and beginning where this ends, by which the particles are disposed to run together into drops; the outermost of all, a sphere of repulsion whereby they repel each other, when removed out of that attraction. [103]

This is an extension of Newton's dictum that where attraction ends there repulsion starts. The repulsion is not only between the particles of air but also between grosser bodies, such as that which enables a fly to walk on water. This favourite instance was repeated, for example, in the first edition of the *Encyclopaedia Britannica*, where was added also the case of a needle that "swims upon water" [104]. Rowning's synthesis differs little from the more fully articulated one developed a few years later by Boscovich.

Rowning's discussion of capillary rise [103] is fully referenced with citations of the works of Hauksbee, Jurin, van Musschenbroek and the French savants (see below), but his conclusions are not wholly in accord with their results. He assumes that the rise is proportional to the wetted area of the tube (notwithstanding Jurin's experiment) and that the size of the sphere of attraction is comparable with the radius of the tube, which is what Hauksbee said, but is contrary to Rowning's own reading of Keill. If this were so, then he acknowledges that tubes of different thicknesses but with the same bore should show different rises, but "no one has as yet been so accurate as to observe it".

Desaguliers turned Newton's conjecture about repulsion between air particles into established fact. He said also that the views expressed in the Queries were not mere conjectures but facts confirmed by "daily experiments and observations" [62]. Cotes made another advance beyond Newton's usual public position on the cause of gravity in his Preface to the second edition of the *Principia*, although Newton then tacitly endorsed it. In the General Scholium Newton committed himself only to the statement: "And to us it is enough that gravity does really exist and act according to the laws we have expressed. ..." [105] Henry Pemberton [106], the editor of the third edition, was the disciple who kept closest to Newton's public view. He relegated the topic of cohesion, however, to the last paragraphs of his own exposition of Newton's work, and wrote there:

From numerous observations of this kind he makes no doubt, that the smallest parts of matter, when near contact, act strongly on each other, sometimes being mutually attracted, at other times repelled. The attractive power is more manifest than the other, for all parts of all bodies adhere by this principle. And the name of attraction, which our author has given to it, has been very freely made use of by many writers, and as much objected to by others. He has often complained to me of having been misunderstood in this matter. What he says upon this head was not intended by him as a philosophical explanation of any appearances, but only to point out a power in nature not hitherto distinctly observed, the cause of which, and the manner of its acting, he thought was worthy of a diligent enquiry. To acquiesce in the explanation of any appearance by asserting it to be a general power of attraction, is not to improve our knowledge in philosophy, but rather to put a stop to our farther search.

FINIS [107]

This careful 'quasi-positivistic' [108] attitude to gravity and cohesion was often impatiently brushed aside by Newton's followers; to them the attractive forces were facts of nature and they did not care how they were effected. It was a cavalier attitude that offended contempory Continental philosophers but which was to pay dividends in the hands of Laplace and his school. Even in Britain it did not always command approval, as we have seen from Pemberton's mild rebuke. Others went further and put it more strongly; *Biographia Britannica* wrote of James Keill carrying his use of attractive forces further than was warranted by "the principles of true philosophy", and added that "he is not the only person, who instead of reflecting honour has thrown a blemish on this point of Newtonian philosophy" [109]. Not only this attitude to the forces but also the wide range of applications of the philosophy came in for criticism. Others in Britain attacked Newton's philosophy *per se*, often on theological grounds, but their influence was small in 'philosophical' circles.

2.3 On the Continent

The question of how Newton's thoughts on cohesion were received on the Continent is easily answered; they were ignored until what were seen as more urgent problems with his physics had been resolved. From the time of the publication of the *Principia* in 1687 he was recognised as one of the leading mathematicians of the day, but his physics was unacceptable to the Cartesians in France and in the Netherlands, and to Liebniz and later to Wolff [110] in Germany.

There were two stumbling blocks. The first was the introduction of the two 'occult' qualities of action at a distance and a vacuum, which was seen as a return to the primitive days before Descartes had filled space with aetherial vortices. The mechanistic philosophy of Descartes had, however, scarcely ousted the scholastic by the time the Continent became fully aware of Newton. D'Alembert [111] was to claim in 1751, with some exaggeration, that "... scholastic philosophy was still dominant there [in France] when Newton had already overthrown Cartesian physics; the vortices were destroyed even before we considered adopting them. It took us as long to get over defending them as it did for us to accept them in the first place." [112]

In one of earliest foreign reviews of the *Principia*, the writer in the *Journal des Sçavans* commended Newton's mathematics but said that he must give us a physics that matched the power of his mechanics [113]. Huygens was equally dismissive in private, even before he had studied the *Principia*. He wrote on 11 July 1687 to Newton's friend Fatio de Duillier: "I should like to see Newton's book. I am happy for him not to be a Cartesian providing that he does not pass on to us suppositions such as that of attraction." [114] Forty years later, Fontenelle, the Secretary of the French Academy, wrote in his *Éloge* for Newton: "Thus attraction and vacuum banished from physicks by Des Cartes, and in all appearance for ever, are now brought back again by Sir Isaac Newton, armed with a power entirely new, of which they were thought incapable, and only perhaps a little disguised." [115] We have seen that Newton shared the Cartesians' disbelief in action at a distance but his honest declaration that he thought it proper to make full use of the inverse-square law of gravitation, even although he could not account for it physically, did not satisfy his Continental critics [116]. Leibniz, in particular, with his strong belief in the continuity of all natural things, could conceive of pull at a distance only as a sequence of pushes. Johann Bernoulli shared the same view [117]. It was the gravitational attraction at which Newton's critics directed their fire; the relatively minor matter of Query 31 was at first ignored in the condemnation of the greater sin. The fullest exposition of Leibniz's opposition is in his correspondence with Samuel Clarke. Here there is much on gravity, on metaphysics and theology, something on mechanics, but only a passing mention of cohesion [118].

The second stumbling block to the acceptance of Newton's physics was the disagreement between the work of Edme Mariotte [119] and others in France and Newton's work on the dissection of white light into colours. It was not until 1716–1717 that Dortous de Mairan [120] and Jean Truchet [121] in France and Desaguliers in England showed decisively that Newton was correct [122]. Nevertheless those who, following Huygens, held to wave theory of light, could not accept his particles of light streaming through a vacuum.

Newton's ideas on cohesion seem to have attracted notice abroad first in the guise of Keill's publications [123] and of Freind's book of chemical lectures. The

Latin edition of this work was reprinted in Amsterdam in 1710 and so became the subject of a highly critical review by Wolff, published anonymously in the Leipzig journal *Acta eruditorum* [124]. Freind's reply to this review is revealing since it shows how soon some of Newton's followers in Britain abandoned their master's cautious stance. He wrote, in obvious exasperation: "Such a principle of attraction they are pleas'd to call a figment; but how any thing shou'd be a figment, which really exists, is past comprehension." [125]

In France the work on cohesion was at first more ignored than criticised [126]. Mariotte [127] had observed the adhesion between floating bodies on the surface of water, and in the early years of the century several sets of observations of capillary rise were reported in the Memoirs of the Academy, but they were less well-designed than those of Hauksbee and Jurin, guided by Newton. Such a comparison is an example of the familiar fact that experiments guided by a well-articulated theory, even if it be not wholly correct, are more useful than those conducted more aimlessly.

Louis Carré [128], assisted by E.-F. Geoffroy [129], measured the rise of water in three tubes of diameter 1/10, 1/6, and 1/3 ligne, and found rises of $2\frac{1}{2}$, $1\frac{1}{2}$ pouces and 10 lignes respectively (12 lignes = 1 pouce $\equiv$ 2.71 cm). These figures, like that quoted by Newton in the *Opticks*, are only about half that expected for clean glass tubes that are perfectly wetted by the water, a discrepancy that shows the difficulty of removing the last traces of grease from the glass. In a partial vacuum they found a slightly larger rise than in the open air [130].

Dufay (or du Fay) [131], who was later to make his name by his electrical researches, studied both the rise of water and the depression of mercury in capillary tubes. Fontenelle notes that he ascribed the depression of mercury to the fact that it did not wet the glass because of a film of air between the liquid and the solid, and so deduced that there would be no depression in a vacuum. Dufay tried to convince himself that this was so by reporting that the meniscus in a Torricellian vacuum was flatter than that in air [132]. Petit, a physician [133], complicated matters by using a narrow tube inside a wider one, so that the water rose in the annular space between them. He believed that the strength of the adherence of water was proportional to the density of the solid wall – a false analogy with gravitation, but one that showed, perhaps, that Newton's ideas were beginning to be treated with respect [134].

These French philosophers made few attempts to account for their findings, writing only in the most general terms of a 'stickiness' (Mariotte, who used the word *viscosité*), or a 'sympathy' (Carré), or an 'adhesion' (Fontenelle), or an 'adherence' (Petit) between the water and the glass, avoiding all mention of the Newtonian 'attraction'. Some years later, Desmarest [135] divided theories of capillarity into three classes: first, those where there is "an unequal pressure of a fluid [i.e. air or an aetherial fluid] which acts with less advantage in the narrow confines of a capillary tube", second, those in which there is an "adherence or innixion [i.e. pressing] of the liquids on the walls of the tubes", and, third, those in which there is a "mutual attraction of the capillary surfaces and the particles that comprise the liquids". In the first class he places Dufay, Dortous de Mairan and Johann and Jakob Bernoulli, in the second, Carré, and in the third, Hauksbee, Jurin and Clairaut.

Fontenelle describes how Dortous de Mairan explained the depression of mercury by the fact that it does not wet the glass but then interpreted this as a consequence of the struggle between the opposing vortices of a subtle magnetic material in the annular space between the mercury and the glass. Even to Fontenelle, a convinced Cartesian, this explanation did not carry conviction [132]. It had what we now think of as a characteristic weakness of many early 18th century theories. They were thought to have done their job if they provided a plausible account of a possible mechanism that did not contradict any known fact, and which satisfied the metaphysical creed of the proposer. It was not held to be necessary that theories should be falsifiable nor that they had predictive power, notwithstanding Boyle's claim that one criterion of a good theory was "That it enable a skilfull naturalist to foretell future phenomena." [136] The need for more searching criticisms of theories became apparent in the second half of the century; it is reflected, for example, in d'Alembert's 'Discours préliminaire' to the *Encyclopédie* of 1751 [112].

The French savants made no mention of the work in London. The paper of Carré and Geoffroy was too early to have been influenced by Hauksbee's work, but Geoffroy, a Fellow of the Royal Society since 1698, was fluent in English and on cordial terms and correspondence with Hans Sloane, a Secretary of the Society until 1713. Dufay made notes on Hauksbee's work [131] which was known to him and his contemporaries through an Italian translation of the first edition, published at Florence in 1716 [137].

Perhaps Geoffroy's most original contribution related to the field of cohesion was his table of 'affinities' of 1718, the first of many such tables compiled in the next eighty years. These showed the comparative strengths of the chemical affinities of one substance for another (usually elements in the modern sense of the word), so that it could be seen at a glance which substance would readily displace another from a chemical combination [138]. Geoffroy accepted a corpuscular theory and spoke in his lectures of water particles being smooth and oval: "An oval figure seems more agreeable to the fluidity and motion of water than a spherical, and likewise to the solidity we observe in ice; the points of contact being too few in spherical bodies to form so strong a cohesion." [139] Although these musings resemble some of those of Freind and others, they probably derive more from Descartes than from Newton. His affinities, or 'rapports' as he calls them, are closer in name to the term 'sociableness' that Newton used in his earliest work before he moved to the more explicit 'attraction' [140]. Geoffroy's translator wrote that, "These affinities gave offence to some particular people, who were apprehensive that they might be only attractions disguised, and so much the more dangerous, as some persons of eminent learning had already cloathed them in seducing forms." [141] Such tables became popular with chemists in the second half of the century. Geoffroy's cautious word 'rapports' was abandoned for the more committing 'affinité' or even 'attraction', although this last word was always more popular with the natural philosophers than with the chemists [142]. Maupertuis [143], when writing 'Sur l'origine des animaux' in 1745, was one of the first to assert that "these *rapports* [of Geoffroy] are nothing but what other more bold philosophers call attraction." [144]

Newton's *Opticks* became available in French in 1720 and 's Gravesande's book on Newtonian physics (see below) was published in Latin at Leiden in 1720–1721 [145] to a hostile review in the Jesuit *Journal de Trévoux* [146]. A few years later Freind's chemical lectures were plundered to make an anonymous book entitled *Nouveau cours de chimie suivant les principes de Newton et de Sthall*, 1723. The first reviewer, in the *Journal des Sçavans*, ascribed it to J.-B. Senac, later the King's physician [147], an ascription that has been accepted [148].

None of these works converted the French to Newtonian physics. The first move in that direction came from a group of whom Voltaire was the eldest and the best known [149]. It comprised himself, his mistress, Émilie, Marquise du Châtelet [150], and the natural philosophers Maupertuis and Clairaut [151]. Voltaire became the first to accept the Newtonian theory of attraction when he was in England in 1727 at the time of Newton's death; they never met but he attended the funeral. It was from his friend Samuel Clarke that he learnt what Newton had achieved [152]. His association with Mme du Châtelet began in 1733. She was the better mathematician, having already had instruction from Maupertuis; he had to struggle to master the principles if never the practice of Newton's work. In the years 1734 to 1738 "the poet definitely became the philosopher." [153] He announced his conversion to his compatriots in his *Letters concerning the English nation* of 1733, which appeared in French the next year as *Lettres philosophiques* [154]. He noted in his 14th Letter that in England attraction prevailed "even in chemistry", and in his 15th, which is 'On attraction', he mentions Newton's ramified structure of matter, but generally he confined himself to gravitational attraction, as he did a few years later in his *Elements of Sir Isaac Newton's philosophy*. At one point in that work he mentions that bodies in contact are "attracted in the inverse cubes of their distances, or even considerately more" [155], but that is in the context of a discussion of the inflection (or diffraction) of light. In the edition of 1741, the first produced under his own control, he adds a final chapter in which he discusses the attraction of small bodies, but he makes no advance on what had already been achieved elsewhere.

Meanwhile Maupertuis had almost taken the plunge. He had been in London in 1728, at the same time as Voltaire, but any Newtonian views that he may then have acquired were soon restrained under the Leibnizian influence of Johann Bernoulli,

the elder, whom he visited in Basel the next year [156]. He was, however, a convinced Newtonian by 1731 when his paper 'De figuris quas fluida rotata ...' was
read at a meeting of the Royal Society on 8 July [157]. On 31 July he wrote
somewhat apologetically to Bernoulli to explain that he was publishing in England
because that was where attraction was taken seriously [156]. This paper was followed by, and contained in, a small book on the shape of the heavenly bodies [157]
which was meant to give the impression of an even balance between Descartes and
Newton, but which, in fact, came down very much on the side of Newton, as he
confessed in a letter to Bernoulli of 10 November 1732 [156]. Bernoulli himself
had, however, become less of a convinced Cartesian by 1735 [158]. Maupertuis
later told Bernoulli that a new theory never convinced the partisans of the old; one
could only hope to convince the bystanders. (Planck was to observe that one had to
wait for the supporters of the old to die [159].) Voltaire studied Maupertuis's book
before he wrote his *Lettres philosophiques*.

In the early part of his *Discours* Maupertuis was at pains to establish that there was
nothing metaphysically inadmissible in the notion, which he ascribed to Newton,
that attraction was an inherent property of matter. In the later chapters he examined
the shape of fluid bodies that gravitate and rotate, under different assumptions about
the dependence on distance of the force betwen any two parts. His study of powers
of the separation other than -2 seems, however, to have been no more than an
academic exercise in generality. He did not, at this stage, have cohesive forces in
view, but he was able to show that, under all reasonable assumptions, a rotating fluid
body would be flatter at the poles, as Newton had claimed, and not at the equator,
as was claimed by the Cartesians, on the basis of what Maupertuis himself showed
by his journey to Lapland to be flawed earlier French evidence of the shape of the
Earth.

Two years later he returned to the question of the attraction of bodies with
powers of the separation other than -2; this time he was interested in applications
to cohesion – Keill and Freind are both mentioned – but again the whole work is an
exercise in applied mathematics rather a serious piece of physics: "I do not examine
if the attraction contradicts or accords with the true philosophy. I treat it here only
as in geometry; that is, as a quality, whatever it may be, of which the phenomena
are calculable. ..." [160] For a solid sphere he reproduces Newton's result for the
inverse-square law and speculates that the particularly simple properties of this law
may have been the reason why God chose it as the force that governs the motions
of the planets. He shows that for a cubic law the force has a term proportional to the
logarithm of the distance of a particle from the nearest point of an attracting sphere,
a result that Newton had stated with less explicit detail in Proposition 91 of the
Principia [161]. This paper attracted the attention of Fontenelle who, as Secretary
of the Academy, reviewed it in the History [162]. He gives there a fair account of

Newtonian theory but without any commitment to support it. He notes that what makes the determination of the cohesive forces "difficult, and perhaps impossible, is that the experiments or the phenomena yield only extremely complicated facts." He finishes somewhat sardonically by noting that "the physicists need have no fear of lack of work to do, but the mathematicians may run out of occupation more quickly."

Mme du Châtelet's opinions changed with time. She was, presumably, a Newtonian when Voltaire was, with her help, writing his *Elements*; in the dedicatory poem he speaks of her as "the pupil, friend of Newton, and of truth". In 1738–1739 she was more of a Leibnizian, in part under of the influence of Samuel König [163] who had learnt his metaphysics from Wolff, and who was introduced into her company at her château at Cirey by Maupertuis [164]. Her *Institutions de physique* appeared first anonymously in 1740 and was wholly Leibnizian in its metaphysics and even its mechanics. The 'principle of sufficient reason' is invoked repeatedly to counter Newton's views. She says that the coherence of matter is "one of the natural effects, the explanation of which has most puzzled [*embarrassée*] the natural philosophers" [165]. (Was there a copy of Chambers's *Cyclopaedia* at Cirey?) Her 16th chapter, 'On Newtonian attraction', records that Newton's disciples invoke forces that fall off as the inverse cube of the separation (or more strongly), that Freind has "put forward a chemistry totally based on this principle" [166], but then she, like Fontenelle, puts her finger on a weak point when she remarks that each new phenomenon seems to need a new force.

She eventually abandoned these Leibnizian "imaginations" and embarked on what is still the only French translation of the *Principia*. This, and her commentary, were finished before her death in childbirth in 1749, after discussions with Clairaut, but they were not published for another seven years [167]. She had had access to Newton's second edition as early as 1737 and was seeking another copy in "a fine edition" in 1739 [168]. The 'Privilege du Roy' of the published book is dated 7 March 1746. Work went on beyond that date and was probably in some disarray in 1747–1748 when Clairaut thought that the motion of the lunar apse was inconsistent with a pure inverse-square law of gravitation (see below). By the time this problem was resolved Mme du Châtelet was approaching her final confinement. When the book did appear it had, at the end of the second volume, a series of exercises on the attracting spheres and spheroids according to different force laws, rather in the manner of Maupertuis, although he is not mentioned, and with a similar lack of physical applications.

The cultural links between Britain and the Netherlands were stronger than those between Britain and France and Newton's ideas were received favourably there during his lifetime [169]. Herman Boerhaave [170] became the professor of botany and medicine at Leiden in 1709 and also the professor of chemistry in 1718. He was a convinced 'corpuscularian' and an admirer of Newton whom he praised particularly for an insistence on the primacy of experiment in the lecture he gave in 1715 on retiring as Rector Magnificus [171]. His *Elementa chemiae* of 1732 showed, however, that he had no particular commitment to or use for Newtonian attraction [172]. He was probably not unsympathetic to the efforts of Newton and his followers; he is known to have had a copy of the 1710 Amsterdam edition of Freind's lectures [173], and it may be significant that Fahrenheit discussed naturally with him the "attraction or adhesion of the particles", a topic that does not occur in Fahrenheit's letters to Leibniz [174]. Boerhaave, in his turn, wrote to Fontenelle praising Newton's work on magnetic and other attractions and on elasticity [175]. Nevertheless he did not ultimately accept Newton's attempt to reduce chemistry to physics and was, perhaps, the most influential writer of his time to insist that chemistry was an autonomous science [176]. Shaw's translation of his *Chemistry* gave it a Newtonian slant that is not in the original; Freind's lectures, for example, appear as a recommended work only in this English edition. A pseudonymous writer in the *Gentleman's Magazine* for 1732 said that Boerhaave's and Freind's "systems and way of reasoning are as different as that of alkali and acid." [177]

Boerhaave's view of heat was also not that of Newton; he rejected the view that it was nothing but the rapid motion of the particles and put forward the hypothesis that it was a material but weightless fluid whose movement constituted the heat. Heat as a weightless but usually static fluid was a view that became increasingly influential as the century wore on, eventually to be subsumed into the caloric theory of Lavoisier and others [178].

It was Boerhaave's younger colleague, W.J.'s Gravesande [179], and Boerhaave's former pupil, Pieter van Musschenbroek [180], who brought Newtonian physics to the Netherlands. Voltaire made the distinction correctly when he wrote in a letter of 1737: "I have come to Leiden to consult Dr Boerhaave about my health and 's Gravesande about Newton's philosophy." [181]

In 1715 the Dutch sent an embassy to London for the coronation of George I, and 's Gravesande, then a young lawyer, was one of the secretaries. He met Newton, became a friend of Keill and Desaguliers, was elected to the Royal Society and, on his return to the Netherlands, became the professor of mathematics and astronomy at Leiden. He declared his colours at once; the second half of his inaugural lecture of 22 June 1717 is devoted to the physics and astronomy of "the celebrated Newton, this great mathematician and restorer of the true philosophy". [182] He lost no time in producing the first Newtonian textbook of physics to be written on the Continent, which was translated into English by Desaguliers [183]. In this book he says that vacua exist "as is proved by the phaenomena" and, following Keill rather than Newton, that a "body is divisible *in infinitum*", since "There are no such things as parts infinitely small; but yet the subtility of the particles of several bodies is such, that they very much surpass our conception." His views on attraction are orthodox Newtonian doctrine with but one slight gesture to the Leibnizians:

By the word *Attraction* I understand, any force by which two bodies tend towards each other; tho' perhaps it may happen by impulse. But that *Attraction* is subject to these laws; That it is very great, in the very contact of the parts; and that it suddenly decreases, insomuch that it acts no more at the least sensible distance; nay, at a greater distance, it is changed into a repellent force, by which the particles fly from each other.

His explanation of the roundness of drops does not sit easily with his views on the range of the attractive forces:

... in attraction, the greater the number is of particles which attract one another between two particles, the greater is the force with which they are carried towards one another; which produces a motion in the drop, till the distance between the opposite points in the surface become everywhere equal; which can only happen in a spherical figure. [184]

This view of the cohesion of drops by the tension in linear arrays of particles becomes more explicit in his treatment of the elasticity of solids, which he ascribes to the stretching of fibres within the body or, at least, that it "may be conceived as consisting of such threads." [185] Stretched threads were then the standard method of explaining the laws governing the rupture of beams [186], but the extension of the idea to liquids was a novelty that was to be used again later in the century.

From the expansion of bodies by heat "it is evident that the particles of which bodies consist, from the action of the fire, acquire a repellent force, by which they endeavour to fly from each other." [187]

If two pieces of cork or two hollow glass beads, or similar bodies that are wetted by water, float on the surface of the water in a glass vessel, then it is seen that they come together and adhere to each other and to the walls of the vessel. At first sight this looks like a simple case of attraction between the bodies or between one of them and the wall, but 's Gravesande explained correctly (as had Mariotte before him [127]) that it was the capillary effect of the distortion of the liquid surface by the floating bodies that was the true cause, not the direct effect of attraction between them [184]. There is a similar coming together of two non-wetting bodies, and a repelling if one is wetted and one is not. Mariotte's and 's Gravesande's explanation did not prevent the naive interpretation being put forward again later in the century.

'S Gravesande's younger colleague, van Musschenbroek, who was first at Utrecht and later at Leiden, was initially more sceptical about attractive forces but was eventually convinced:

That attraction obtains in all bodies whatever I am sufficiently assured by a multiplicity of experiments. I do not advance this as an hypothesis, nor maintain it out of prejudice, or in complaisance to any party: for formerly I exploded it as a fiction, as many learned men have done. But a multitude of experiments since made upon bodies, repeated examinations of the phenomena, and serious and continued meditations on the subject, have now convinced me of the truth of this principle of attraction. . . . But what this attractive force is, how it inheres, in what manner it operates upon other bodies, and in what proportion of the distance it constantly acts, we cannot by any means conceive clearly. [188]

He is not convinced by the argument of Keill and 's Gravesande that matter is infinitely divisible because a geometric figure has this property; the matter is one of physics not of mathematics. He returns instead to Newton's concept of the particles being composite structured entities, composed of different arrangements of unknowable "first elements".

He gives more attention than most of his contemporaries to the physical properties that result from attraction, such as the forces between magnets and the phenomena of capillarity, subjects to which he devoted two long dissertations packed with new experimental results [189]. The magnetic work did not have any decisive outcome but the capillary work was more accurate than anything that had gone before. He must have cleaned his tubes carefully since he found rises of water much greater than those found previously. His eight series of experiments repeated much of the work of Hauksbee, Jurin and Petit, but in his first series he found that the rise in different tubes of the same diameter but of different lengths was a little greater in the longer tubes, thus showing, he believed, that the attraction of the whole length of the tube was the cause of the rise [190].

In his textbook he retains throughout a healthy scepticism about the depth of our understanding of cohesive forces, "but here we want sure and accurate experiments", and of the underlying structure of matter, it is "an ample field for making experiments that we must leave to posterity" [191]. His is perhaps one of the most balanced account of the attractive cohesive forces in the century between Newton's *Opticks* and the revival of the subject by Young and Laplace.

Thus Newton's concepts of corpuscular impenetrable matter, of the existence of vacua, of attractive forces acting at a distance through these vacua (however they be caused) and, more tentatively, of repulsive forces between the particles of air, made their way slowly in France but were accepted more readily in the Netherlands. Germany and Switzerland never fell under the spell of Boyle, Locke and Newton, but followed Descartes or Leibniz. Russia was essentially a German–Swiss outpost in the years following the founding of the Academy at St Petersburg in 1725–1726 [192]. The Cartesian exposition of capillarity there by Bilfinger [193] attracted criticism from Jurin, whose paper [194] was published with liberal footnotes by Bilfinger; perhaps inevitably their disagreement spread to the field of gravitation. Ten years later Josias Weitbrecht [195] adopted a more Newtonian stance at St Petersburg. Like Keill, he had thirty theorems on the attraction of bodies and the rise of water in capillary tubes. He committed himself to no definite statement about the range of these forces except to say that it was very short [*brevissimus*] between water and glass. He saw that this supposition led to a problem if the tube was wider than the range of the forces but solved this by supposing that the cylindrical layer of water next to the wall was attracted to the glass and raised by it, and that this cylinder then acted on the next layer of water inside it, and so raised that.

In Italy even the Copernican system was suspect until about 1740 when the more liberal Pope Benedict XIV came into office. Newtonianism soon followed, mainly in the form of of the Latin editions of 's Gravesande and van Musschenbroek [196].

2.4 A science at a halt

Newton's views of interparticle forces, as expressed in Query 31, are now known to have been substantially correct, although not, of course, written in the language of modern physics. Under his supervision, Hauksbee and Jurin had established with qualitative correctness and reasonable accuracy, all the important laws of capillarity. The Keills, Freind and Hales had tried to extend his ideas into other areas of physics, chemistry, botany and physiology. These attempts had met with varying success, but an extension into geology was a step too far – attraction is not the power that causes "the ascent of water to the tops of high mountains" [197]. Desaguliers had had some perceptive thoughts about the elasticity of solids, and he and Rowning had proposed substituting a short-ranged repulsive force for the more qualitative concept of impenetrability. Only Newton's tentative theory of the repulsion of static particles of air was to prove seriously amiss. But after all these advances and intellectual ferment the study of the cohesion of matter fell out of the main stream of scientific enquiry. After about 1735 little new was done for the next seventy years, and much of what was done was the work of those not of the first rank [198]. Pemberton in England, and Voltaire and du Châtelet in France had little to say about this aspect of Newton's work, and Maclaurin in Scotland restricted himself to a few words [199]. Desaguliers and van Musschenbroek were more interested, but it was only Robert Helsham in Dublin who went so far as to open his course of lectures with two on cohesion before turning to electricity and gravitation [200].

In the first half of the 18th century Newtonianism meant, first, a commitment to experiment as the true source of knowledge of the physical world, second, the eschewing in public of metaphysical 'systems' (other than a belief in a corpuscular structure of matter), third, his laws of mechanics, fourth, the gravitational theory, fifth, his theory of colours and a corpuscular theory of light, and finally, the existence of short-ranged attractive forces between the particles of matter. The gravitational force and, when the Newtonians thought of them, the cohesional forces also, were usually treated as deductions from observations and many cared little, or regarded as unknowable, what was the mechanical source of these forces or whether they were inherent to matter. David Hume summed up this point of view in 1739:

Nothing is more requisite for a true philosopher, than to restrain the intemperate desire of searching into causes, and having establish'd any doctrine upon a sufficient number of experiments, rest contented with that, when he sees a farther examination would lead him into obscure and uncertain speculations. [201]

Freind and a few others went further, and Daniel Bernoulli, in a letter to Euler of 4 February 1744, said that God could well have "imprinted in matter a universal attraction" [202], but most would have subscribed to Whewell's ruling of a hundred years later, that gravity was "a property which we have no right to call *necessary* to matter, but every reason to suppose *universal*." [203]

   After about 1740 aetherial explanations began to multiply, but much of the motivation for these lay in the wish to explain the more fashionable phenomena of electricity, magnetism and heat, rather than the neglected cohesive forces [204]. The obvious distinction and even antagonism between the Newtonians and the followers of Descartes and of Leibniz became less marked as the century advanced, with many taking their views from more than one camp. It is, however, convenient to retain the names as useful labels to identify the metaphysical bias of each natural philosopher.

   The undeniable success of the gravitational theory led to its more rapid acceptance than that of the doctrine of the cohesive forces, but there was a moment of doubt in 1747. Euler, then in Berlin, had had a problem with the Moon's orbit [205] and now he, Clairaut and d'Alembert, in Paris, all tried, independently, to calculate the annual change in the position of the apses of its orbit, and all obtained an answer that was only half the observed value [206]. It was Clairaut who, in a paper read to the Academy on 15 November 1747, tried boldly to remove the discrepancy by adding a correction term to the inverse-square law of attraction. He supposed that the force of gravitation might vary with separation r as $(ar^{-2} + br^{-4})$, where a was proportional to the product of the masses of the bodies, but b was a new coefficient, still to be determined [207]. He supposed that the second term might be related to the cohesive and capillary forces, but added in a footnote that if it were to have an effect at the distance of the Moon it might prove to be too strong for the purpose and to lead to too great a gravitational force at surface of the Earth. Euler had already written to him on 30 September to point out that such a term was also incompatible with the regular motion of Mercury [208]. On 6 January 1748 Euler admitted that Newton's law seemed to be at fault, "but I have never thought of correcting the theory by making changes in the expression for the forces" [209]. D'Alembert wrote on 16 June to Gabriel Cramer [210] in Geneva, a friend and correspondent of all the parties, to say that he thought the force between the Earth and the Moon did not depend only on their distance apart and he wondered if a magnetic force might be involved. Nevertheless he was reluctant to criticize Newton in public [211].

There was a further complication when Pierre Bouguer [212], who was to make his name in photometry, revised a prize essay that he had submitted to the Academy in 1734. The second edition of this work [213], for which Clairaut was the assessor appointed by the Academy, was published in 1748. Bouguer considered rays emanating from a spherical body. If the rays maintained their strength as they moved out then their increasing separation would lead to an inverse-square law, but if they became more feeble as they spread then the force would fall off more rapidly. He believed that Newton, and after him Keill and Freind, had argued that "an infinite number of phenomena which strike the eyes of naturalists" require an inverse-cube law, so he simply added this to produce $(ar^{-2} + br^{-3})$: "We cannot use any other expression, as soon as we embrace the principles of Mr Newton, fully understood." Bouguer suggested, without calculation, that his inverse-cube term might solve Clairaut's problem with the motion of the Moon. Clairaut also considered such a term, and in a letter to James Bradley even toyed with a series of inverse powers of the separation [214].

If Euler had his doubts about such proposals, Buffon [215] was outraged by this tampering with the inverse-square law and there was a rapid exchange of notes between him and Clairaut in the Memoirs of the Academy [216]. Clairaut was probably the better mathematician and Buffon did not try to refute him directly but resorted to metaphysical arguments. For him gravity was a single effect and so needed only a single algebraic term; each term in a series had to correspond to a *force réelle* or a *qualité physique*. If there were to be two terms, what was to determine the relative sizes of the coefficients? He clearly did not accept the common French view of Newtonian doctrine that it required the strength of the cohesive forces to be proportional to the product of the densities of the attracting bodies. Clairaut patiently rebutted Buffon's arguments; for him metaphysics was not the right weapon to bring to the field, and it must therefore have been particularly galling for him when, in his penultimate note, he had to admit that, after all, the inverse-square law sufficed. He and his colleagues had not taken their calculations to a high enough degree of approximation; once this was done the anomaly disappeared. His withdrawal, he wrote to Cramer, had caused "something of a scandal" [217].

This episode confirmed in Buffon's mind the conviction that the cohesive forces were also inverse square, and that the apparent change to higher inverse powers at short distances arose from the shapes of the particles. Only for spheres does the inverse-square law between the particles lead to the same law between larger bodies down to the point of contact; for cubes, cylinders, etc., the law would change. He attempted no calculations, however. His definitive statement on the subject is to be found bizarrely prefaced to a volume of his *Histoire naturelle* that deals with a range of animals from the giraffe to the hamster. There, in italics, he writes:

All matter is attracted to itself in inverse ratio of the squares of the distance, and this general law does not seem to vary in particulate attractions, except by reason of the shape of the constituent particles of each substance, since this shape enters as a factor into each distance. [218]

He argues that if we knew, for example, that the apparent law of attraction was inverse cubic then we should be able to reason backwards and deduce the shape of the particles. The chemist Guyton de Morveau [219] was a friend and follower of Buffon and shared his views on this subject. He attempted, as an example of the effect of shape, to calculate the force between two tetrahedra, each composed of an array of ten close-packed spheres. He had an unusual and, what was surely even then, a heterodox view of how to sum the interactions of the spheres. He supposes that the 'attraction' of one sphere for another at a certain separation is a, and so "since we know that the action is reciprocal, it follows that the two particles will be attracted one towards the other with a force $2a$". Each tetrahedron has one apical particle, three in the next layer, and six in the base. He assumes that when an apical particle interacts with three or six in the other tetrahedron the force is to be counted four or seven times, that is, as $(1 + 3)$ or $(1 + 6)$. When two trios interact the force is counted six times, or $(3 + 3)$, etc. In spite of his remark about reciprocity he sees the force as a property residing in each particle, and not as a mutual property of a pair of particles. His final numbers have little meaning, but his conclusion that two tetrahedra approaching tip-to-tip follow a different law from those approaching base-to-base is sound [220].

Buffon's view was shared later by the Dutch natural philosopher J.H. van Swinden [221], by the Swedish chemist Torbern Bergman [222], and the French *physicien* Antoine Libes [223], who ranked Buffon's contribution to the field as highly as Newton's [224]. The conviction that all the underlying forces were inverse square was strengthened by the discovery, first, that magnetic poles, and, later, that electric charges follow this law, although Buffon did not try to include electric forces in his original scheme of things. As long as the only magnets available were natural lodestones it had proved impossible to find the 'true' law of magnetism; Hauksbee, Jurin and van Musschenbroek all tried and failed. Artificial magnets became available from the middle of the century, thanks first to the efforts of Gowin Knight [225] and the astronomer John Michell [226], whose short *Treatise of artificial magnets* was published in 1750 [227]. He inferred, and others were then able to show, that the total force between two magnets was explicable as an inverse-square force between well-characterised poles, a result that was soon followed by the more important discovery of Cavendish and Coulomb that the same law held for electric charges [228].

Diderot [229] had also worried about the identity or otherwise of the gravitational and cohesive forces. In 1754 he wrote that "all phenomena, whether of weight, elasticity, attraction, magnetism, or electricity, are only different facets of the same affection". At the same time, he stated explicity what others had tacitly assumed, that the presence of a third body has no effect on the force between the first two [230]. That the attractive forces were a property of a pair of particles was so widely accepted that it comes as a surprise to find Guyton de Morveau dissenting. It was a view that was spelled out more clearly in the next century; Maxwell ascribes the first explicit statement to Gauss [231]. In 1874 Wilhelm Weber [232] set it out formally; single particles have only the properties of mass and permanence, pairs of particles have these properties plus those of mutual attraction and repulsion, and groups of three or more particles have no properties that are not found in the constituent pairs. This was put forward in the context of his attempt to interpret electrodynamics and magnetism in terms of an action-at-a-distance model, but it only put formally what many had assumed for more than a hundred years.

In an anonymous article in the *Journal de Trévoux* of 1761 Diderot discussed more fully the difficulty of deciding between one universal inverse-square law and a possible multiplicity of laws for cohesive forces. He came down in favour of the former, but his attempt to show that such a law would lead to strong attractions between close spheres is fallacious [233].

Kant [234], arguing on metaphysical grounds, decided that the forces between the parts of matter could be both attractive and repulsive. The former was Newton's inverse-square law and the latter, he believed, was inverse-cubic. To these he added that heat contributed an inverse first-power force of repulsion between contiguous parts, a notion that clearly derives from Newton's hypothesis for gases. Heat as a source of repulsion was an idea that became more formally established in the work of Laplace and his followers. Kant argued that the concept of impenetrability was an occult one that should be banned, and the assumption of repulsive forces was one that promised the chance of future explanations. He had little problem with the idea of action-at-a-distance [235].

D'Alembert distinguishes between a passive force of adhesion that acts only between the points of particles actually in contact, and an active force that pulls them together from a distance. He regards the second as the more important and notes that it would lead to the particles compressing a liquid "from the outside inwards"; at one point he seems to be coming close to what we now call Laplace's equation for the excess pressure within a drop [236]. He was one of the few writers on hydrodynamics who had anything useful to say on cohesion and capillarity. Euler had little interest, and although Daniel Bernoulli speaks of the mutual attraction of particles of mercury and its capillary depression, he was content to follow his uncle Jakob in ascribing the capillary rise of water to a lower density, and so a lower pressure, of 'aero-aetherial' particles within the tube than above the level surface of the water outside it [237]. He did, however, think of himself as a good Newtonian [202]. The macroscopic approach to hydrodynamics was based on the concept of *pressure* in a fluid, initially a scalar entity, which was to be subsumed into the wider concept of a tensorial *stress* in an elastic body. In this way the subjects could advance on firm foundations, but the lack of enquiry into the microscopic forces that underlay pressure and stress left many dissatisfied, and was to be the cause of much argument in the next century.

The chemist P.J. Macquer [238] was one of those who appeared to conflate the two phenomena of gravity and chemical attraction. In his *Dictionnaire de chymie* he wrote that "the causticity of a body is nothing but its dissolving power, or its disposition to combine with other bodies; and this disposition is nothing other than the attraction, which is one and the same thing as gravity." But in his article on 'Gravity' [*Pesanteur*] he showed that he interpreted this term widely, discussing many aspects of physical and chemical association. He wrote that: "The law that gravitation follows at small distances does not yet appear to have been well determined." [239] He was clearly willing to entertain laws other than the inverse-square.

Thus there was a range of broadly Newtonian views in the middle and second half of the 18th century. There were those who believed that cohesive forces were the same as gravitational and that they maintained their inverse-square character down to the smallest distances, any apparent departure from this law being ascribed to the non-spherical shape of the attracting particles. There were those who believed that cohesive forces were gravitational but thought that the inverse-square law changed into something steeper at short distances, and there were those who thought that the two forces were distinct. Some of the last class thought that they might be related to electric or magnetic forces. Only the first two classes necessarily believed that the strength of the forces was proportional to the product of the masses or densities of the attracting bodies, but some of the last class implicitly assumed it.

These Newtonian philosophers thought it proper to try to find the mathematical form of the law of attraction but only a few went further and speculated on the causes or underlying mechanism. Such speculation was more in the tradition of Descartes than of Newton and had a long tradition in France. As early as 1680, Claude Perrault [240], an architect and physician, supposed that air was composed of three kinds of particles of decreasing size, the *partie grossière*, the *partie subtile*, and the extremely small *partie etherée*. Those of the second kind pressed against solids and were responsible for their adhesion since they could not insert themselves into the gap between two solid blocks until the gap was as wide as their diameter [241]. Boyle had had a similar triple set of particles, and Newton had, in his early days, speculated that cohesion might arise from the aether being less dense between particles [242], but only a few of the 18th century philosophers followed up these ideas.

Dortous de Mairan used supplementary particles to explain cohesion but in a different way from his predecessors. His *Dissertation sur la glace* had originally been submitted to the Academy of Bordeaux for a prize offered in 1716, but was revised substantially for a new edition in 1749 [243]. He, like many later writers in French, uses the term *parties intégrantes* for the massy corpuscles, and the *matière subtile*, or the *molécules* of this matter, for the smaller particles of an aether. He suggested that these moved more slowly between the massy particles than they did in free space, rather as a wind does in a forest than over the open ground outside it, and he thought that the cohesion was due to the lowering of the pressure consequent on this motion.

In 1758 the Academy of Sciences at Rouen offered a prize for an essay on the improvement of Geoffroy's scheme of chemical affinities, and for finding "a physico-mechanical scheme" that would explain them. There were at least four entries, one of which "deals fully with the first part of the question, but says nothing about the second; [its author] does not even believe, in spite of the approval of the Academy, that the discovery of the mechanism is possible." This author was J.P. de Limbourg, a physician from Theux, near Liège [244]. In his essay he stresses the analogy between chemical affinities and cohesive forces, cites with approval Newton, van Musschenbroek and Nollet, but leaves open the question of whether the forces are to be ascribed to "the sole decree of the Creator, or depend on some internal principle that which acts by pulling one [body] towards another, or if it is only the effect of heat, or of the air, or of some other more subtle matter." [245] His approach is in the tradition of Newtonian chemistry [246].

A second entry tackled boldly the question of mechanism with a proposal that was more sophisticated and apparently more convincing than that of Dortous de Mairan; it came from G.-L. Le Sage [247]. His father had been a French Protestant refugee in England early in the century who had moved to Geneva, where the younger Le Sage became a pupil of Cramer. His proposal was essentially the same as one put forward by Bouguer in 1734 and again in 1748 [248], and is close to that of the young Newton in a letter to Boyle [242], but Le Sage was unaware of these. His proposal is remembered today as Le Sage's theory of gravitation but its first appearance was as a theory of cohesion. The Academy awarded prizes both to de Limbourg for his Newtonian chemistry and to Le Sage for his Cartesian physics.

In his essay [249] Le Sage envisages the particles of matter as hollow spheres with arrays of holes in their walls (Fig. 2.2) which, as in Dortous de Mairan's model, are subject to bombardment by a dense cloud of rapidly moving tiny bodies. These he calls *corpuscules ultramondains*, since they are not acted on by the gravitational

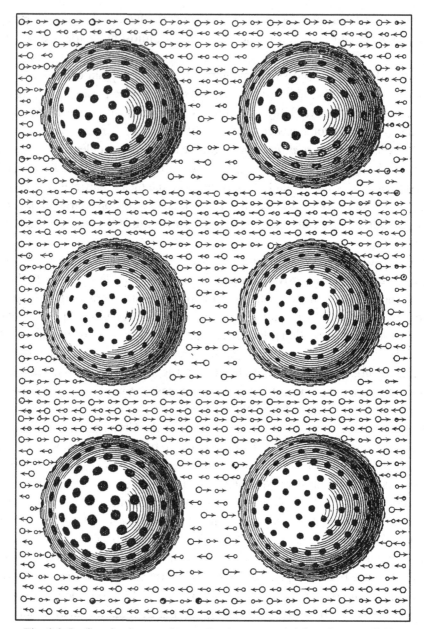

Fig. 2.2 Le Sage's picture of attraction between particles of matter [249].

field. The apparent attraction of the particles of matter is now a consequence of each of a nearby pair partially shielding the other from this bombardment. (This achievement of an attraction by means of forces that are themselves only repulsive is what we now call a depletion force, or an entropic attraction, and it has re-appeared in the second half of the 20th century; see Section 5.6.) By adjusting the size and number of holes he can adjust the reduction of density between each pair and so explain different intensities of attraction, so that, for example, the mutual attraction of two particles of water (*top*) or two of oil (*middle*) exceeds that of water and oil (*bottom*). Such a theory, like Dortous de Mairan's explanation of why mercury does not wet glass, does everything that was expected of it in the first half of the century; it was plausible, apparently consistent with known facts and with the laws of mechanics, but it was not falsifiable and had no predictive power.

Le Sage had known that the same model could be used to explain gravitation and this became the focus of a second exposition more than twenty years later [250]. The bombarding particles are now called *atomes gravifiques*. It was this second application of his theory that attracted attention in the next century, and it was as a theory of gravitation, not of cohesion, that it is now remembered [251]. A similar theory of cohesion was proposed in Russia in 1760 by M.V. Lomonosov [252], a pupil of Wolff, and Le Sage's theory was commended by De Luc [253], but both soon faded from the main stream of physical thought. Le Sage himself may have had second thoughts for in some philosophical notes published after his death he refers disparagingly to "the hypotheses of vortices, and all other hypotheses by which physics has been disfigured for a century"; they are but "chimeric fictions" [254].

In what seems to have been little more than a mathematical jeu d'esprit, Le Sage did propose an all-embracing law of attraction a few years after his Rouen essay [255]. He supposed that the attractive force varied not as the inverse square of the separation r, but as the inverse of the 'triangular numbers', $\frac{1}{2}r(r-1)$, where the diameter of the particles is taken to be unity. The apparent power by which the force changes with distance is $(1-2r)/(r-1)$; that is, the force changes as r^{-2} at infinite separation, so satisfying the gravitational law, as r^{-3} at $r=2$, as r^{-4} at $r=\frac{3}{2}$, etc., becoming infinitely steep as r approaches 1. The force changes sign at $r=1$ and is therefore repulsive at shorter distances. It was ingenious, not without faults in our eyes [256], but neither he nor anyone else seems to have taken it seriously at the time.

Many French clerics remained perceptive critics of Newtonian doctrine which they associated with Voltaire and the Enlightenment. Their wrath fell particularly on Pierre Sigorgne, a professor at the Collège de France until his dismissal in 1749 for criticising the King [257]. An uncritical Newtonian, he added the cohesive force to the gravitational, to give again the "general law of attraction" as $(ar^{-2} + br^{-3})$, and in his book of 1747 he noted, as Newton had done, that there was then an infinite force on a particle in contact with a sphere [258]. He has a long chapter on 'Capillary tubes: how this effect comes from attraction'. He is familiar with what has already been done, and explains Jurin's observation that the rise of the water is determined only by the diameter of the tube at the height of the liquid surface by supposing that, in a tube of conical shape, the sloping walls contribute to the suspension of the liquid. He has a series of propositions on the relative strength of the forces between water, glass and mercury that seems to derive from the work of Keill and Freind. In his later years he took on himself the job of interpreting chemistry in a Newtonian fashion that was no longer fashionable [257].

Sigorgne's first critic was Giacinto Gerdil, a Savoyard, and a Barnabite priest, the professor of philosophy at Turin, who later became a cardinal [259]. He had already, in 1747, complained about those who ascribed to Newton the view that gravity was inherent to matter, and he had invoked microscopic vortices to explain capillary rise [260]. In 1754 he returned to the attack, criticizing Keill and Sigorgne, and making the powerful point that the infinite force on contact that followed from an inverse-cubic (or higher) law was incompatible with the fact that bodies can be pulled apart [261]. This point was seized upon in an anonymous and neutral review of Gerdil's book in the *Journal des Sçavans* [262]; it was one that the Newtonians were unable to counter and so generally ignored.

Gerdil's most original contribution was the set of experiments he made with mercury in metal tubes which are reported in a further dissertation [263]. He opens with a sentence that shows the importance of capillary rise in 18th century attempts to understand cohesion: "Nothing is more commonplace in the eyes of the vulgar than the phenomena of capillary tubes; nothing more astonishing in the eyes of a philosopher." He has tubes of gold, silver and tin, of which the gold have internal diameters of $\frac{1}{2}$ and $\frac{1}{3}$ ligne, and the others, by implication, are similar. He notes that the densities of the metals are in the increasing order of tin (less than 7), silver (11), mercury (14), and gold (18). He believed that Newtonian theory required that the strength of the attraction be proportional to the product of the densities of the materials. He argued therefore that mercury should rise in gold tubes but fall in those of silver and tin. His results were not so simple. Mercury at first fell in the gold tubes, but after a short time some of it became incorporated into the gold and then it showed a small rise. The silver and tin tubes behaved similarly. It is probable that there was a thin contaminating film on the inner surfaces of the tubes that prevented the immediate amalgamation of the solid metal with mercury, but it is easy to see why Gerdil thought that he had refuted Newtonian doctrine.

He makes other points, many of which were concerned with the rate of rise or fall of the liquids. He observes, for example, that water rises more slowly in a long tube than in a short one of the same diameter. We can see that it is more difficult to expel the air from a long tube and so account for his observation, but the assumption of a naive theory of attraction, and perhaps a memory of van Musschenbroek's results with long and short tubes, could lead one to expect the opposite result. His own attempts to explain capillarity centre on differences in the pressure of air or of some other subtle fluid inside and outside the tube, and on the internal friction [*frottement*] in mercury. He dismisses the fact that the same rise is found in a vacuum as in air by saying that even the best pumps cannot remove all the air. He carried his opposition to Newtonianism to great lengths and it may be that he was the author of two pseudonymous and fraudulent papers (in the names of Coultaud and Mercier) that alleged that the apparent weight of a body increased with its height above sea level. The fraud was unmasked by Le Sage who knew the area of the Alps where the experiments were supposed to have been made [264].

Sigorgne came under attack also from Aimé-Henri Paulian, a Jesuit who was professor of physics at the Collège d'Avignon [265]. It was his aim to establish peace between the Cartesians and the Newtonians and he did this by adopting a stance that was common on the Continent around the middle of the century; in celestial physics Descartes was mistaken and Newton was correct, but in the physics of everyday matter Newton's ascription of all phenomena to 'attractions' was wrong, and he listed Sigorne's 22 propositions without any attempt to endorse them [266]. His principal criticism depends, as with Gerdil, on the failure of the assumption that the attraction is proportional to the product of the densities; an assumption that fits the trio water, glass and mercury, but which fails with metal tubes. His own explanation of capillary rise is to ascribe it to small asperities on the inner walls of the tubes which can support the particles. He invokes also the viscosities of the liquids, but uses these more to explain the dynamics of the rise rather than its occurrence.

Gerdil's work was also cited with approval by Bonaventure Abat [267], a Franciscan friar in Marseille whose *Amusemens philosophiques* [268] contain a long and effective criticism of Newtonian attraction as an explanation of capillarity. Abat, like Sigorgne and Bouguer, believed that Newtonian theory required that the cohesive forces fell off as the inverse cube of the separation but his principal criticisms are independent of this gratuitously precise assumption. He divides liquids into two classes, humid and dry, noting that a given liquid can fall into one class or the other according to the nature of the solid with which it is in contact. Humid liquids wet the surface of the solid and rise in capillary tubes, dry liquids do not wet the walls and fall. (The word 'humid' had been used previously in this context by G.E. Hamberger in his *Elementa physices* of 1727 [269].) Abat finds that water falls in the fine quills from the wings of sea-birds and of a partridge, but after a few hours this fall is reversed and the water rises, a change that he says is quite inexplicable on any theory of attraction. Gerdil's experiments with mercury in gold tubes showed the same behaviour, presumably, although Abat does not make the point, because the gold is initially unwetted by the mercury but becomes wetted when amalgamation sets in [270]. Moreover if one invokes attraction to explain why two drops of a liquid coalesce on contact, how does one explain why two bubbles in a liquid behave in just the same way? [271]

The criticisms of Gerdil, Paulian and Abat have substance; they show the difficulties that a simple theory of attraction can lead to in the absence of a clear concept of surface tension and an understanding of how this tension arises from the attractive forces. Their works were critical rather than constructive, for they made no systematic attempts to develop an alternative theory. Even the Newtonians were content for many years to leave their explanations in the qualitative form of Hauksbee and Jurin. Clairaut was the first to try to give the theory an adequate mathematical form. His book of 1743, *Théorie de la figure de la Terre*, deals mainly with the recently controversial subject of the flattening of the Earth at the poles, and so with the hydrostatics of sea level at different latitudes. Into this he inserts, rather incongruously, a chapter 'On the rise and fall of a liquid in a capillary tube' [272]. He writes:

In this research I shall consider the particles of fluid as perfectly smooth and infinitesimally small by comparison with the diameter of the tube. I shall suppose the material of the tube to be perfectly homogeneous and the surface perfectly smooth. Moreover I shall use the same function of the distance to express the attraction of the material of the tube as the attraction of the particles of the fluid, distinguishing these attractions only by their coefficients or intensities. . . . Finally, I suppose that the function of the distance that expresses the law of attraction, both of the glass and the water, be given, and that it has been established.

After this precise and promising start his analysis quickly goes astray. He assumes that all the liquid in a tube of, say, one-twentieth of an inch in diameter is within the attractive range of the glass walls, an assumption tentatively made by Hauksbee in spite of the evidence of his own experiments. Clairaut's analysis has other faults: an arbitrary choice of the points in the liquid where these forces act, and a neglect of inconvenient terms. He is unable to show that the rise is inversely proportional to the diameter of the tube, and he does not ask why the rise in a tube is the same as that between parallel plates at a separation equal to the radius of the tube, an omission for which he was later criticised by Laplace. His principal result which he arrives at by a route that seems to be adjusted to lead him to the answer he wanted, is that liquids rise only if the attraction of their particles by the glass is half or more of that between the particles of the liquid. This result is correct, on his premises, and a simple route to it, in the spirit of the later work of Young, is given in the Appendix to this chapter.

A few years later another attempt at a theory of the behaviour of liquids in contact with solids arose from work on the properties of water in bulk. János-Andràs Segner was a Hungarian who was professor of mathematics and physics at Göttingen from 1735 to 1755, when he moved to Halle [273]. Around 1750 he invented the improved water-wheel or turbine that now bears his name. This invention led to a long correspondence with Euler who worked out the theory of the device, and in one of these letters Segner touched on the loosely related topic of the shape of liquid drops [274]. On 23 April 1751 he had been enrolled into the newly-founded Royal Society of Göttingen and he celebrated his election with a long paper in the first volume of its Proceedings on the shape of a sessile drop of liquid, that is, of one resting on a flat surface [275]. He knew of Clairaut's work only by repute and says that he had been unable to get a copy of his book. He introduced for the first time the explicit notion of a surface tension but unfortunately believed that it acted only if the shape of the surface departed from circular, when he thought that the hypothetical filaments in the surface would be extended. Thus for a sessile drop, whose shape is determined by the interplay of surface and gravitational (or bulk) forces, his filaments exert forces only in the vertical sections since, by symmetry, the horizontal sections are circular. His calculations of the tensions in these filaments follows from his own observations on drops of mercury and from van Musschenbroek's on water. He commits himself to no opinion on the range of the forces responsible for the tension. The idea of filaments in tension may have been derived from 's Gravesande's work; he cites only Clairaut and van Musschenbroek but probably knew of 's Gravesande's work also. His notion of a surface tension was a valuable one but it was flawed; it was to be another forty years before it was to be formulated more correctly by Monge and Young.

The lack of progress throughout the 18th century is shown by the way that the same topics were repeatedly brought forward, often in ignorance of what had gone before, and often with errors that had already been refuted. A striking illustration is that of the adhesion of two glass balls floating on the surface of water. Mariotte and, more particularly, 's Gravesande in 1720, had explained that this phenomenon was a secondary consequence of the distortion of the liquid surface and not a primary effect of attraction between the floating bodies. Nevertheless many still plumped wrongly for the naive explanation; they form an interesting list: Helsham in a posthumous book of 1739 [276], Hjortsberg in 1772 [277, 278], the first edition of *Encyclopaedia Britannica* in 1773 [104], and Atwood in his lectures at Cambridge in 1784 [279]. By the end of the century matters were improving; Godart [280] gets it right in 1779, as do Bennet [281] and Banks [282] in 1786, Monge [283] in an influential paper in 1789, an anonymous article in the *Philosophical Magazine* of 1802 [284], and Cavallo [285] in his popular exposition of physics in 1803 [286]. Banks is the only one who refers to 's Gravesande's work.

A second subject that became fashionable again in the later part of the century was what was usually called 'Dr Taylor's experiment', that is, Brook Taylor's measurement of the force needed to lift a floating strip of wood from the surface of water [45]. Père Bertier [287] revived interest in the experiment in about 1764, when he showed again that the excess force was proportional to the area of contact and independent of the mass of the floating body [288]. He was an Oratorian, and like many of the clergy, had little sympathy with Newtonian attraction, preferring to invoke an "invisible fluid" to explain the weak adhesion of slabs of marble in vacuo, arguing for the presence of such a fluid from the stronger adhesion in the presence of a tangible fluid such as air. He had earlier tried to measure the interaction between suspended needles of wood, iron, or paper, and other bodies brought near them [289]. Like the fictitious Coultaud and Mercier, he also claimed to have shown that weight of a body increased with its altitude [264]. Louis XV is said to have called this Cartesian *physicien* 'le père aux tourbillons', while Rousseau enjoyed his good humour in spite of his pedantry [290].

Taylor's experiment was repeated also by G.F. Cigna [291] in 1772. He was the professor of anatomy at Turin and, with the support of his confrère Lagrange [292], then in Berlin, he held that what he was measuring was the adherence caused by the pressure of the overlying air [293]. He confirmed this conclusion by repeating the experiment with a glass slide coated with grease, when he still found an apparent attraction although it was known that water and grease do not attract. Guyton de Morveau rebutted Cigna's conclusion by noting that different surfaces give different attractions, so the effect cannot be due solely to the atmospheric pressure; moreover, the effect, like the rise of liquids in tubes, persists in a vacuum [294]. E.-F. Dutour [295] and Père Bésile [296] took up the subject and by putting one liquid inside a narrow tube claimed to be able to measure the force of adhesion between two liquid surfaces. The most comprehensive single set of results was obtained by F.C. Achard in Berlin; he was later a pioneer of the sugar-beet industry [297]. He studied, often at more than one temperature, most combinations of 30 liquids and 20 solids. He emphasised the importance of keeping the plate truly horizontal, of removing all bubbles of air from below the plate, and of adding the last weights in small increments as the point of detachment is approached. He found that the adhesive force did not scale with the densities of the solid or liquid but must depend on the shapes and number of the points of contact between the constituent particles of each partner, and he tried to estimate these in terms of those of his standard pair, water and glass [298].

In October 1768 the astronomer J.J. Lalande [299], stung by the frequent opposition to attractive forces, made a passionate defence of capillary effects as a source of information about cohesion. He wrote:

It seems to me that we have here the considerable advantage of becoming well-informed about the general attraction of matter, a subject in dispute for too many years. Capillary tubes place in our hands a tangible clue to the generality of that law which is the key to physics, the greatest power in Nature, and the prime mover of the Universe. [299]

He dismisses the objections of Gerdil and his followers and the theories of Hauksbee and Jurin. His own view is not original, being essentially that of Clairaut, as he acknowledges, but his paper is of value for his explicit discussion of the range of the attractive forces. He wrote: "Some may think, perhaps, that if the sphere of attraction of the glass is very small, for example a quarter of a ligne, then it [the liquid] should ascend the tube only for a quarter of a ligne." This clearly was not the case and this range, about 0.1 mm, is probably already greater than most Newtonians would have chosen. Its size accounts for the physical (as distinct from the mathematical) flaws in Clairaut's work. Lalande knew of Hauksbee's experiment with tubes of different wall thicknesses but says, correctly, that this showed only that the forces were shorter in range than the thinnest wall used. If it were practicable to use a tube with a wall thickness of less than a quarter of a ligne, then a smaller rise might be seen. There is nothing wrong with Clairaut's and Lalande's reasoning on this point but they should not have assumed that the liquid in the centre of the tube was within the range of the forces from the glass since a rise is found in a tube of an internal diameter of 5 ligne or more. Lalande's paper marks, perhaps, the last flourish of the French Newtonian era in the treatment of capillarity. His sentiments on the importance of this "key to physics" were to be revived forty years later by Laplace who shared Newton's devotion to the attractive forces.

In the middle of the century there appeared two very different books on cohesion. The first, by Gowin Knight, is a long obscure exposition of his views on attraction and repulsion, in part Cartesian but mainly Newtonian. It is replete with Propositions, Corollaries, etc., and often seems to be a caricature of the *Principia* [300]. As a contribution to this field it is evidence only of a subject that is beyond its intellectual prime. His treatment of magnetism is of more value for here he had done some important original work.

The second book is altogether more serious. In 1758 the Jesuit priest Rudjer Bošković [301], or Roger Boscovich as his name is usually transcribed in English, had published in Vienna the first edition of his *Theoria philosophiae naturalis*. He was from Ragusa (Dubrovnik) but spent most of his life in Italy, Austria and France. He was not satisfied with the Vienna edition of his book and a second version, prepared under his supervision, was published at Venice in 1763; it is this edition that is now taken to be the authoritative source of his theory [302].

His declared aim was to reconcile Newtonian attractions at a distance with Leibniz's doctrine of continuity of cause and effect. To achieve this he postulated a force between the particles of matter that is a continuous function of their separation r. At the largest separations the force is attractive and varies as r^{-2}, that is, it is gravitational; at intermediate distances the force undergoes several oscillations from attraction to repulsion and back again as r diminishes. This range of r is where the force accounts for cohesion and related properties: "the alternation of the arcs, now repulsive, now attractive, represent[s] fermentations and evaporations of various kinds, as well as sudden conflagrations and explosions." [303]

He admits, however, that "There are indeed certain things that relate to the law of forces of which we are altogether ignorant, such as the number and distances of the intersections of the curve with the axis, the shape of the intervening arcs, and other things of that sort." [304] At short distances his curve becomes steeply repulsive and tends to a positive infinite value as r goes to zero. This feature is the most original aspect of his work (although Gowin Knight had had similar ideas); he dispensed with particles of rigid impenetrability and replaced them with massy points that repelled each other ever more strongly as their separations diminished. It is in this aspect that his work goes beyond that of Rowning, of which he probably knew nothing [305], since Rowning had retained hard central cores in his repelling particles. Boscovich emphasises that his system of particles can never form a hard body, there must always be some compressibility. He writes: "It is usual to add a third class of bodies [to soft and elastic ones], namely such as are called hard; and these never alter their shape at all; but these also, according to general opinion, never occur in Nature; still less can they exist in my theory." [306] This was an unsettled question at the time. In 1743 the Abbé Nollet had cited the 17th century experiments of the Accademia del Cimento which appeared to show that water was incompressible, but warned that the work was inconclusive and said that he thought that all bodies were compressible in some degree [307]. D'Alembert had, however, no qualms about taking the experiments at face value [308]. The matter was settled in 1762 when John Canton succeeded in measuring the coefficient of compressibility of water with what we can now see was remarkable accuracy [309]. The fact that both water and solids transmit sound at (presumably) finite speeds [310] is evidence of their compressibility, but when was this inference first drawn? It is in Brisson's *Dictionnaire* of 1781 but may have been noted earlier [311].

Boscovich takes the standard Newtonian stance on the question of the meaning of the forces:

The objection is frequently brought forward against mutual forces that they are some sort of mysterious qualities or that they necessitate action at a distance.... I will make just one remark, namely that is quite evident that these forces exist, that an idea of them can be easily formed, that their existence is demonstrated by direct reasoning, and that the manifold results that arise from them are a matter of continual ocular observation. [312]

Boscovich's theory, like Rumford's cannon-boring experiment, acquired a greater significance in the 19th century than it had for most of his contempories. He was not on the closest terms with many European mathematicians and philosophers. He was, for example, a contempory of Euler, and they were interested in many of the same problems, but in over 3000 letters to and from Euler, Boscovich receives only four passing mentions; many lesser men are more strongly represented [208]. He was a friend of Lalande but antagonised Lagrange, d'Alembert and Laplace. Lagrange, in a letter to d'Alembert, refers scathingly to 'la briga fratesca', the intrigue of monks, when discussing Boscovich [313]. The earliest to take a more positive view of Boscovich's theory were Joseph Priestley [314], the astronomer John Michell (Priestley's neighbour in Yorkshire), and the Scottish philosophers [315]. When David Brewster published the textbook of his fellow Scot, John Robison, who had died in 1805, the section on Boscovich's theory nominally occupied over a hundred pages of the first volume [316]. In the early 19th century the concept of Boscovichian particles was sometime used rhetorically in opposition to Daltonian atoms [317], and Kelvin used the noun 'Boscovichianism' as late as 1905 for the doctrine of an atom as a point source of force [318]. Today we are at ease with the idea of chemically indestructible atoms that are, nevertheless, the source of force fields, but in the 18th and early 19th centuries these were apparently opposing views. Either atoms were hard, inelastic, massy, and indestructible, or else they were a source of fields with, possibly, a massy point at their centres. The latter view fitted more with the prevailing field theories of matter and its interaction that were held by Faraday and many other British physicists in the 19th century.

In some unpublished papers read at Bath in 1780, William Herschel adapted Boscovich's model to the interaction of the particles of matter with those of light [319]. His particles had an inner zone of attraction in which the cohesive forces act; these are "in the inverse ratio of some very high power of the distance"; this zone is surrounded by one of repulsion, which governs the reflection of light, by one of attraction for refraction, then by another of repulsion for diffraction, and finally by the attractive zone in which gravity acts. He wisely did not try to include Newton's repulsion of gas particles into this scheme, and he insisted that he differed from Boscovich in requiring a small hard core in his particles.

The encyclopaedias that became a notable feature of the intellectual life of the 18th century reflect the changes of opinion with time and place. The French ones were the most important since the scientific articles in them were written by the leading *savants* of the day. The British and German were less influential until, in the early 19th century, the successive editions of *Encyclopaedia Britannica* attracted the leading physicists as contributors. Harris's *Lexicon* [16] of 1704 was followed by the *Cyclopaedia* of Ephraim Chambers in 1728 [89]. He was a convinced if not very perceptive Newtonian. A few years later a twenty-five year old Leipzig bookseller, J.H. Zedler [320] embarked on the first volume of his *Universal Lexicon* [321]. By an accident of the alphabet the articles on 'Attractio', 'Capillares tubi', and 'Cohaesio' all appeared in the early years. The first deals mainly with gravitation; the second is a long and even-handed description of the experimental work published up to 1733 and is agnostic on the cause of the attachment of water to glass. The third is more sympathetic to Newtonianism, quoting freely both from Query 31 and from Hamberger's *Elementa physices* [269], indeed, it is possible that

Hamberger was the author, although he is generally held to have been more of a Leibnizian.

The most influential work of the century was the great French encyclopaedia of Diderot, d'Alembert and their colleagues, the first volume of which was published in 1751. In his 'Discours préliminaire', d'Alembert summarised the position in France in the middle of the century by saying that Descartes who had previously had disciples without number was now reduced to apologists [322]. The original proposal for the work had been for a translation of Chambers's *Cyclopaedia*, and this influence is apparent particularly in the articles written by d'Alembert himself [323]. In that on 'Attraction' he shows a Newtonian bias by having 24 theorems on short-ranged attractions [324]. He is not willing to claim that they can explain all of chemistry, although he suggests that such an explanation is "less vague" than any alternative. On 'Cohesion' [325], he opens by running together the first two sentences of Chambers's article, so altering their meaning: "In all times the cause of cohesion has puzzled philosophers in all systems of physics". An apt summary of the 18th century, this phrase was still being used, without attribution, as late as 1800 [311]. In other articles, notably that on 'Capillaire' [326], it is clear that although he is wholly convinced of the correctness of the gravitational theory he has not quite the same confidence in the attractive forces of cohesion. Diderot was less of a Newtonian than d'Alembert, and the many articles on chemistry by G.-F. Venel of Montpellier are firmly non-Newtonian in tone [327].

The *Encyclopédie méthodique* which started to appear in 1784 was a revision and extension of the *Encyclopédie* of 1751. Its 'method' was the division of knowledge into its constituent areas, with groups of volumes on mathematics, on physics, chemistry, pharmacy, metallurgy, etc. Its lack of method is evident, however, in the repetition of articles. There are, for example, different articles on 'Attraction' in the mathematics, physics and chemistry volumes, and on 'Adhésion' in physics and chemistry. The mathematical article on 'Attraction' is by d'Alembert; it was published posthumously and is little changed from that of 1751 [328]. The physics article on 'Attraction Newtonienne' [329] appears to be by Monge, one the authors of this set of volumes. It describes 'Taylor's experiment' and Monge's work on the attraction of floating bodies. Both in this article and in the chemical volumes there is a distinction between adhesion and cohesion, which was not new [330], but which is here given more than usual emphasis. Adhesion is the sticking together of bodies brought together, as in Taylor's experiment, while cohesion is that which prevents the breaking into their parts of solid and liquid bodies. It is 'stronger' than adhesion.

The first volume of the chemistry series started to appear in 1786; it was written by Guyton de Morveau. His article on 'Adhérence, Adhésion' [331] came out in 1789. In it he describes at length his own and Achard's repetitions of Taylor's experiment and, as befits a chemist, he concentrates on those in which mercury is in contact with another metallic surface. He regrets that Achard had not chosen his pairs of substances with chemistry more in mind. Fourcroy pointed out that some of the precise figures for the strength of adhesion of mercury to other metals could be in error since amalgamation could have changed the weight of the disc [332]. Fourcroy was responsible for the second volume on chemistry in 1792, but the article on 'Attraction' is still by Guyton [333]. He is now in a position to recognize the importance of Coulomb's proof of the inverse-square law for electrical attraction and so is convinced that attractive forces really exist, but still confesses to the difficulty of envisaging a mechanism without "impulsion". Not for the first nor for the last time do we hear the dictum that a body cannot act where it is not. His long article on 'Affinité' has already been cited for his calculation of the force of attraction of two tetrahedra each composed of mutually gravitating particles. Later in the article he admits that there are difficulties in the assumption of a pure inverse-square law, but is uncertain what to suggest. He believes that adhesion and chemical affinity are closely related, notwithstanding the fact that there is strong adhesion between bodies such as water and glass which have no chemical affinity. (We have seen that Newton had not differentiated between physical and chemical attraction in his Query 31, and the position was little changed at the end of the century; the chemical aspects have been discussed more fully elsewhere [140, 142].) The stronger effect of cohesion is, for Guyton, a different phenomenon and may follow a different law. This distinction, with him as with others, remained a purely verbal one; nothing useful flowed from it. The chemical article on 'Cohésion' did not appear until 1805 when the subject was treated by Fourcroy who regarded cohesion as something to be overcome before chemical action could start [334]. 'Tubes capillaires' were reached in the physics series only in 1822, when the whole subject had been transformed by Laplace.

2.5 Conclusion

It is hard to discern any real progress after the work of Newton and his immediate heirs. Those who thought that adhesion and cohesion were the result of short-ranged forces of attraction between exceedingly small particles were to be proved correct, but they failed to find any convincing mechanism by which such forces brought about their most spectacular effect, capillary rise. Desmarest, a convinced Newtonian, summed it up in 1754: "It is not sufficient to say, in a vague way, that attraction is the cause of the suspension of water in capillary tubes; one must explain how the attraction acts, and there lies the difficulty." [335] Those who objected to the invocation of attractive forces acting at a distance made valid criticisms of the Newtonians' efforts, and often had a keener sense of the importance of the distinction between wetting and non-wetting systems. Their own explanations, when they offered them, were less convincing even than those they opposed. D'Alembert said of one of their efforts that "an explanation so vague condemns itself" [336].

It is not hard to find reasons for this failure to take the subject forward. Firstly, many natural philosophers put forward mechanically impossible schemes to explain cohesion, and the general understanding of mechanics was inadequate to cope with these deficiencies. It was well into the middle of the century before the distinction between the vector conservation of momentum and the more restricted scalar conservation of kinetic energy (to use the modern terms) was satisfactorily resolved [337]. Secondly, the modern abstraction of a perfectly hard but nevertheless elastic particle (the plaything of those who have used computers to simulate the dynamics of fluids for the last forty years) was held to be self-contradictory, since elasticity implied deformation and this implied parts that could move with respect to each other, and these hard atoms had no parts [338]. A hard but elastic body is a concept that can only be reached as the mathematical limit of a continuously varying or Boscovichian force. To us, taking such a limit is a natural step, but it is an interesting comment on the present neglect of metaphysics among the practitioners of 'normal' science that no one now using a model of hard elastic spheres would ask if it raised any formal problem. Thirdly, there was in the 18th century no use of a potential field from which the vector force could be derived by taking the gradient at each point. Such a field is not necessary for handling molecular dynamics but its use greatly simplifies the calculations. It was not known to those working on attractive forces. During the course of the century these and other deficiencies were made good, and 'Newtonian mechanics' was put into the form that we now associate with that phrase by the efforts of the Bernoullis, Euler, d'Alembert, Lagrange and Laplace.

At the conceptual level the most obvious gap in the thinking of the natural philosophers was the absence of a clear idea of surface tension. It seems so natural an idea to us, and one that follows from so many elementary observations, that it difficult for us to see why the idea moved forward so slowly from 's Gravesende's 'threads' to Segner's notion that there was a tension but only in surfaces of changing curvature. At the end of Monge's paper published in 1789 on the forces between floating bodies we find the first formulation of a surface tension that is "constant in all directions" [283]. He did not, however, exploit this idea in the way that Young did a few years later, although he is known to have been interested in capillarity since as early as 1783 [339].

We can see in retrospect how these theoretical deficiencies held the subject back, but there were other less direct hindrances. The most obvious is, perhaps, that the leading philosophers of the day had other, and in their view, more important things to do, and there were in the early part of the century few to do them [340]. Astronomy retained throughout the century and beyond its place as the most prestigious branch of applied mathematics. Electricity and magnetism were the rising physical subjects where new and spectacular experiments were pouring forth. The science of heat, hovering uncertainly between physics and chemistry, was a field in which there was great progress in establishing the basic facts; the distinction between heat and temperature was resolved quantitatively, scales of temperature were established, and specific and latent heats were recognised and measured. There was no agreement on the interpretation of this wealth of new work; there were one- and two-fluid theories of electricity, heat as a movement of particles fell out of fashion and heat as subtle fluid came in, and 'imponderable fluids' were to be found in many fields. Different opinions were held at all times [341].

One of the metaphysical debates of the early years was resolved in Newton's favour. Gravity was allowed to act at a distance and it was agreed not to pursue the unprofitable question of how it acted. Pragmatically this was the right way forward and for two hundred years physicists were content to accept action-at-a-distance as a *de facto* feature of gravitational forces. The proper but sterile worries of the Cartesians were of little interest to most physicists until the 20th century – Maxwell being one of the exceptions. By implication and analogy the same point of view came eventually to be accepted also for the interparticulate forces and, because of the short distances involved, it generally suffices even today to treat intermolecular forces as acting instantaneously at a distance. The first correction for the finite speed of propagation was not made until 1946 (see Section 5.4).

Newton and Freind had tried to bring chemistry within the purview of corpuscular physics, and their ideas were taken up by some French chemists in the second half of the 18th century, but this was not to be the way forward for many years to come. Chemistry had first to establish itself as a reputable and independent science. Boerhaave in Leiden and Cullen in Glasgow [342] were both good Newtonians in that they believed in a corpuscular structure of matter and they put experiment and deduction from it before metaphysical systems, but both were adamant that chemistry was an autonomous branch of science. Even Peter Shaw, Boerhaave's Newtonian translator, came to this view in his own writings [343]. This stance was justified when Lavoisier and Dalton, in their different ways, put chemistry on the path it was to follow so successfully in the 19th century. Both were interested in physical problems but their chemistry owed nothing to Newton. A few chemists, such as Guyton de Morveau and lesser known men such as Hjortsberg and Sigorgne, kept alive the link between adhesion and chemical affinity. Guyton's views influenced Berthollet and Laplace, and, through them, Gay-Lussac and Dulong, who were to contribute experimental and theoretical work that we recognise as physical chemistry, but this branch did not establish itself as a strong and continuing field in the first half of the 19th century. It was eclipsed particularly by the rise of organic chemistry. This failure is linked to what has been called the fall of Laplacian physics, which will be discussed in the next chapter.

So the study of cohesion failed to prosper in the 18th century under the internal difficulties of its own subject matter and the external competition of other more exciting branches. "Everything has its fashions, even philosophy has its own", wrote Réaumur in 1749 [344], and cohesion became an unfashionable subject for many of the leading figures of the day. Euler, the most productive mathematician of his time, is an extreme example. Only one of his 234 *Letters to a German princess* is on cohesive attraction and he dismisses it with the words:

Were there a single case in the world, in which two bodies attracted each other, while the intermediate space was not filled with subtle matter [*matière subtile*], the reality of attraction might very well be admitted; but as no such case exists, we have, consequently, reason to doubt, nay, even to reject it. [345]

A contrast, and a fitting end to the 18th century, is provided by two papers, one in 1802 by John Leslie [346], who was soon to be elected Professor of Natural Philosophy at Edinburgh, and a more important one in 1804 by Thomas Young [347], until recently at the Royal Institution in London. Leslie opens with a robust defence of action-at-a-distance, noting that Laplace had recently "proved" that gravity acts instantaneously and riding roughshod over the metaphysical squeamishness of those who had difficulty with this idea [348]. He treats interparticle force in the same way as Boscovich, adding that it "is indifferent whether we consider the elementary portions of matter as points, atoms, particles or molecules. Their magnitude, if they have any, never enters into the estimate." He laments that much of the work in this field has "been left to the culture of a secondary order of men", and then proceeds to give his own explanation of capillary rise. He insists that it can only be lateral forces between the particles of glass and water that are responsible for the vertical rise, and then tries to explain this paradox by emphasising the spreading of water on a glass plate, whatever its orientation, as a consequence of the force on the particles in the layers of water not immediately next to the wall and their consequent movement to places where they can be in positions closer to the glass. He thus comes nearer than his contemporaries to using the concept of potential energy. He makes no firm statement on the range of the forces but his mechanism seems to require one that is comparable with the radius of the tube. In this he shows no advance on Hauksbee, Clairaut and Lalande, but he is able to produce a plausible argument for the rise in a tube being equal to that between plates at a separation equal to that of the radius of the tube, and he correctly explains Jurin's results with tubes of variable diameter by noting that the pressure depends only on the height and not on any other dimension. But without a clear idea of surface tension he could go no further.

Two years later, on 20 December 1804, Thomas Young read a paper to the Royal Society in which he brought together in a masterly way the ideas that lay behind the work of Clairaut, Monge and Leslie [349]. He criticises Segner's notion of a tension only in surfaces of variable curvature and recognises that Monge had said that there was a tension whatever the shape of the surface. He couples this idea with the assertion that there is a fixed angle of contact between any given pair of liquid and solid, an assertion which he describes (probably correctly [350]) as "one observation, which appears to be new, and which is equally consistent with theory and with experiment". He uses these two facts to produce the first satisfactory phenomenological treatment of capillary rise. He writes:

It is well known, and it results immediately from the composition of forces, that where a line is equably distended, the force that it exerts, in a direction perpendicular to its own, is directly as its curvature; and the same is true of a surface of simple curvature; but where the curvature is double, each curvature has its appropriate effect, and the joint force must be as the sum of the curvatures in any two perpendicular directions. For this sum is equal, whatever pair of perpendicular directions may be employed, as is easily shown by calculating the versed sines of two equal arcs taken at right angles in the surface.

(The versed sine of an angle θ is $(1 - \cos \theta)$. This theorem had been proved by Euler [351].) If now he could have overcome his well-known aversion to using explicit algebraic expressions and equations he could have written this result in the form of the equation usually ascribed to Laplace [352], namely that the difference of pressure, Δp, across a surface of tension σ and principal curvatures R_1 and R_2, is

$$\Delta p = \sigma \left(R_1^{-1} + R_2^{-1} \right).$$

It follows that if the combined effect of gravity and a fixed angle of contact of the liquid with the solid wall produce a curved surface, then the pressure in the liquid under this curved surface must be lower or higher than that in the liquid at a point remote from the wall. The liquid will therefore rise or fall until the difference of hydrostatic pressure compensates for this surface-tension-induced difference. The change of height is proportional to the curvature. He writes that "the curvature must be every where as the ordinate [i.e. height]; and where it has double curvature, the sum of the curvatures in different directions must be as the ordinate."

These two results, first Young's assertion of constancy of the angle of contact, and secondly, the Young–Laplace equation for the difference of pressure across a curved surface in tension, are what we need in principle to solve all the problems of capillarity. Some of the more obvious are tackled by Young in the rest of his paper. His exposition is, however, "unduly concise and obscure", as even his friendly editor and biographer is compelled to admit [353], or, in part, faulty, as a more hostile critic claims [354]. He was, as another biographer puts it, a mathematician "of an older school" [355]. Nevertheless he makes a fair attempt at treating the rise

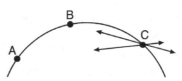

Fig. 2.3 The forces exerted by particles A and B on particle C, at the surface of a drop, according to the ideas of Thomas Young.

of water in tubes and between parallel plates, of the mutual attraction of floating bodies, of Newton's 'oil-of-oranges' experiment, of 'Dr Taylor's experiment', and of Clairaut's result that a liquid would neither rise nor fall if the liquid–solid attraction is half that of the liquid–liquid. He showed for the first time why the rise in a fine tube is inversely proportional to the diameter, and equal to that between parallel plates at a separation equal to the radius of the tube. The paper ends with what we now call Young's equation (see the Appendix to this chapter); if the surface of the liquid meets the solid wall at an angle θ, and if the tensions of the solid–gas, liquid–gas, and solid–liquid surfaces are σ_{sg}, σ_{lg} and σ_{sl}, then

$$\sigma_{sg} = \sigma_{sl} + \sigma_{lg} \cos\theta.$$

All these results depend on the existence of a surface tension. To what does he ascribe this tension? Here his account becomes less satisfactory. He assumes that there is a constant force of attraction that extends to an unspecified distance. He takes from Newton the idea that the pressure of a gas arises from a repulsive force that is "in simple inverse ratio of the distance of the particles from each other", but he ignores Newton's necessary restriction that such a force can act only between immediate neighbours if it is not to lead to wholly unacceptable physical consequences. With these two forces he explains the inward force on a particle on the convex surface of a liquid, as follows (Fig. 2.3). Particles A and B exert equal attractive forces on C, as shown by the arrows to the left. The repulsive force from B is stronger than that from A, as shown by the arrows to the right, and the net effect on C is a force acting towards the interior of the drop. This argument suggests that the unbalanced force and so the tension might vary with the curvature or even vanish at a planar surface, but since he does not discuss the effects of the particles in the interior of the drop there is no way of settling these points.

Two years after this paper appeared in the *Philosophical Transactions*, Young published the lectures that he had prepared earlier for the Royal Institution. He repeats his 1805 paper but also includes Lecture 49, 'On the essential properties of matter', and Lecture 50, 'On cohesion' [356]. These repeat many of the arguments of the earlier paper but he now notices Newton's restriction of the repulsive force to neighbouring particles, and realises that the small compressibility of water implies a much stronger repulsive force than one that varies as the reciprocal of the separation. Both here and in the articles he wrote for the *Encyclopaedia Britannica* [357] he is less dogmatic about the form of the forces than he had been earlier, but as late as 1821 he was still maintaining "that the mean sphere of action of the repulsive force is more extended than that of the cohesive", a conclusion which, he admits, is "contrary to the tendency of some other modes of viewing the subject" [358].

The 1807 *Lectures* include also an account of the modulus of elasticity of solids, but his definition differs from what we now call 'Young's modulus'. He throws no further light on the cohesion of solids other than to repeat an earlier assertion that "lateral forces" are called into play [359].

The papers of Leslie and, more particularly, of Young mark the limits of Newtonian or 18th century science in handling the problem of cohesion. Young's work was, and would have been seen by his predecessors as a triumphant success. The next advance came at once; it required Laplace's combination of physical insight and a mathematical grasp which grew from the resurgence of French mathematics at the turn of the century. This was guided first by the teaching of such men as Coulomb and Monge, and then by the new institutions for higher education in mathematics and engineering that were fostered in revolutionary France.

Appendix

Clairaut tried to show in 1743 that a liquid would neither rise nor fall in a capillary tube if the force of cohesion between two of its particles were twice that between one of them and one in the wall. His attempt can scarcely be called a proof; it is more a sketch of an argument that looks as if it were designed to lead to a result that he had already reached intuitively. A simple derivation, more in the spirit of Young than of Clairaut, runs as follows.

Let a_{ij} be a measure of the strength of the cohesive force between a particle of species i and one of species j. A measure of the affinity of a liquid of pure i for one of pure j might be the difference $(2a_{ij} - a_{ii} - a_{jj})$. If this is zero or positive the two will mix freely since the balance of forces is either neutral or favourable. If the difference is negative then complete mixing will not occur since there is a penalty to be paid on replacing ii and jj contacts by ij contacts. The more negative the difference the greater will be the tension σ at the boundary between the two liquids. Let us therefore put

$$\sigma_{ij} = k(a_{ii} + a_{jj} - 2a_{ij}) \geq 0, \tag{A.1}$$

where k is a constant that is assumed to be the same for all substances. Consider now three phases in equilibrium as shown in Fig. 2.4. Phase 1 is a solid with a vertical wall. If the point of contact of the fluid phases 2 and 3 with the solid is not to move, then by resolving the forces vertically (Young's argument) we have

$$\sigma_{13} - \sigma_{12} - \sigma_{23} \cos\theta = 0, \tag{A.2}$$

or

$$a_{33} - a_{22} - 2a_{13} + 2a_{12} - (a_{22} + a_{33} - 2a_{23})\cos\theta = 0. \tag{A.3}$$

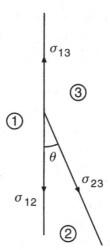

Fig. 2.4 Young's description of three phases, 1 to 3, meeting along a horizontal line (shown here in section as a point) at which the three surface tensions, σ_{12}, σ_{13} and σ_{23}, are in balance.

We now suppose that phase 2 is a liquid and phase 3 is air or a vacuum, so that $a_{13} = a_{23} = a_{33} = 0$, or

$$a_{12} = \tfrac{1}{2}a_{22}(1 + \cos\theta) = a_{22}\cos^2(\theta/2). \tag{A.4}$$

Thus if $a_{12} = 0$, the liquid has no attraction for the wall and $\theta = \pi$, or the wall is not wetted by the liquid, which would therefore fall in a capillary tube. Mercury in a glass tube comes quite close to this limit. If $a_{12} = \tfrac{1}{2}a_{22}$, then $\theta = \pi/2$, or the liquid surface is perpendicular to the wall and the liquid neither rises nor falls in a capillary tube (Clairaut's result). If $a_{12} = a_{22}$ then $\theta = 0$, or the wall is fully wetted by the liquid, since its particles have as strong an attraction for the wall as they have for each other. The liquid then rises in the tube. Water in a clean glass tube reaches this limit.

These results are plausible but they have no strict validity since eqn A.1 is only a crude representation of the relation between the forces and the surface tension.

One cannot resolve the forces horizontally; the force $\sigma_{23}\sin\theta$ has to be balanced by an elastic deformation of the solid that is outside the scope of this simple description.

Notes and references

1 I. Newton (1642–1727) I.B. Cohen, DSB, v. 10, pp. 42–103; R.S. Westfall, *Never at rest, a biography of Isaac Newton*, Cambridge, 1980; A.R. Hall, *Isaac Newton, adventurer in thought*, Oxford, 1992.
2 I. Newton, *Philosophiae naturalis principia mathematica*, London, 1687; 2nd edn, ed. R. Cotes, Cambridge, 1713; 3rd edn, ed. H. Pemberton, London, 1726; English translation of the 3rd edn by A. Motte, *The mathematical principles of natural philosophy*, 2 vols., London, 1729 (facsimile reprint, London, 1968). Quotations are from Motte's translation unless an earlier edition is needed when the variant readings in the edition of A. Koyré, I.B. Cohen and A. Whitman, 2 vols., Cambridge, 1972, have been used. Motte's translations have been checked against those of the 3rd edn by

I.B. Cohen and A. Whitman; I. Newton, *The Principia: Mathematical principles of natural philosophy*, Berkeley, CA, 1999.

3 I. Newton, *Opticks: or, a treatise on the reflexions, refractions, inflexions and colours of light*, London, 1704; Latin translation by S. Clarke, London, 1706; 2nd English edn, London, 1717; 3rd English edn, London, 1721; 4th English edn, London, 1730 (reprinted in 1979, New York, with a Preface by I.B. Cohen). Quotations are from the reprint of 1979.

4 K. Lasswitz, 'Der Verfall der 'kinetischen Atomistik' im siebzehnten Jahrhundert', *Ann. Physik* **153** (1874) 373–86; W.B. Hardy, 'Historical notes upon surface energy and forces of short range', *Nature* **109** (1922) 375–8; E.C. Millington, 'Studies in cohesion from Democritus to Laplace', *Lychnos* (1944–1945) 55–78; 'Theories of cohesion in the seventeenth century', *Ann. Sci.* **5** (1954) 253–69. These articles and others make clear the obvious fact that the subject did not start with Newton but I have chosen to take his great contribution as the point to open this history.

5 Newton, ref. 2, v. 1, Author's Preface (no pagination). This Preface from the first edition was retained in both the later ones.

6 Newton, ref. 2, v. 2, pp. 77–9, Book 2, Proposition 23. This passage is reprinted by S.G. Brush, *Kinetic theory*, 3 vols., Oxford, 1965–1972, v. 1, pp. 52–6.

7 A.R. Hall and M.B. Hall, *Unpublished scientific papers of Isaac Newton*, Cambridge, 1962, p. 307. The translation is the Halls'.

8 Hall and Hall, ref. 7, p. 333. This proposed Conclusion of 1687 contains much that was to appear twenty years later in the *Opticks*.

9 R. Hooke (1635–1702) R.S. Westfall, DSB, v. 6, pp. 481–8. Hooke's pamphlet of 1661 on capillarity, his first publication, is reproduced in facsimile in R.T. Gunther, *Early science in Oxford*, v. 10, pp. 1–50, printed privately, 1935.

10 R. Cotes (1682–1716) J.M. Dubbey, DSB, v. 3, pp. 430–3.

11 I.B. Cohen, *Introduction to Newton's 'Principia'*, Cambridge, 1971, p. 240.

12 *The Correspondence of Isaac Newton*, 7 vols., Cambridge, 1959–1977, ed. variously by H.W. Turnbull, J.F. Scott, A.R. Hall and L. Tilling, v. 5, p. 384.

13 Hall and Hall, ref. 7, pp. 348–64.

14 Newton, ref. 2, v. 1, pp. 293–4. See also J.S. Rowlinson, 'Attracting spheres: Some early attempts to study interparticle forces', *Physica A* **244** (1997) 329–33.

15 J. Harris (*c*.1666–1719) R.H. Kargon, DSB, v. 6, pp. 129–30.

16 J. Harris, *Lexicon technicum: or, an universal English dictionary of arts and sciences*, London, v. 1, 1704, v. 2, 1710; art. 'Attraction' in vs. 1 and 2. See also D. McKie, 'John Harris and his *Lexicon technicum*', *Endeavour* **4** (1945) 53–7; L.E. Bradshaw, 'John Harris's *Lexicon technicum*', pp. 107–21 of 'Notable encyclopedias of the seventeenth and eighteenth centuries', ed. F.A. Kafker, *Studies on Voltaire and the eighteenth century*, v. 194, 1981. Harris's distinction between the physical and the mathematical was made by Newton himself in the 8th Definition at the opening of the *Principia*.

17 Koyré *et al.*, ref. 2, v. 2, p. 576; R. Palter, 'Early measurements of magnetic force', *Isis* **63** (1972) 544–58; J.L. Heilbron, *Elements of early modern physics*, Berkeley, CA, 1982, pp. 79–89.

18 Newton, ref. 2, v. 2, p. 225, Book 3, Proposition 6, Cor. 5.

19 Palter, ref. 17.

20 B. Taylor (1685–1731) P.S. Jones, DSB, v. 13, pp. 265–8. Taylor was a Secretary of the Royal Society from 1714 to 1718.

21 F. Hauksbee (*c*.1666–1713) H. Guerlac, DSB, v. 6, pp. 169–75; 'Francis Hauksbee: expérimenteur au profit de Newton', *Arch. Int. d'Hist. Sci.* **16** (1963) 113–28,

reprinted in his *Essays and papers in the history of modern science*, Baltimore, MD, 1977, pp. 107–19; J.L. Heilbron, *Physics at the Royal Society during Newton's presidency*, Los Angeles, CA, 1983; M.B. Hall, *Promoting experimental learning: Experiment and the Royal Society, 1660–1727*, Cambridge, 1991, pp. 116–39.

22 F. Hauksbee, 'An account of experiments concerning the proportion of the power of the load-stone at different distances', *Phil. Trans. Roy. Soc.* **27** (1712) No. 335, 506–11; [B. Taylor] 'An account of an experiment made by Dr. Brook Taylor assisted by Mr. Hawkesbee, in order to discover the law of magnetical attraction', *ibid.* **29** (1715) No. 344, 294–5; 'Extract of a letter from Dr. Brook Taylor, F.R.S. to Sir Hans Sloan, dated 25. June, 1714. Giving an account of some experiments relating to magnetism', *ibid.* **31** (1721) No. 368, 204–8. The date 1714 is a misprint for 1712. The experiments in the last two papers were probably made before those in the first.

23 The fullest transcription of these notes from Cambridge University Library, Add. Ms. 3970 is in I.B. Cohen, 'Hypotheses in Newton's philosophy', *Physis* **8** (1966) 163–84. Other versions, differing in detail, are in J.E. McGuire, 'Force, active principles, and Newton's invisible realm', *Ambix* **15** (1968) 154–208; Westfall, ref. 1, pp. 521–2; and A.R. Hall, *All was light: An introduction to Newton's Opticks*, Oxford, 1993, p. 141.

24 A. Koyré, 'Les queries de l'Optique', *Arch. Int. d'Hist. Sci.* **13** (1960) 15–29; Hall, ref. 23, p. 141ff.

25 Newton, ref. 3, p. 375ff.

26 B.J.T. Dobbs, 'Newton's alchemy and his theory of matter', *Isis* **73** (1982) 511–28; *The Janus face of genius: the role of alchemy in Newton's thought*, Cambridge, 1991; J.E. McGuire and M. Tamny, *Certain philosophical questions: Newton's Trinity notebook*, Cambridge, 1983, pp. 275–95.

27 Newton, ref. 3, pp. 267–9.

28 D. Gregory (1659–1708) D.T. Whiteside, DSB, v. 5, pp. 520–2; W.G. Hiscock, *David Gregory, Isaac Newton and their circle: extracts from David Gregory's memoranda, 1677–1708*, printed privately, Oxford, 1937, pp. 29–30; A. Thackray, 'Matter in a nut-shell', *Ambix* **15** (1968) 29–53; *Atoms and powers: an essay on Newtonian matter-theory and the development of chemistry*, Cambridge, MA, 1970, p. 57; Hall, ref. 1, App. A.

29 I. Newton, 'De natura acidorum', printed, in part, in English in the Introduction to Harris, ref. 16, v. 2, reprinted by I.B. Cohen, ed., *Isaac Newton's papers and letters on natural philosophy and related documents*, 2nd edn, Cambridge, MA, 1978, pp. 255–8, and in full in both Latin and English in Newton's *Correspondence*, ref. 12, v. 3, pp. 205–14.

30 C. Huygens (1629–1695) H.J.M. Bos, DSB, v. 6, pp. 597–613; H.A.M. Snelders, 'Christiaan Huygens and the concept of matter', pp. 104–25 of *Studies on Christiaan Huygens*, ed. H.J.M. Bos, M.J.S. Rudwick, H.A.M. Snelders and R.P.W. Visser, Lisse, Netherlands, 1980. For Hooke's repetition of the experiment, see p. 108.

31 H. Guerlac, 'Newton's optical aether', *Notes Rec. Roy. Soc.* **22** (1967) 45–57; J.L. Hawes, 'Newton's revival of the aether hypothesis and the explanation of gravitational attraction', *ibid.* **23** (1968) 200–12.

32 F. Hauksbee, 'An experiment made at Gresham-College, shewing that the seemingly spontaneous ascention of water in small tubes open at both ends is the same *in vacuo* as in the open air', *Phil. Trans. Roy. Soc.* **25** (1706) No. 305, 2223–4. The dates on which his experiments were performed are recorded in the Journal Book of the Society.

33 Hardy, ref. 4.

34 F. Hauksbee, 'Several experiments touching the seeming spontaneous ascent of water', *Phil. Trans. Roy. Soc.* **26** (1709) No. 319, 258–65, 265–6.

35 J. Jurin (1684–1750) DNB; Pogg., v. 1, col. 1213–4 ; A.A. Rusnock, ed. *The correspondence of James Jurin (1684–1750): Physician and Secretary to the Royal Society*, Amsterdam, 1996, pp. 3–61.

36 J. Jurin, 'An account of some experiments shown before the Royal Society; with an enquiry into the cause of the ascent and suspension of water in capillary tubes', *Phil. Trans. Roy. Soc.* **30** (1718) No. 355, 739–47; 'An account of some new experiments, relating to the action of glass tubes upon water and quicksilver', *ibid.* **30** (1719) No. 363, 1083–96. He reprinted his papers on capillarity and other subjects, with some additional notes, in *Dissertationes physico-mathematicae*, London, 1732.

37 Guerlac, ref. 21, 1963.

38 F. Hauksbee, 'An account of an experiment touching the direction of a drop of oil of oranges, between two glass planes, towards any side of them that is nearest press'd together', *Phil. Trans. Roy. Soc.* **27** (1711) No. 332, 395–6; 'An account of an experiment, concerning the angle requir'd to suspend a drop of oyl of oranges, at certain stations, between two glass planes, placed in the form of a wedge', *ibid.* **27** (1712) No. 334, 473–4; 'A farther account of the ascending of drops of spirit of wine between two glass planes twenty inches and a half long; with a table of the distances from the touching ends, and the angles of elevation', *ibid.* **28** (1713) No. 337, 155–6. Hauksbee's experiments up to 1709 are collected in his book *Physico-mechanical experiments on various subjects*, London, 1709. A second edition of 1719 contains a Supplement recording those carried between 1709 and his death in 1713. The second edition was reprinted with an Introduction by D.H.D. Roller, New York, 1970. The books have additional material not in the original papers.

39 *De vi electrica*, Newton's *Correspondence*, ref. 12, v. 5, pp. 362–9. There is a modern version of the calculation in Heilbron, ref. 21, p. 69.

40 Hauksbee, ref. 38, 2nd edn, 1719, pp. 194–217.

41 B. Taylor, 'Concerning the ascent of water between two glass planes', *Phil. Trans. Roy. Soc.* **27** (1712) No. 336, 538.

42 F. Hauksbee, 'An account of an experiment [... some farther experiments] touching the ascent of water between two glass planes, in an hyperbolick figure', *Phil. Trans. Roy. Soc.* **27** (1712) No. 336, 539–40; **28** (1713) No. 337, 153–4.

43 H. Ditton (1675–1715) DNB.

44 H. Ditton, *The new law of fluids or, a discourse concerning the ascent of liquors, in exact geometrical figures, between two nearly contiguous surfaces; ...*, London, 1714.

45 Taylor, ref. 22 (1721). See also Thackray, ref. 28, p. 79.

46 Jurin, ref. 36 (1719). See also A. Quinn, 'Repulsive force in England, 1706–1744', *Hist. Stud. Phys. Sci.* **13** (1982) 109–28.

47 Newton, ref. 3, pp. 395–7.

48 *Four letters from Sir Isaac Newton to Doctor Bentley containing some arguments in proof of a deity*, London, 1756; reprinted in Newton's *Correspondence*, ref. 12, v. 3, pp. 233–41, 244–5, 253–6, and in facsimile by Cohen, ref. 29, pp. 279–312.

49 See M.B. Hesse, *Forces and fields: the concept of action at a distance in the history of physics*, London, 1961, p. 49.

50 J. Locke (1632–1704) M. Cranston, DSB, v. 8, pp. 436–40; J. Locke, *An essay concerning human understanding*, 4th edn, London, 1700, Book II, chap. 8, § 11. Kelvin was still using the expression in 1893 in his Presidential Address to the Royal Society, *Proc. Roy. Soc.* **54** (1893) 377–89, see 382.

51 Translated from the Latin in Newton's *Correspondence*, ref. 12, v. 4, pp. 265–8.

52 Guerlac, ref. 31; Hawes, ref. 31; Thackray, ref. 28, pp. 26–32; G. Buchdahl, 'Gravity and intelligibility: Newton to Kant', in *The methodological heritage of Newton*, eds.

R.E. Butts and J.W. Davis, Oxford, 1970, pp. 74–102; P. Heimann and J.E. McGuire, 'Newtonian forces and Lockean powers: Concepts of matter in eighteenth-century thought', *Hist. Stud. Phys. Sci.* **3** (1971) 233–306; Z. Bechler, 'Newton's law of forces which are inversely as the mass: a suggested interpretation of his later efforts to normalise a mechanistic model of optical dispersion', *Centaurus* **18** (1974) 184–222; E. McMullin, *Newton on matter and activity*, Notre Dame, Ind., 1978; Westfall, ref. 1, in which see the entry in the index under 'system of nature: aetherial hypotheses'; Hall, ref. 23, pp. 146ff.

53 G.W. Leibniz (1646–1716) J. Mittelstrass, E.J. Aiton and J.E. Hofman, DSB, v. 8, pp. 149–68.

54 E.J. Aiton, *The vortex theory of planetary motions*, London, 1972.

55 Hiscock, ref. 28.

56 *Biographia Britannica*, London, v. 3, 1750; v. 4, 1757. John Freind, v. 3, pp. 2024–44; David Gregory, v. 4, pp. 2365–72; John Keill, v. 4 pp. 2801–8; James Keill, v. 4, pp. 2809–11; John Desaguliers, 2nd edn, v. 5, 1793, pp. 120–5.

57 John Keill (1671–1721) D. Kubrin, DSB, v. 7, pp. 275–7.

58 F. Rosenberger, *Isaac Newton und seine physikalischen Principien*, Leipzig, 1895, Buch II, 'Die Bildung der Newton'schen Schule', esp. pp. 344ff.; R. Gunther, ref. 9, v. 11, 1937; A.V. Simcock, *The Ashmolean Museum and Oxford science, 1683–1983*, Oxford, 1984; L.S. Sutherland and L.G. Mitchell, eds., *The history of the University of Oxford*, v. 5, *The eighteenth century*, Oxford, 1986, see the chapters: L.S. Sutherland, 'The curriculum', pp. 469–91, A.G. MacGregor and A.J. Turner, 'The Ashmolean Museum', pp. 639–58, and G.L'E. Turner, 'The physical sciences', pp. 659–81; C. O'Meara, *Oxford chemistry, 1700–1770*, an unpublished dissertation for Part 2 of the Chemistry Finals examination at Oxford, 1987.

59 "Aug. 9, 1720. Sr. Is. Newton went to Oxford with Dr. Keil, he having not been there before.", *Family memoirs of the Rev. William Stukeley, M.D.*, Surtees Soc., Durham, 1880, v. 73, p. 61; also in similar words in *Memoirs of Sir Isaac Newton's life by William Stukeley, . . . 1752, . . .*, ed. A.H.White, London, 1936, p. 13.

60 A. Guerrini and J.R. Shackleford, 'John Keill's 'De operationum chymicarum ratione mechanica'', *Ambix* **36** (1989) 138–52.

61 J.T. Desaguliers (1683–1744) A.R. Hall, DSB, v. 4, pp. 43–6.

62 J.T. Desaguliers, *A course of experimental philosophy*, London, v. 1, 1734, v. 2, 1744, Preface to v. 1.

63 E.G. Ruestow, *Physics at seventeenth and eighteenth-century Leiden: Philosophy and the new science in the University*, The Hague, 1973; P.R. de Clercq, *The Leiden cabinet of physics*, Leiden, 1989.

64 John Keill, *Introductio ad veram physicam . . .*, Oxford, 1702; *An introduction to natural philosophy: or, philosophical lectures read in the University of Oxford, Anno Dom. 1700*, London, 1720.

65 John Keill, 'In qua leges attractiones aliaque physices principia traduntur', *Phil. Trans. Roy. Soc.* **26** (1708) No. 315, 97–110. Nineteen of the thirty theorems were translated into English in v. 2 of Harris's *Lexicon*, ref. 13, art. 'Particle'. There is a complete translation of the theorems in *The Philosophical Transactions of the Royal Society . . . abridged*, ed. C. Hutton, G. Shaw and R. Pearson, London, 1809, v. 5, pp. 417–24, and a summary in J.R. Partington, *A history of chemistry*, London, v. 2, 1961, pp. 478–9.

66 R.E. Schofield, *Mechanism and materialism: British natural philosophy in an age of reason*, Princeton, NJ, 1970, chap. 3; 'The Counter-Reformation in eighteenth-century science – last phase', in *Perspectives in the history of science and technology*, ed.

D.H.D. Roller, Norman, OK, 1971, pp. 39–54, and comments by R.J. Morris, pp. 55–60, and R. Siegfried, pp. 61–6.

67 J. Freind (1675–1728) M.B. Hall, DSB, v. 5, pp. 156–7; *Philosophical Transactions . . . abridged*, ref. 65, v. 4, p. 423.

68 For Newton's and others' views on the divisibility of matter, see E.W. Strong, 'Newtonian explications of natural philosophy', *J. Hist. Ideas* **18** (1957) 49–93.

69 C.A. Coulomb (1736–1806) C.S. Gillmor, DSB, v. 3, pp. 439–47; *Coulomb and the evolution of physics and engineering in eighteenth-century France*, Princeton, NJ, 1971.

70 C.A. Coulomb, 'Recherches théoretiques et expérimentales sur la force de torsion, et sur l'elasticité des fils de métal. . . . Observations sur les loix de l'elasticité et de la cohérence', *Mém. Acad. Roy. Sci.* (1784) 229–69. This memoir is discussed by Gillmor, ref. 69, 1971, pp. 150–62, and by C.A. Truesdell, 'The rational mechanics of flexible or elastic bodies, 1638–1788', in *Leonhardi Euleri opera omnia*, Leipzig, Berlin, etc., 1911 onwards, 2nd Series, v. 11, part 2, Zürich, 1960, pp. 396–401, 405–8.

71 James Keill (1673–1719) F.M. Valadez, DSB, v. 7, pp. 274–5; James Keill, *An account of animal secretion . . .* , London, 1708.

72 S. Hales (1677–1761) H. Guerlac, DSB, v. 6, pp. 35–48; D.G.C. Allan and R.E. Schofield, *Stephen Hales, scientist and philanthropist*, London, 1980.

73 S. Hales, *Vegetable staticks; or, an account of some statical experiments on the sap in vegetables . . .* , London, 1727; reprinted with an Introduction by M.A. Hoskin, London, 1969; H.L. Duhamel de Monceau, *La physique des arbres*, Paris, 1758, Part 1, pp. 74–8.

74 J. Freind, *Praelectiones chymicae, in quibus omnes fere operationes chymicae ad vera principia . . . redigunter . . .* , London, [1709]; *Chymical lectures: In which almost all the operations of chymistry are reduced to their true principles and the laws of Nature*, London, 1712; Partington, ref. 65, pp. 479–82.

75 Freind, ref. 74, 1712, p. 8.

76 Schofield, ref. 66, 1971, pp. 44–5.

77 Freind, ref. 74, 1712, pp. 95–102.

78 Freind, ref. 74, 1712, p. 147.

79 Freind, ref. 74, 1712, p. 149.

80 *Remarks and collections of Thomas Hearne*, ed. C.E. Doble and others, Oxford, 11 vols., 1885–1921, v. 1, pp. 88–90, 122–3, entries for 21 November and 10 December 1705.

81 Dobbs, ref. 26, 1991, pp. 193–4; P. Casini, 'Newton: the classical Scholia', *Hist. Sci.* **22** (1984) 1–58.

82 J. Gascoigne, *Cambridge in the age of the Enlightenment*, Cambridge, 1989, pp. 68, 142–5.

83 S. Clarke (1675–1729) J.M. Rodney, DSB, v. 3, pp. 294–7.

84 R. Smith (1689–1768) E.W. Morse, DSB, v. 12, pp. 477–8.

85 R. Smith, *A compleat system of opticks*, 2 vols., Cambridge, 1738; see v. 1, p. 89.

86 W. Whiston (1667–1752) J. Roger, DSB, v. 14, pp. 295–6; J.E. Force, *William Whiston, honest Newtonian*, Cambridge, 1985.

87 W.J.H. Andrewes, ed., *The quest for longitude*, Cambridge, MA, 1996, pp. 116, 128, 142–4.

88 I.B. Cohen, *Franklin and Newton*, Philadelphia, PA, 1956, pp. 243–61, esp. pp. 255–7; L. Stewart, *The rise of public science: rhetoric, technology, and natural philosophy in Newtonian Britain, 1660–1750*, Cambridge, 1992.

89 E. Chambers (*c*.1680–1740) DNB. E. Chambers, *Cyclopaedia: or, an universal dictionary of arts and sciences*, London, 1728, 2 vols.; L.E. Bradshaw, 'Chambers' *Cyclopaedia*', in Kafker, ref. 16, pp. 123–40.

90 H. Guerlac, 'The Continental reputation of Stephen Hales', *Arch. Int. d'Hist. Sci.* **4** (1951) 393–404.

91 J.T. Desaguliers, 'An account of a book entitl'd Vegetable Staticks . . . by Stephen Hales', *Phil. Trans. Roy. Soc.* **34** (1727) No. 398, 264–91; **35** (1727) No. 399, 323–31.

92 J.T. Desaguliers, 'An attempt to solve the phaenomenon of the rise of vapours, formation of clouds and descent of rain', *Phil. Trans. Roy. Soc.* **36** (1729) No. 407, 6–22.

93 J.T. Desaguliers, 'Some thoughts and conjectures concerning the cause of elasticity', *Phil. Trans. Roy. Soc.* **41** (1739) No. 454, 175–85.

94 H. Beighton (1686?–1743) DNB; Pogg., v. 1, col. 136; Stewart, ref. 88.

95 J.-A. Nollet (1700–1770) J.L. Heilbron, DSB, v. 10, pp. 145–8; J.-A. Nollet, *Leçons de physique expérimentale*, Paris, 1743–1748, v. 4, p. 73.

96 D. Bernoulli (1700–1782) H. Straub, DSB, v. 2, pp. 36–46; D. Bernoulli, *Hydrodynamica, sive, De viribus et motibus fluidorum commentarii*, Strasbourg, 1738; English translation by T. Carmody and H. Kobus, *Hydrodynamics by Daniel Bernoulli*, New York, 1968. See p. 16 of either edition. The work was substantially complete by 1733.

97 R. Watson (1737–1816) E.L. Scott, DSB, v. 14, pp. 191–2; R. Watson, 'Experiments and observations on various phaenomena attending the solution of salts', *Phil. Trans. Roy. Soc.* **60** (1770) 325–54.

98 H. Hamilton (1729–1805) DNB; Pogg., v. 1, col. 1009; H. Hamilton, 'A dissertation on the nature of evaporation and several phaenomena of air, water, and boiling liquors', *Phil. Trans. Roy. Soc.* **55** (1765) 146–81.

99 B. Franklin (1706–1790) I.B. Cohen, DSB, v. 5, pp. 129–39. See also Cohen, ref. 88; B. Franklin, 'Physical and meteorological observations, conjectures, and suppositions', *Phil. Trans. Roy. Soc.* **55** (1765) 182–92.

100 J. Rowning (1701?–1771) R.E. Schofield, DSB, v. 11, pp. 579–80.

101 J. Rowning, *A compendious system of natural philosophy* . . . , Part 1, 3rd edn, London, 1738, p. 12. The dates of publication of the four parts of this book are confusing, but Part 1 seems to have appeared in 1735 and Part 2 in 1736. Rowning's work is sometimes associated with that of Robert Greene (1678?–1730) whose *The principles of the philosophy of the expansive and contractive forces*, . . . , Cambridge, 1727, seems, from its title, to promise a similar treatment. Greene's work, however, lacks the clarity of Rowning's; his biographer in the DNB went so far as to describe it as "a monument of ill-digested and mis-applied learning".

102 Rowning, ref. 101, pp. 13–14.

103 Rowning, ref. 101, Part 2, 1st edn, pp. 5–6, see also pp. 56–72.

104 *Encyclopaedia Britannica*, London, 1773, art. 'Mechanics'.

105 Newton, ref. 2, v. 2, p. 392. This form of words was specifically endorsed by Newton in a letter to Cotes of 28 March 1713, see *Correspondence*, ref. 12, v. 5, pp. 396–9. See also D. Bertolini Meli, *Equivalence and priority: Newton versus Leibniz*, Oxford, 1993, chap. 9, pp. 191–218.

106 H. Pemberton (1694–1771) R.S. Westfall, DSB, v. 10, pp. 500–1.

107 H. Pemberton, *A view of Sir Isaac Newton's philosophy*, London, 1728, pp. 406–7. A similar view was expressed by C. Maclaurin, *An account of Sir Isaac Newton's philosophical discoveries*, London, 1748, pp. 108–11.

108 The adjective is Cohen's: I.B. Cohen, *The Newtonian revolution*, Cambridge, 1980, p. 131. Newton as a positivist was an identification proposed by Léon Bloch in 1907; for a discussion, see H. Metzger, *Attraction universelle et religion naturelle, chez quelques commentateurs anglais de Newton*, Paris, 1938, pp. 13–19; the chapter 'Léon Bloch et Hélène Metzger: 'La quête de la pensée newtonienne'', by M. Blay in *Études sur Hélène Metzger*, ed. G. Freudenthal, Leiden, 1990, pp. 67–84; and Dobbs, ref. 26, 1991, pp. 188, 211.

109 *Biographia Britannica*, ref. 56, art. 'James Keill'.

110 C. Wolff (1679–1754) G. Buchdahl, DSB, v. 14, pp. 482–4; T. Frängsmyr, 'The mathematical philosophy', chap. 2, pp. 27–44 of *The quantifying spirit in the 18th century*, ed. T. Frängsmyr, J.L. Heilbron and R.E. Rider, Berkeley, CA, 1990.

111 J. Le R. d'Alembert (1717–1783) J.M. Briggs, DSB, v. 1, pp. 110–17; T.L. Hankins, *Jean d'Alembert, science and the Enlightenment*, Oxford, 1970.

112 English translation of d'Alembert's 'Discours préliminaire' to the *Encyclopédie* of 1751 in R.N. Schwab, *Preliminary discourse to the Encyclopedia of Diderot*, Indianapolis, Ind., 1963, p. 88.

113 *Journal des Sçavans*, 2 August 1688, 128. Extracts from this review are given by A. Koyré, *Newtonian studies*, London, 1965, p. 115; by Cohen, ref. 11, pp. 156–7; ref. 29, pp. 428–9; and ref. 108, pp. 96–9. This review had been preceded by a favourable one in French by John Locke in the Amsterdam journal *Bibliothèque Universelle* and by a neutral one in the *Acta eruditorum*; see J.T. Axtell, 'Locke's review of the *Principia*', *Notes Rec. Roy. Soc.* **20** (1965) 152–61.

114 C. Huygens to N. Fatio de Duillier, Letter 2473 in *Oeuvres complètes de Christiaan Huygens*, The Hague, 1901, v. 9, pp. 190–1. The letter is quoted also by M.B. Hall, 'Huygens' scientific contacts with England', in Bos *et al.*, ref. 30, pp. 66–82, see p. 79.

115 B. le B. de Fontenelle (1657–1757) S. Delorme, DSB, v. 5, pp. 57–63; [Fontenelle] 'Éloge de M. Neuton', *Hist. Acad. Roy. Sci.* (1727) 151–72; English translation, *The elogium of Sir Isaac Newton*, London, 1728, p. 15, reprinted in facsimile by Cohen, ref. 29, pp. 444–74, and in an another translation in A.R. Hall, *Isaac Newton: eighteenth-century perspectives*, Oxford, 1999, pp. 59–74.

116 Koyré, ref. 113, Appendices A to E, pp. 115–72; H. Guerlac, *Newton on the Continent*, Ithaca, NY, 1981.

117 Johann Bernoulli (1667–1748) E.A. Fellmann and J.O. Fleckenstein, DSB, v. 2, pp. 51–55.

118 H.G. Alexander, ed., *The Leibniz–Clarke correspondence*, Manchester, 1956, pp. 66, 92, 115–18. See also F.E.L. Priestley, 'The Clarke–Leibniz controversy', in Butts and Davis, ref. 52, pp. 34–56, and A.R. Hall, *Philosophers at war: the quarrel between Newton and Leibniz*, Cambridge, 1980, pp. 159–67.

119 E. Mariotte (?–1684) M.S. Mahoney, DSB, v. 9, pp. 114–22.

120 J.J. Dortous de Mairan (1678–1771) S.C. Dostrovsky, DSB, v. 9, pp. 33–4; H. Guerlac, 'The Newtonianism of Dortous de Mairan', in Guerlac, ref. 21, 1977, pp. 479–90; E.McN. Hine, 'Dortous de Mairan, the 'Cartonian'', *Studies on Voltaire and the eighteenth century*, v. 266, pp. 163–79, 1989; 'Dortous de Mairan and eighteenth century 'Systems theory'', *Gesnerus* **52** (1995) 54–65.

121 Jean Truchet (1657–1729), known as Father Sebastien; see Newton's *Correspondence*, ref. 12, v. 7, pp. 111–18, and A.R. Hall, 'Newton in France; a new view', *Hist. Sci.* **13** (1975) 233–50.

122 Guerlac, ref. 21, 1977, pp. 78–163.

123 See the letter from Wolff to Leibniz of 14 December 1709, in *Briefwechsel zwischen Leibniz und Christian Wolf*, . . . , ed. C.I. Gerhardt, Halle, 1860, pp. 111–12, in which Wolff criticises Keill's views on the structure of matter and the motion of bodies in fluids under the action of gravity. Keill and Wolff disputed in *Acta eruditorum* about these topics and about the existence of a vacuum: 'Johannis Keill . . . Epistola ad clarissimum virum Christianum Wolfium . . .', *Acta eruditorum* (1710) 11–15; C.W., 'Responsio ad epistolam viri clarissimi Johannis Keill . . .', *ibid.*, 78–81. See also J. Edleston, *Correspondence of Sir Isaac Newton and Professor Cotes: including letters of other eminent men*, London, 1850, pp. 211–13.

124 [C. Wolff] 'Praelectiones chymicae, . . . a Johanne Freind . . .', *Acta eruditorum* (1710) 412–16. This review was reprinted, in Latin, in the English edition of Freind's lectures, ref. 74, pp. 161–71. For Wolff's authorship, see the letter from William Jones to Roger Cotes of 15 November 1711, in Newton's *Correspondence*, ref. 12, v. 5, pp. 204–5, and Heilbron, ref. 17, pp. 41–2, who notes Wolff's letters to Leibniz of 6 June and 16 July 1710, in Gerhardt, ref. 123, pp. 119–22.

125 J. Freind, 'Praelectionem chymicarum vindiciae, in quibus objectiones, in *Actis Lipsiensibus* anno 1710, mense septembri, contra vim materiae attractricem allatae, diluuntur', *Phil. Trans. Roy. Soc.* **27** (1711) No. 331, 330–42. This reply was translated in the English edition of his lectures, ref. 74, pp. 172–200, and attracted, in turn, further criticism from Wolff, 'Responsio ad imputationes Johannis Freindii in Transactionibus Anglicanis', *Acta eruditorum* (1713) 307–14, reprinted in *Opuscula omnia Actis eruditorum Lipsiensibus inserta* . . . , Venice, 1743, v. 5, pp. 160–6.

126 P. Brunet, *Introduction des théories de Newton en France au xviiie siècle*, Paris, 1931, v. 1 [all published]; Aiton, ref. 54, chap. 8, pp. 194–208; R.L. Walters and W.H. Barber, Introduction to *Eléments de la philosophie de Newton*, v. 15 of *Complete works of Voltaire*, Oxford, 1992.

127 E. Mariotte, *Traité du mouvement des eaux et des autres corps fluides*, new [3rd] edn, Paris, 1700, pp. 116–26. The first edition was published in 1686; the second, published in 1690, is reprinted in *Oeuvres de Mr Mariotte*, 2 vols., Leiden, 1717, v. 2, pp. 321–476. An English translation by J.T. Desaguliers was published with Newtonian glosses to counteract Mariotte's Cartesian explanations: *The motion of water, and other fluids, being a treatise of hydrostaticks*, London, 1728, see pp. 84–6, and for the glosses, pp. 279–90.

128 L. Carré (1663–1711) [Fontenelle] 'Éloge' in *Hist. Acad. Roy. Sci.* (1711) 102–7. There is a list of the Academy éloges in C.B. Paul, *Science and immortality: the éloges of the Paris Academy of Sciences (1699–1791)*, Berkeley, CA, 1980, pp. 111–26; see also the bibliography of biographies in R. Hahn, *The anatomy of a scientific institution: The Paris Academy of Sciences, 1666–1803*, Berkeley, CA, 1971, pp. 330–73. For Carré, see also Pogg., v. 1, col. 383–4, C. Hutton, *A mathematical and philosophical dictionary*, 2 vols., London, 1795, 1796, v. 1, pp. 245–6, and *Dictionnaire de biographie française*, 1956, v. 7, col. 1228–9.

129 É.-F. Geoffroy (1672–1731) W.A. Smeaton, DSB, v. 5, pp. 352–4; Hahn, ref. 128, p. 346.

130 L. Carré, 'Expériences sur les tuyaux capillaires', *Mém. Acad. Roy. Sci.* (1705) 241–54; see also [Fontenelle] 'Sur les tuyaux capillaires', *Hist. Acad. Roy. Sci.* (1705) 21–5.

131 C.-F. de C. Dufay (1698–1739) J.L. Heilbron, DSB, v. 4, pp. 214–17; Hahn, ref. 128, p. 344.

132 [Fontenelle] 'Sur l'ascension des liqueurs dans les tuyaux capillaires', *Hist. Acad. Roy. Sci.* (1724) 1–14.

133 F.P. du Petit (1644–1741) J. Dortous de Mairan, *Éloges des Academiciens de l'Academie Royale des Sciences morts dans les années 1741,1742 et 1743*, Paris, 1747, pp. 1–36. The original éloge is in *Hist. Acad. Roy. Sci.* (1741) 169–79; see also Pogg., v. 2, col. 415, and Hahn, ref. 128, p. 362.

134 F.P. du Petit, 'Nouvelle hypothèse par laquelle on explique l'élevation des liqueurs dans les tuyaux capillaires, et l'abaissement du mercure dans les mêmes tuyaux plongés dans ces liquides', *Mém. Acad. Roy. Sci.* (1724) 94–107; see also Fontenelle, ref. 132.

135 N. Desmarest (1725–1815) K.L. Taylor, DSB, v. 4, pp. 70–3; N. Desmarest, 'Discours historique et raisonné', the preface to F. Hauksbee, *Expériences physico-méchaniques sur différens sujets*, 2 vols., Paris, 1754, v. 1, pp. cxliv–v, and also v. 2, pp. 165–233. This translation was made by François de Brément (1713–1742) and edited by Nicholas Desmarest, who is now remembered as a geologist. His notes and commentary are longer than Hauksbee's text. He includes an éloge for de Brément by Dortous de Mairan (v. 1, pp. viii–xx) and in the second volume (pp. 165–306) an 'Histoire critique' of theories of capillarity up to 1750. 'Innixion' is a word apparently invented by Hauksbee.

136 Quoted by R.S. Westfall, *The construction of modern science*, New York, 1971, p. 115, from a manuscript of Robert Boyle. Hesse, ref. 49, p. 99, says that these modern criteria for the usefulness of theories were an innovation of the second half of the 17th century, but they were not widely accepted for another hundred years.

137 Desmarest, ref. 135, v. 1, p. xlii.

138 É.-F. Geoffroy, 'Table des differents rapports observés en chimie entre differentes substances', *Mém. Acad. Roy. Sci.* (1718) 202–12. J.L. Gay-Lussac reprinted this table in his review, 'Considérations sur les forces chimiques', *Ann. Chim. Phys.* **70** (1839) 407–34. There is an English translation in *Science in context* **9** (1996) 313–20.

139 É.-F. Geoffroy, *A treatise of the fossil, vegetable, and animal substances, that are made use of in physick*, trans. G. Douglas, London, 1736, pp. 10–11.

140 M.M. Pattison Muir, *A history of chemical theories and laws*, New York, 1907, chap. 14, 'Chemical affinity', pp. 379–430; A.M. Duncan, 'Some theoretical aspects of eighteenth-century tables of affinity', *Ann. Sci.* **18** (1962) 177–94, 217–32; 'The functions of affinity tables and Lavoisier's list of elements', *Ambix* **17** (1970) 28–42; W.A. Smeaton, 'E.F. Geoffroy was not a Newtonian chemist', *ibid.* **18** (1971) 212–14; M. Goupil, *Du flou au clair? Histoire de l'affinité chimique: de Cardan à Prigogine*, Paris, 1991, pp. 134–9; U. Klein, *Verbindung und Affinität. Die Grundlegung der neuzeitlichen Chemie an der Wende vom 17. zum 18. Jahrhundert*, Basel, 1994; 'É.F. Geoffroy's table of different 'rapports' observed between different chemical substances – a reinterpretation', *Ambix* **42** (1995) 79–100. Newton's early use of the term 'sociableness' is from his letter to Boyle of 28 February 1678/9, first printed by T. Birch in 1744 in his edition of Boyle's works, and reprinted in Newton's *Correspondence*, ref. 12, v. 2, pp. 288–96, see p. 292.

141 Douglas, ref. 139, p. xii. His words are taken almost directly from Fontenelle's 'Éloge de M. Geoffroy', *Hist. Acad. Roy. Sci.* (1731) 93–100, see 99. Fontenelle had used similar words in his Éloge on Newton.

142 A. Duncan, *Laws and order in eighteenth-century chemistry*, Oxford, 1996, pp. 94–102, 110–19.

143 P. L.M. de Maupertuis (1698–1759) B. Glass, DSB, v. 9, pp. 186–9; P. Brunet, *Maupertuis*, 2 vols., Paris, 1929; M.L. Dufrenoy, 'Maupertuis et le progrès scientifique', *Studies on Voltaire and the eighteenth century*, v. 25, pp. 519–87, 1963;

D. Beeson, 'Maupertuis: an intellectual biography', *ibid.*, v. 299, 1992; Hahn, ref. 128, pp. 358–9.

144 'Sur l'origine des animaux' is a section of his *Vénus physique* of 1745, reprinted in *Oeuvres de Mr. Maupertuis*, new edn, Lyon, 4 vols., 1756, v. 2, pp. 1–133, see p. 88. The first edition was published in Berlin in 1753.

145 See Brunet, ref. 126, pp. 84, 97.

146 *Journal de Trévoux* (1721) 823–57, reprinted in more correct form in pp. 1761–96.

147 *Journal des Sçavans* (1724) 29–33; J.-B. Senac (*c.*1693–1770) W.A. Smeaton, DSB, v. 12, pp. 302–3; Hahn, ref. 128, p. 366.

148 See, for example, P.J. Macquer to T. Bergman, 22 February 1768: "[it] is a work of his youth which he has never acknowledged", in *Torbern Bergman's foreign correspondence*, ed. G. Carlid and J. Nordström, Stockholm, 1965, v. 1, pp. 229–31; Brunet, ref. 126, p. 110; Thackray, 1970, ref. 28, pp. 94–5; Duncan, ref. 142, pp. 78–81. Partington, ref. 65, v. 3, pp. 58–9 gives an abstract of Senac's book.

149 F.M.A. de Voltaire (1694–1778) C.C. Gillispie, DSB, v. 14, pp. 82–5; I.O. Wade, *The intellectual development of Voltaire*, Princeton, NJ, 1969; R. Vaillot, *Avec Madame du Châtelet,1734–1749*, v. 2 of *Voltaire et son temps*, ed. R. Pomeau, Oxford, 1988.

150 G.-É. le T. de B. Marquise du Châtelet (1706–1749) R. Taton, DSB, v. 3, pp. 215–17; R. Vaillot, *Madame du Châtelet*, Paris, 1978; C. Iltis, 'Madame du Châtelet's metaphysics and mechanics', *Stud. Hist. Phil. Sci.* **8** (1977) 29–48; M. Terrall, 'Émilie du Châtelet and the gendering of science', *Hist. Sci.* **33** (1995) 283–310.

151 A.-C. Clairaut (1713–1765) J. Itard, DSB, v. 3, pp. 281–6; P. Brunet, 'La vie et l'oeuvre de Clairaut', *Rev. d'Hist. Sci.* **4** (1951) 13–40, 109–53; **5** (1952) 334–49; **6** (1953) 1–15.

152 W.H. Barber, 'Voltaire and Samuel Clarke', *Studies on Voltaire and the eighteenth century,* v. 179, pp. 47–61, 1979.

153 Wade, ref. 149, p. 253.

154 F.M.A. Voltaire, *Letters concerning the English nation*, London, 1733; M. de V . . . , *Lettres philosophiques*, Amsterdam, 1734; new edn by G. Lanson and A.M. Rousseau, 2 vols., Paris, 1964.

155 F.M.A. Voltaire, *The elements of Sir Isaac Newton's philosophy*, trans. J. Hanna, London, 1738, p. 89; the French edition was published in Amsterdam the same year. For a critical version of the 1741 edition, see Walters and Barber, ref. 126. The book received a favourable but not flattering review in the *Journal des Sçavans* (1738) 534–41. Vaillot, ref. 150, p. 152, ascribes the review to Mme du Châtelet. I thank Michael Hoare for a discussion of this book.

156 Beeson, ref. 143, pp. 62–88.

157 P.L. de Maupertuis, 'De figuris quas fluida rotata induere possunt, problemato duo; cum conjectura de stellis quae aliquando prodeunt vel deficiunt; & de annulo Saturni', *Phil. Trans. Roy. Soc.* **37** (1732) No. 422, 240–56; English version in the abridged edition, ref. 65, v. 7, pp. 519–28. It appeared in French in Chapters 7 and 8 of P.L.M. Maupertuis, *Discours sur les différentes figures des astres: . . . avec une exposition abrégée des systèmes de M. Descartes et de M. Newton*, Paris, 1732, reprinted in *Oeuvres*, ref. 144, v. 1, pp. 79–170, which, in turn, was translated into English in John Keill, *An examination of Dr. Burnet's Theory of the Earth . . . , To the whole is annexed a Dissertation on the different figures of celestial bodies . . . by Mons. de Maupertuis*, 2nd edn, Oxford, 1734. Maupertuis and Clairaut took part in the French expedition to Lapland in 1736 to determine the shape of the Earth, see Beeson, ref. 143, pp. 88–134.

158 Brunet, ref. 126, pp. 272–93.
159 M. Planck, *Scientific autobiography and other papers*, London, 1950, pp. 33–4.
160 P.L.M. Maupertuis, 'Sur les loix d'attraction', *Mém. Acad. Roy. Sci.* (1732) 343–62; Rowlinson, ref. 14.
161 Newton, ref. 2, v. 1, pp. 302–5.
162 [Fontenelle] 'Sur l'attraction newtonienne', *Hist. Acad. Roy. Sci.* (1732) 112–17.
163 J.S. König (1712–1757) E.A. Fellmann, DSB, v. 7, pp. 442–4.
164 R.L. Walters, 'Chemistry at Cirey', *Studies on Voltaire and the eighteenth century*, v. 58, pp. 1807–27, 1967; Iltis, ref. 150. Cirey (-sur-Blaise) is near Joinville, to the south-east of Paris.
165 G.-É. du Châtelet, *Institutions physiques*, new edn, Amsterdam, 1742, pp. 217–18; first published anonymously as *Institutions de physique* in 1740, see W.H. Barber, 'Mme du Châtelet and Leibnizianism; the genesis of the *Institutions de physique*', in *The age of enlightenment. Studies presented to Theodore Besterman*, Edinburgh, 1967, pp. 200–22; L.G. Janik, 'Searching for the metaphysics of science: the structure and composition of Madame Du Châtelet's Institutions de physique, 1737–1740', *Studies on Voltaire and the eighteenth century*, v. 201, pp. 85–113, 1982.
166 Du Châtelet, ref. 165, pp. 329–50. The phrase quoted was used by, and may be quoted from Maupertuis, ref. 160.
167 "Imaginations" was Voltaire's word used in his 'Preface historique', p. vi, to Madame la Marquise du Châtelet, *Principes mathématiques de la philosophie naturelle*, Paris, 1756, 1759. The edition of 1756 was 'l'édition préliminaire' and that of 1759, 'l'édition définitive', according to R. Taton, 'Madame du Châtelet, traductrice de Newton', *Arch. Int. d'Hist. Sci.* **22** (1969) 185–210. See also J.P. Zinsser, 'Translating Newton's *Principia*: The Marquise du Châtelet's revisions and additions for a French audience', *Notes Rec. Roy. Soc.* **55** (2001) 227–45.
168 T. Besterman, *Les lettres de la Marquise du Châtelet*, 2 vols., Geneva, 1958, see v. 1, p. 329, Letter 186; see also Walters and Barber, ref. 126. She already had copies of Newton's *Optique* and the physics texts of 's Gravesande and van Musschenbroek.
169 Schofield, ref. 66, chap. 7, pp. 134–56.
170 H. Boerhaave (1668–1738) G.A. Lindeboom, DSB, v. 2, pp. 224–8; *Herman Boerhaave, the man and his work*, London, 1968; H. Metzger, *Newton, Stahl, Boerhaave et la doctrine chimique*, Paris, 1930; Partington, ref. 65, v. 2, chap. 20. There are short biographies, by the editors, of some Dutch scientists in *A history of science in the Netherlands*, ed. K. van Berkel, A. van Helden and L. Palm, Leiden, 1999; for Boerhaave, see pp. 419–24.
171 Lindeboom, ref. 170, 1968, pp. 100, 268–70.
172 H. Boerhaave, *Elementa chemiae*, 2 vols., Leiden, 1732; 2nd corr. edn, Paris, 1733; *A new method of chemistry*, trans. P. Shaw, 2nd edn, London, 1741. Shaw and E. Chambers had originally translated an unauthorised and inaccurate version of Boerhaave's lectures, which they published in 1727. The first authorised English translation, *Elements of chemistry*, was by Timothy Dallow in 1735.
173 E. Cohen, *Herman Boerhaave en zijne beteeknis voor de chemie*, Ned. Chem. Ver., [Utrecht, 1918], p. 44.
174 P. van der Star, *Fahrenheit's letters to Leibniz and Boerhaave*, Leiden, 1983.
175 G.A. Lindeboom, ed., *Boerhaave's correspondence*, Leiden, 1964, v. 2, p. 15.
176 Metzger, ref. 170; F. Greenaway, 'Boerhaave's influence on some 18th century chemists', in *Boerhaave and his time*, ed. G.A. Lindeboom, Leiden, 1970, pp. 102–13.
177 *Gentleman's Magazine* **2** (1732) 1099–100, quoted by Schofield, ref. 66, p. 154.

178 Metzger, ref. 170, pp. 55ff.; R. Fox, *The caloric theory of gases from Lavoisier to Regnault*, Oxford, 1971.
179 W.J. 's Gravesande (1688–1742) A.R. Hall, DSB, v. 5, pp. 509–11; Van Berkel *et al.*, ref. 170, pp. 450–3; P. Brunet, *Les physiciens hollandais et la méthode expérimentale en France au xviiie siècle*, Paris, 1926; Ruestow, ref. 63, chap. 7; F.L.R. Sassen, 'The intellectual climate in Leiden in Boerhaave's time', in Lindeboom, ref. 176, pp. 1–16; J.N.S. Allamand, *Oeuvres philosophiques et mathématiques de Mr. G.J. 's Gravesande*, 2 vols., Amsterdam, 1774, v. 1, pp. ix–lix; for Allamand (1713–1787), see Pogg., v. 1, col. 31–32.
180 P. van Musschenbroek (1692–1761) D.J. Struik, DSB, v. 9, pp. 594–7; C. de Pater, 'Petrus van Musschenbroek (1692–1761), a Dutch Newtonian', *Janus* **64** (1977) 77–87; *Petrus van Musschenbroek (1692–1761), een Newtoniaans natuuronderzoeke*, Thesis, Utrecht, 1979; Van Berkel *et al.*, ref. 170, pp. 538–40.
181 Letter quoted by Brunet, ref. 126, pp. 117–18, and by Thackray, ref. 28, p. 83. Voltaire may not have succeeded in meeting Boerhaave, see Lindeboom, ref. 170, 1968, pp. 365–6.
182 Allamand, ref. 179, v. 2, pp. 311–28.
183 W.J. 's Gravesande, *Physices elementa mathematica, experimentis conformata; sive introductio ad philosophiam Newtonianam*, 2 vols., Leiden, 1720, 1721; *Mathematical elements of natural philosophy, confirmed by experiments; or, an introduction to Sir Isaac Newton's philosophy*, trans. J.T. Desaguliers, 2 vols., London, 1720, 1721.
184 'S Gravesande, ref. 183, English edn, v. 1, pp. 8–16.
185 'S Gravesande, ref. 183, English edn, v. 1, Book 1, chap. 26, 'Of the laws of elasticity'.
186 I. Todhunter and K. Pearson, *A history of the theory of elasticity and of the strength of materials*, Cambridge, 1886, v. 1, chap. 1; Truesdell, ref. 70.
187 'S Gravesande, ref. 183, English edn, v. 2, p. 20.
188 P. van Musschenbroek, *The elements of natural philosophy*, trans. J. Colson, 2 vols., London, 1744, v. 1, pp. vi–viii. Colson was the Lucasian Professor at Cambridge.
189 P. van Musschenbroek, *Physicae experimentales et geometricae ...*, Leiden, 1729, 'Dissertatio physica experimentalis de magnete', pp. 1–270; 'Dissertatio physica experimentalis de tubis capillaribus vitreis', pp. 271–353. For his magnetic experiments, see also 'De viribus magneticis', *Phil. Trans. Roy. Soc.* **33** (1722–1725) No. 390, 370–8, and for those on capillarity, de Pater, ref. 180, 1979, pp. 227–313.
190 Van Musschenbroek, ref. 188, v. 1, p. 221.
191 Van Musschenbroek, ref. 188, v. 1, pp. 18–36, 197–234, see pp. 203 and 204.
192 V. Boss, *Newton and Russia: The early influence, 1698–1796*, Cambridge, MA, 1972; R. Calinger, 'Leonhard Euler: the first St Petersburg years (1727–1741)', *Historia mathematica* **23** (1996) 121–66.
193 G.B. Bilfinger (1693–1750) Pogg., v. 1, col. 189–90; Aiton, ref. 54, pp. 155, 168–71; Boss, ref. 192, pp. 105–15; *Neue Deutsche Biographie*, 1955, v. 2, pp. 235–6; G.B. Bülffinger, 'De tubulis capillaribus, dissertatio experimentalis', *Comment. Acad. Sci. Imp. Petropol.* **2** (1727) 233–87.
194 J. Jurin, 'Disquisitiones physicae de tubulis capillaribus', *Comment. Acad. Sci. Imp. Petropol.* **3** (1728) 281–92.
195 J. Weitbrecht (1702–1747) Pogg., v. 2, col. 1291–2; Boss, ref. 192, p. 145; J. Weitbrecht, 'Tentamen theoriae, qua ascensus aquae in tubis capillaribus explicatur', *Comment. Acad. Sci. Imp. Petropol.* **8** (1736) 261–309; see also Desmarest, ref. 135, v. 2, pp. 233ff., and J.C. Fischer, *Geschichte der Physik*, Göttingen, 1803, v. 4, pp. 69–71.

196 P. Casini, 'Les débuts du newtonianisme en Italie, 1700–1740', *Dix-huit. Siècle* **10** (1978) 85–100; C. de Pater, 'The textbooks of 's Gravesande and van Musschenbroek in Italy', pp. 231–41 of *Italian scientists in the Low Countries in the xviith and xviiith centuries*, ed. C.S. Maffioli and L.C. Palm, Amsterdam, 1989.

197 S. Switzer (1682?–1745) DNB; S. Switzer, *An introduction to a general system of hydrostaticks and hydraulicks, philosophical and practical*, 2 vols., London, 1729, v. 1, pp. 50–2 and Plate 1.

198 A useful list of books and papers on cohesion and capillarity (and many other subjects) is in T. Young, *A course of lectures on natural philosophy and the mechanical arts*, 2 vols., London, 1807, v. 2, pp. 87–520.

199 C. Maclaurin (1698–1746) J.F. Scott, DSB, v. 8, pp. 609–12; 'An account of the life and writings of the author', in Maclaurin, ref. 107, pp. i–xx.

200 R. Helsham (1682?–1738) DNB; Pogg., v. 1, col. 1061; R. Helsham, *A course of lectures in natural philosophy*, Dublin, 1739. A 1999 reprint of the 1767 edition of this book by Trinity College, Dublin, is preceded by a life of the author.

201 D. Hume, *A treatise of human nature* [v. 1, London, 1739], quoted from the 2nd edition, ed. L.A. Selby-Bigge and P.H. Nidditch, Oxford, 1978, p. 13.

202 Letter of D. Bernoulli to Euler of 4 February 1744, quoted, in German, by W. Thomson, 'On the ultramundane corpuscles of Le Sage', *Proc. Roy. Soc. Edin.* **79** (1872) 577–89, and *Phil. Mag.* **45** (1873) 321–32, and, in English, by J.T. Merz, *A history of European thought in the nineteenth century*, Edinburgh, 1896, v. 1, pp. 351–2.

203 W. Whewell, *Astronomy and general physics considered with reference to natural theology*, London, 1833, p. 222.

204 P.M. Heimann, 'Ether and imponderables' in *Conceptions of ether. Studies in the history of ether theories, 1740–1900*, ed. G.N. Cantor and M.J.S. Hodge, Cambridge, 1981, pp. 61–83; Heimann and McGuire, ref. 52.

205 L. Euler (1707–1783) A.P. Youschkevitch, DSB, v. 4, pp. 467–84; L. Euler, 'Sur le mouvement des noeuds de la Lune et sur la variation de son inclinaison à l'Écliptique'. This paper was read to the Academy in Berlin on 5 October 1744, but I have not traced any formal publication. There is a report on it in *Hist. Acad. Roy. Sci. Berlin* (1745) 40–4, publ. 1746. The calculation of the Moon's orbit appears in his long correspondence with Christian Goldbach in St Petersburg on 20 September 1746, see P.-H. Fuss, *Correspondance mathématique et physique de quelque célèbres géomètres du xviii ième siècle*, St Petersburg, 1843, v. 1, pp. 397–400.

206 Brunet, ref. 126; L. Hanks, *Buffon avant 'L'Histoire naturelle'*, Paris, 1966, p. 115; Hankins, ref. 111, pp. 32–7; E.G. Forbes, 'Introduction' to *The Euler–Mayer correspondence (1751–1755), a new perspective on eighteenth-century advances in the lunar theory*, London, 1971; P. Chandler, 'Clairaut's critique of Newtonian attraction: Some insights into his philosophy of science', *Ann. Sci.* **32** (1975) 369–78; Heilbron, ref. 17, pp. 52–3; J. Roger, *Buffon, un philosophe au Jardin du Roi*, Paris, 1989, pp. 85–91, English trans., *Buffon, a life in natural history*, Ithaca, NY, 1997, pp. 53–8.

207 A.C. Clairaut, 'Du système du monde, dans les principes de la gravitation universelle', *Mém. Acad. Roy. Sci.* (1745) 329–64. The paper was read on 15 November 1747, but the volume was not published until 1749.

208 L. Euler, *Briefwechsel*, ed. A.P. Juskevic, V.I. Smirnov and W. Habicht, Basel, 1975, v. 1 of 4th Ser of *Opera omnia*, ref. 70; Abstracts of letters between Euler and d'Alembert and Clairaut. Letters 26–7, 418–23, see esp. 420 of 30 September 1747.

209 Euler, ref. 208, Letter 422, quoted by G. Maheu, 'Introduction à la publication des lettres de Bouguer à Euler', *Rev. d'Hist. Sci.* **19** (1966) 206–24, see 222.

210 G. Cramer (1704–1752) P.S. Jones, DSB, v. 3, pp. 459–62.
211 *Correspondance inédite de d'Alembert* . . . , an extract from *Bullettino di bibliografia e di storia delle scienze mathematiche e fisiche*, an 18 (1885) 507–70, 605–50, modern reprint, Bologna, n.d. Letter of 16 June 1748, "Je serois faché d'ailleurs d'attirer à Newton le coup de pied de l'âne, ..."
212 P. Bouguer (1698–1758) W.E.K. Middleton, DSB, v. 2, pp. 343–4; Aiton, ref. 54, pp. 219–27; Hahn, ref. 128, p. 336.
213 P. Bouguer, *Entretien sur la cause d'inclination des orbites des planètes*, 2nd edn, Paris, 1748, pp. 48–61; 1st edn, 1734. This work was reprinted in v. 2 of *Recueil des pièces qui ont remporté les prix de l'Académie Royale des Sciences 1721–1772*, Paris, various dates.
214 S.P. Rigaud, ed., *Miscellaneous works and correspondence of the Rev. James Bradley, D.D., F.R.S.*, Oxford, 1832, pp. 451–4, Letter from Clairaut, 19 August 1748.
215 G.-L. Leclerc, Comte de Buffon (1707–1788) J. Roger, DSB, v. 2, pp. 576–82, and ref. 206; Hahn, ref. 128, pp. 337–8; Hanks, ref. 206.
216 Clairaut's paper, ref. 207, was followed by those of Buffon, (1745) 493–500, 551–2, 580–3, and replies from Clairaut, (1745) 577–8, 578–80, 583–6.
217 P. Speziali, 'Une correspondance inédite entre Clairaut et Cramer', *Rev. d'Hist. Sci.* **8** (1955) 193–237, Letter of 26 July 1749, pp. 226–8.
218 G.-L. Leclerc, Comte de Buffon, *Histoire naturelle, générale et particulaire*, Paris, 1765, v. 13, pp. i–xx, 'De la Nature: Seconde vue', see p. xiii. See also Thackray, ref. 28, pp. 205ff.
219 L.B. Guyton de Morveau (1737–1816) W.A. Smeaton, DSB, v. 5, pp. 600–4; Hahn, ref. 128, pp. 347–8.
220 L.B. Guyton de Morveau, art. 'Affinité' in *Encyclopédie méthodique; Chimie*, Paris, 1786, v. 1, pp. 535–613, see pp. 546–7.
221 J.H. van Swinden (1746–1823) W.D. Hackmann, DSB, v. 13, 183–4; Heilbron, ref. 17, pp. 63–4.
222 T.O. Bergman (1735–1784) W.A. Smeaton, DSB, v. 2, pp. 4–8; T. Bergman, *A dissertation on elective attractions*, London, 1785, pp. 2–3. The original edition of 1775 was in Latin; Thomas Beddoes was the translator.
223 A. Libes (1752?–1832) Pogg., v. 1, col. 1449–50; *Grande Larousse*, Paris, 1866, v. 10, p. 475.
224 A. Libes, *Traité élémentaire de physique*, 4 vols., Paris, 1801, v. 2, pp. 1–40, see p. 3.
225 G. Knight (1713–1772) DNB; Pogg., v. 1, col. 1279–80; P. Fara, *Sympathetic attractions: Magnetic practices, beliefs, and symbolism in eighteenth-century England*, Princeton, NJ, 1996, esp. pp. 36–46; Heimann and McGuire, ref. 52.
226 J. Michell (1724?–1793) Z. Kopal, DSB, v. 9, pp. 370–1; C.L. Hardin, 'The scientific work of the Reverend John Michell', *Ann. Sci.* **22** (1966) 27–47.
227 J. Michell, *Treatise of artificial magnets*, Cambridge, 1750; trans. into French by Père Rivoire, see A.-H. Paulian, *Dictionnaire de physique*, Paris, 1761, v. 1, p. xix; Under 'Attraction', v. 1, pp. 171–80, Paulian discusses only gravitational attraction. Palter, ref. 17.
228 Heilbron, ref. 17, pp. 79–89.
229 D. Diderot (1713–1784) C.C. Gillispie, DSB, v. 4, pp. 84–90.
230 D. Diderot, *Pensées sur l'interpretation de la nature*, 1754, reprinted in *Oeuvres complètes de Diderot*, Paris, 1981, v. 9, pp. 3–111. The quotation is from B.L. Dixon, 'Diderot, philosopher of energy: the development of his concept of physical energy, 1745–1769', *Studies on Voltaire and the eighteenth century*, v. 255, p. 60, 1988.
231 J.C. Maxwell, art. 'Atom', *Encyclopaedia Britannica*, 9th edn, London, 1875.

232 W. Weber, 'Ueber des Aequivalent lebendiger Kräfte', *Ann. Physik. Jubelband* (1874) 199–213, see § 2. M.N. Wise claims, I think wrongly, that Laplace would not have subscribed to this codification; M.N. Wise, 'German concepts of force, energy and the electromagnetic ether, 1845–1880', chap. 9 of Cantor and Hodge, ref. 204.

233 [Diderot] 'Reflexions sur une difficulté proposée contre la manière dont les Newtoniens expliquent la cohésion des corps, et les autres phénomènes qui s'y rapportent', *Journal de Trévoux* (1761) 976–98; *Oeuvres,* ref. 230, v. 9, pp. 333–51. The attribution to Diderot was made at least as early as 1831.

234 I. Kant (1724–1804) J.W. Ellington, DSB, v. 7, pp. 224–35.

235 [I. Kant] *Kant's Prolegomena and Metaphysical foundations of natural science,* trans. E.B. Bax, London, 1883, pp. 171–98. The first German edition was published in 1786. For the metaphysics of Kant, see P.M. Harman, 'Kant: the metaphysical foundations of physics', chap. 4, pp. 56–80, of *Metaphysics and natural philosophy,* Brighton, 1982, and W. Clark, 'The death of metaphysics in Enlightened Prussia', chap. 13, pp. 423–73 of *The sciences in Enlightened Europe,* ed. W. Clark, J. Golinski and S. Schaffer, Chicago, 1999.

236 J. d'Alembert, *Traité de l'équilibre et du mouvement des fluides,* Paris, 1744, chap. 4, pp. 36–47.

237 Bernoulli, ref. 96, pp. 19, 20, 27 and 86 of the Latin edition, or pp. 20, 21, 32 and 96 of the English translation.

238 P.J. Macquer (1718–1784) W.A. Smeaton, DSB, v. 8, pp. 618–24; Hahn, ref. 128, p. 357.

239 P.J. Macquer, *Dictionnaire de chymie,* 2 vols., Paris, 1766, see v. 1, pp. 239–40 and v. 2, pp. 184–99. An English translation of the second edition was published in 1777.

240 C. Perrault (1613–1688) A.G. Keller, DSB, v. 10, pp. 519–21; Hahn, ref. 128, p. 362.

241 C. Perrault, *Essais de physique,* Paris, 1680. This work was reprinted as the first part of *Oeuvres diverses de physique et de méchanique de Mrs C. et P. Perrault,* Leiden, 1721, v. 1, and was probably more easily accessible in the 18th century in this form.

242 For Boyle, see the article 'Air' in Chambers, ref. 89, and for Newton, see his letter to Boyle of 28 February 1678/9, ref. 140.

243 J. Dortous de Mairan, *Dissertation sur la glace, ou explication physique de la formation de la glace, et de ses divers phénomènes,* Paris, 1749, chap. 4, pp. 22–9. See also Hine, ref. 120.

244 J.P. de Limbourg (1726–1811) Pogg., v. 1, col. 1462; *Biographie nationale . . . de Belgique,* Brussels, 1892, v. 12, col. 198–201; Goupil, ref. 140, pp. 139–43. The quotations are from the Academy's report on the competition which is prefaced to Limbourg's essay: *Dissertation . . . sur les affinités chymiques,* Liège, 1761. Liège was then an independent principality and bishopric.

245 De Limbourg, ref. 244, pp. 40–1.

246 Duncan, ref. 142, pp. 70–2.

247 G.L. Lesage (1724–1803) J.B. Gough, DSB, v. 8, pp. 259–60.

248 Bouguer, ref. 213, 'Des principes de physique qu'on pourrait substituer aux attractions', pp. 61–6.

249 G.L. Le Sage, *Essai de chymie méchanique,* [Rouen, 1758?, Geneva, 1761]. The Geneva edition is a revision; see his letter to Euler of 20 March 1761, Letter 2065 of Euler, ref. 208. Euler was critical of the 'ultramundane particles', see his letter to Le Sage of 16 April 1763, Letter 2068. The Geneva copy of the *Essai* in the Royal Society has no printed title-page, and so no date, but is preceded by hand-written pages by Le Sage in which he sets out some of the particulars of the Rouen prize and adds notes on his essay. He has added further notes in the margins of some pages.

The explanation of the Figure is in chap. 4, pp. 35–48. There is a summary of the essay in the *Journal des Sçavans*, 1762, 734–8.

250 G.L. Le Sage, 'Lucrèce Newtonien', *Nouv. Mém. Acad. Roy. Sci. Berlin* (1782) 404–32, and also as a pamphlet of 1784, the year of publication of the Berlin memoir. There is a partial translation in Thomson, ref. 202.

251 See Thomson, ref. 202 and Maxwell, ref. 231.

252 M.V. Lomonosov (1711–1765) B.M. Kedrov, DSB, v. 8, pp. 467–72; M.V. Lomonosov, *On the solidity and liquidity of bodies*, a pamphlet in Latin published by the Academy of Sciences in St Petersburg in 1760, and translated into English by H.M. Leicester, *Mikhail Vasil'evich Lomonosov on the corpuscular theory*, Cambridge, MA, 1970, pp. 233–46; Boss, ref. 192, pp. 165–99.

253 J.A. De Luc (or Deluc) (1727–1817) R.P. Beckinsale, DSB, v. 4, pp. 27–9; J.A. De Luc, 'Sur la cohésion et les affinités', *Observ. sur la Phys.* **42** (1793) 218–37.

254 G.L. Le Sage, 'Quelques opuscules relatifs à la méthode', in P. Prevost, *Essais de philosophie ou étude de l'esprit humain*, Geneva, 1805, v. 2, pp. 253–336, see p. 299.

255 G.L. Le Sage, 'Loi, qui comprend, malgré sa simplicité, toutes les attractions et repulsions, chacune entre des limites conformes aux phénomènes', *Journal des Sçavans* (1764) 230–4.

256 Rowlinson, ref. 14.

257 P. Sigorgne (1719–1809) M. Fichman, DSB, v. 12, pp. 429–30.

258 P. Sigorgne, *Institutions newtoniennes*, Paris, 1747, v. 2, chaps. 13 and 14.

259 G.S. Gerdil (1718–1802) Pogg., v. 1, col. 877–8; *Dictionnaire de biographie française*, 1982, v. 15, pp. 1282–3; *Dizionario biografico degli Italiani*, Rome, 1999, v. 53, pp. 391–7.

260 G.S. Gerdil, *De l'immatérialité de l'âme demontrée contre M. Locke*, Turin, 1747, reprinted in *Opere edite ed inedite*, Rome, 1806, v. 3, pp. 1–265, see pp. 102 and 239.

261 G.S. Gerdil, *Discours ou dissertation de l'incompatibilité de l'attraction et ses differentes loix avec les phénomènes*, Paris, 1754, reprinted in *Opere*, ref. 260, 1807, v. 5, pp. 181–256, see pp. 198–206.

262 *Journal des Sçavans* (1754) 515–25.

263 Gerdil, ref. 261, 1807, pp. 257–328, *Dissertation sur les tuyaux capillaires*.

264 J. Evans, 'Fraud and illusion in the anti-Newtonian rear guard: the Coultaud–Mercier affair and Bertier's experiments', *Isis* **87** (1996) 74–107.

265 A.-H. Paulian (1722–1801) Pogg., v. 2, col. 379.

266 [A.-H. Paulian] *Système général de philosophie extrait des ouvrages de Descartes et de Newton*, 4 vols., Avignon, 1769, see v. 2, pp. 102–11. The writer is identified as the author of the *Dictionnaire de physique*, ref. 227. See also art. 'Tube capillaire' in v. 3, pp. 411–24 of the *Dictionnaire*.

267 B. Abat. Little is known about this man. N.-L.-M. Desessarts, *Les siècles littéraires de la France*, Paris, 1800, v. 1, p. 4, and A. Cioranescu, *Bibliographie de la litterature française du dix-huitième siècle*, Paris 1969, v. 1, p. 215, list only the one work and give no dates of birth or death. The sources listed by H. and B. Dwyer, *Archives biographiques françaises*, London, 1993, add only that he was a member of the Academy at Barcelona and of the Royal Society of Montpellier. His work on capillarity is discussed by Fischer, ref. 195, and that on spherical mirrors by J.E. Montucla (in fact, J.J. de Lalande), *Histoire des mathématiques*, Paris, 1802, v. 3, pp. 552–6.

268 B. Abat, *Amusemens philosophiques sur diverses parties des sciences, et principalement de la physique et des mathématiques*, Amsterdam, 1763.

J.J. de Lalande says that it was printed at Marseille, see his 'Lettre sur les tubes capillaires', *Journal des Sçavans* (1768) 723–43, see 738. Abat's book is reviewed in this journal, (1764) 222–30; (1765) 333–41.

269 G.E. Hamberger (1697–1755) Pogg., v. 1, col. 1007–8; *Neue Deutsche Biographie*, Berlin, 1966, v. 7, pp. 579–80. G.E. Hamberger, *Elementa physices,* rev. edn, Jena, 1735, chap. 3, 'De cohaesione corporum', pp. 72–167, see p. 111. The first edition was published in 1727. There is an account of this book by R.W. Home in his Introduction to *Aepinus's Essay on the theory of electricity and magnetism*, trans. P.J. Connor, Princeton, NJ, 1979, see pp. 9–14.

270 Abat, ref. 268, 'Amusemen 11', pp. 497–557.

271 Abat, ref. 268, 'Amusemen 12', pp. 559–64.

272 A. Clairaut, *Théorie de la figure de la Terre*, Paris, 1743, pp. 105–28. This chapter is analysed by J.J. Bikerman, 'Capillarity before Laplace: Clairaut, Segner, Monge, Young', *Arch. Hist. Exact Sci.* **18** (1978) 103–22.

273 J.-A. Segner (1704–1777) A.P. Youschkevitch and A.T. Grigorian, DSB, v. 12, pp. 283–4. His hydrodynamic work and his correspondence with Euler is discussed by H.W. Kaiser, *Johann Andreas Segner: der 'Vater der Turbine'*, Leipzig, 1977, chap. 3, pp. 45–61. Euler's papers on this topic are in his *Opera omnia*, ref. 70, 2nd Series, v. 15.

274 Euler, ref. 208, Letter 2447, 12 December 1751.

275 I.A.S[egner]., 'De figuris superficierum fluidarum', *Comment. Soc. Reg. Sci. Gottingensis* **1** (1751) 301–72. Bikerman, ref. 272, has analysed the mathematics of this paper and his account is followed here.

276 Helsham, ref. 200, pp. 15–16.

277 Lars Hjortsberg (1727–1789) was 'Docent' in chemistry at Uppsala from 1753. I am indebted to Dr A. Lungren of Uppsala for the dates of his birth and death.

278 [L.] Hiotzeberg, 'Sur la cause de l'attraction des corps' (1772), in F. Rozier, *Introduction aux observations sur la physique, sur l'histoire naturelle et sur les arts*, 2 vols., Paris, 1777, v. 1, pp. 527–33, with an editorial comment on p. 534. These two volumes contain most of the papers in the first 18 issues of his important journal, *Observations sur la physique . . .* , which started afresh with volume 1 in 1773 and became the *Journal de physique* in 1794. Thomas Young and others cite it as *Rozier's Journal*. See D. McKie, 'The 'Observations' of the Abbé François Rozier (1734–1793)', *Ann. Sci.* **13** (1957) 73–89; J.E. McClellan, 'The scientific press in transition: Rozier's Journal and the scientific societies in the 1770s', *ibid.* **36** (1979) 425–49.

279 G. Atwood (1745–1807) E.M. Cole, DSB, v. 1, pp. 326–7; G. Atwood, *An analysis of a course of lectures on the principles of natural philosophy, read in the University of Cambridge*, London, 1784, p. 1. There is a similar error in the first experiment of his *Description of the experiments, intended to illustrate a course of lectures in natural philosophy*, London, 1776.

280 [-] Godard, 'Amusement philosophique sur quelques attractions et répulsions qui ne sont qu'apparentes', *Observ. sur la Phys.* **13** (1779) 473–80. He is described as 'Médecin à Vervier', and is almost certainly G.-L. Godart (1717 or 1721–1794), a doctor in Verviers, near Liège, and so a neighbour of de Limbourg, ref. 244. His other publications are medical; *Biographie nationale . . . de Belgique*, Brussels, 1883, v. 7, col. 831–3.

281 A. Bennet (1749–1799) Pogg., v. 1, col. 143–4; D.C. Witt, DNB, *Missing persons*, Oxford, 1993, p. 56; A. Bennet, 'Letter on attraction and repulsion', *Mem. Lit. Phil. Soc. Manchester* **3** (1790) 116–23, read 11 October 1786.

282 [J.] Banks, 'Remarks on the floating of cork balls in water', *Mem. Lit. Phil. Soc. Manchester* **3** (1790) 178–92, read 6 December 1786. The identification of this 'Mr Banks' with John Banks (1740–1805), the author of books on mechanics and mills, is confirmed by J.D. Reuss, *Das gelehrte England … 1770 bis 1790, Nachtrag und Fortsetzung, 1790 bis 1803*, Berlin, 1804, Part 1, pp. 48–9. John Banks was a student at Kendal Dissenting Academy and then a peripatetic lecturer in N.W. England from about 1775; see A.E. Musson and E. Robinson, *Science and technology in the Industrial Revolution*, Manchester, 1969, pp. 107–9; Thackray, ref. 28, p. 254, and sources cited by R.V. and P. J. Wallis, *Index of British mathematicians, Part III, 1701–1800*, Newcastle upon Tyne, 1993, p. 7.

283 G. Monge (1746–1818) R. Taton, DSB, v. 9, pp. 469–78; G. Monge, 'Mémoire sur quelques effets d'attraction ou de répulsion apparente entre les molécules de matière', *Mém. Acad. Roy. Sci.* (1787) 506–29, published in 1789.

284 Friend to physical enquiries, 'On the apparent attraction of floating bodies', *Phil. Mag.* **14** (1802) 287–8.

285 T. Cavallo (1749–1809) J.L. Heilbron, DSB, v. 3, pp. 153–4.

286 T. Cavallo, *The elements of natural or experimental philosophy*, 4 vols., London, 1803, see v. 2, chap. 5, pp. 116–49, esp. p. 120. His understanding of the cause of capillary rise had, however, not advanced beyond the point reached by Jurin.

287 J.-É.Bertier (1702–1783) The *Index biographique des membres et correspondents de l'Académie des Sciences*, Paris, 1954, gives his names as Étienne-Joseph. Pogg., v. 1, col. 168–9, reverses the order of the names and gives his date of birth as 1710. The *Grande Larousse* of 1866, has J.-É. but spells his surname 'Berthier'. The Royal Society, which uses both spellings, describes him on the certificate proposing his election in 1768 as Joseph Stephen. He is sometimes confused with G.-F. Berthier (1704–1782), the Jesuit who edited of the *Journal de Trévoux*. See also P. Costabel, 'L'Oratoire de France et ses collèges', chap. 3, pp. 66–100 of *Enseignement et diffusion des sciences en France au xviiie siècle*, ed. R. Taton, Paris, 1964.

288 J-É. Bertier, *Principes physiques, pour servir de suite aux Principes mathématiques de Newton*, 4 vols., Paris, 1764–1770, see v. 3, pp. 304–68.

289 Untitled note in *Hist. Acad. Roy. Sci.* (1751) 38–9.

290 J.J. Rousseau, *Confessions*, in *Oeuvres complètes*, Paris, 1959, v. 1, pp. 504–5.

291 G.F. Cigna (1734–1790) Pogg., v. 1, col. 445; *Dizionario biografico degli Italiani*, Rome, 1981, v. 25, pp. 479–82.

292 J.L. Lagrange (1736–1813) J. Itard, DSB, v. 7, pp. 559–73; Hahn, ref. 128, p. 351.

293 G.F. Cigna, 'Dissertation sur les diverses élévations du mercure dans les baromètres de différens diamètres', in Rozier, ref. 278, v. 2, pp. 462–73.

294 L.B. Guyton de Morveau, 'Sur l'attraction ou la répulsion de l'eau et des corps huileux, pour vérifier l'exactitude de la méthode par laquelle le Docteur Taylor estime la force d'adhésion des surfaces, et détermine l'action du verre sur le mercure des baromètres', *Observ. sur la Phys.* **1** (1773) 168–73, 460–71. See also Thackray, ref. 28, pp. 211–14.

295 E.-F. Dutour (or Du Tour) (1711–1784) Pogg., v. 1, col. 633; *Dictionnaire de biographie française*, Paris, 1970, v. 12, p. 927. E.-F. Dutour, 'Expériences sur les tubes capillaires', *Observ. sur la Phys.* **11** (1778) 127–37; **14** (1779) 216–24; 'Expériences relatives à l'adhésion', *ibid.* **15** (1779) 234–52; **16** (1780) 85–117; **19** (1782) 137–48, 287–98.

296 Père Bésile de l'Oratoire. I know nothing of this man. He is not in any of the sources listed by the Dwyers, ref. 267, nor apparently in any history of the Paris Oratory. [-] Bésile, 'Expériences relatives à la cohésion des liquides', *Observ. sur la Phys.* **28**

(1786) 171–87; **30** (1787) 125–302. There follow some anonymous papers, 'Épreuves relatives à l'adhésion', *ibid.* **29** (1786) 287–90, 339–46, whose author is identified only as 'M.'

297 F.C. Achard (1753–1821) Pogg., v. 1, col. 7; *Neue Deutsche Biographie*, Berlin, 1953, v. 1, pp. 27–8.

298 F.C. Achard, 'Mémoire sur la force avec laquelle les corps solides adherent aux fluides . . .', *Nouv. Mém. Acad. Roy. Sci. Berlin* (1776) 149–59; *Chymisch-physische Schriften*, Berlin, 1780, pp. 354–67 and ten Tables.

299 J.J. LeF. de Lalande (1732–1807) T.L. Hankins, DSB, v. 7, pp. 579–82; Hahn, ref. 128, p. 352; De Lalande, ref. 268, see p. 724.

300 G. Knight, *An attempt to demonstrate that all the phenomena in Nature may be explained by two simple active principles, attraction and repulsion; wherein the attractions of cohesion, gravity, and magnetism, are shewn to be one and the same, and all the phenomena of the latter are more particularly explained*, London, 1748. See also Duncan, ref. 142, pp. 84–5, and Fara, ref. 225.

301 R.J. Bošković (1711–1787) Z. Marković, DSB, v. 2, pp. 326–32; L.L. Whyte, ed., *Roger Joseph Boscovich, S.J., F.R.S., 1711–1787: studies of his life and work on the 250th anniversary of his birth*, London, 1961.

302 R.J. Boscovich, *Theoria philosophiae naturalis*, 2nd edn, Venice, 1763; English trans. by J.M. Child, *A theory of natural philosophy*, Chicago, 1922, which is the version cited here.

303 Boscovich, ref. 302, § 459, p. 325.

304 Boscovich, ref. 302, § 102, p. 95.

305 Rowning wrote in English which Boscovich did not then read. He later visited Britain for seven months and met Bradley, Michell and others; see Whyte, ref. 301 and Schofield, ref. 66, pp. 237, 242.

306 Boscovich, ref. 302, § 266, p. 205.

307 Nollet, ref. 95, v. 1, pp. 114–22.

308 J. d'Alembert, art. 'Compression' in *Encyclopédie ou dictionnaire raisonné des sciences, des arts et des métiers*, ed. D. Diderot and J. d'Alembert, Paris, 1751–1780, v. 3, 1753, pp. 775–6.

309 J. Canton (1718–1772) J.L. Heilbron, DSB, v. 3., pp. 51–2; J. Canton, 'Experiments to prove that water is not incompressible', *Phil. Trans. Roy. Soc.* **52** (1762) 640–3.

310 F. Hauksbee, 'An account of an experiment, touching the propagation of sound, passing from a sonorous body into the common air, in one direction only', *Phil. Trans. Roy. Soc.* **26** (1709) No. 321, 369–70; 'An account of an experiment touching the propagation of sound through water', *ibid.* 371–2.

311 M.-J. Brisson (1723–1806) R. Taton, DSB, v. 2, pp. 473–5; M.-J. Brisson, *Dictionnaire raisonné de physique*, 2 vols., Paris, 1781, art. 'Cohésion', v. 1, pp. 357–8, and art. 'Compressibilité', v. 1, p. 371. See also the 2nd edn in 6 vols., Paris, 1800, art. 'Cohésion', v. 2, pp. 210–6. This work was based on the *Encyclopédie* of Diderot and d'Alembert. The speed of sound in a solid is difficult to measure. According to Biot, there were some attempts in Britain and Denmark but the earliest that gave a quantitative result was an observation of E.F.F. Chladni, from the frequency of the longitudinal vibrations of rods, that in "certain solid bodies" the speed was 16 to17 times that in air. Biot's own results, from the study of pipes up to a kilometre long, was that in iron the speed was $10\frac{1}{2}$ times that in air; J.-B. Biot, 'Expériences sur la propagation du son à travers les corps solides et à travers l'air, dans les tuyaux très-alongés', *Mém. Phys. Chim. Soc. d'Arcueil* **2** (1809) 405–23.

312 Boscovich, ref. 302, § 101, p. 95.

313 *Oeuvres de Lagrange*, Paris, 1882, v. 13, pp. 274–81, see p. 278; see also, v. 14, 1892, pp. 66–8, Lagrange to Laplace.

314 J. Priestley (1733–1804) R.E. Schofield, DSB, v. 11, pp. 139–47.

315 J. Priestley, *The history and present state of discoveries relating to vision, light, and colours*, 2 vols., London, 1772, v. 1, pp. 390–3, and for a fuller account five years later: *Disquisitions relating to matter and spirit*, London, 1777, pp. 11–23; L.P. Williams, 'Boscovich and the British chemists', pp. 153–67, and R.E. Schofield, 'Boscovich and Priestley's theory of matter', pp. 168–72, in Whyte, ref. 301. Priestley maintained that Michell arrived independently of Boscovich at the idea of replacing the hard core of particles by a continuous repulsion. See also J. Golinski, *Science as public culture: Chemistry and the Enlightenment in Britain, 1760–1820*, Cambridge, 1992, chaps. 3 and 4; R. Olson, 'The reception of Boscovich's ideas in Scotland', *Isis* **60** (1969) 91–103; Heimann and McGuire, ref. 52.

316 J. Robison (1739–1805) H. Dorn, DSB, v. 11, pp. 495–8; J. Robison, *A system of mechanical philosophy*, 4 vols., Edinburgh, 1822, v. 1, pp. 267–368. (From p. 306 onwards the connection with Boscovich becomes tenuous.) This section is based on the article on Boscovich that Robison wrote for the *Supplement* of 1801 to the 3rd [1797] edn of *Encyclopaedia Britannica*.

317 See e.g. F.A.J.L. James, 'Reality or rhetoric? Boscovichianism in Britain: the cases of Davy, Herschel and Faraday' in *R.J. Boscovich, Vita e attività scientifica*, ed. P. Bursill-Hall, Rome, 1993, pp. 577–85.

318 Lord Kelvin, 'Plan of an atom to be capable of storing an electrion with enormous energy for radio-activity', *Phil. Mag.* **10** (1905) 695–8. For Kelvin's changing views on Boscovich, see D.B. Wilson, *Kelvin and Stokes: A comparative study in Victorian physics*, Bristol, 1987.

319 W. Herschel (1738–1822) M.A. Hoskin, DSB, v. 6, pp. 328–36; *The scientific papers of Sir William Herschel*, 2 vols., London 1912, 'Observations on Dr Priestley's desideratum: "What becomes of light?" ', v. 1, pp. lxv–lxxii, 'Additions to observations on Dr Priestley's desideratum etc.', pp. lxii–lxxiv, 'On the central powers of the properties of matter', pp. lxxv–lxxvii.

320 J.H. Zedler (1706–1763) *Allgemeine Deutsche Biographie*, Leipzig, 1898, v. 44, pp. 741–2.

321 J.H. Zedler, *Grosses vollständiges Universal Lexicon aller Wissenschaft und Künste* . . . , 64 vols., Halle and Leipzig, 1732–1750, with four supplementary volumes, 1751–1754; P. E. Carels and D. Flory, 'Johann Heinrich Zedler's *Universal lexicon*', in Kafker, ref. 16, pp. 165–96.

322 D'Alembert, ref. 112, p. 80.

323 P. Quintili, 'D'Alembert "traduit" Chambers: les articles de mécanique, de la *Cyclopaedia* à l'*Encyclopédie*', *Studies on Voltaire and the eighteenth century*, v. 347, pp. 685–7, 1996.

324 *Encyclopédie*, ref. 308, 1751, v. 1, pp. 846–56.

325 *Encyclopédie*, ref. 308, 1753, v. 3, pp. 605–7.

326 *Encyclopédie*, ref. 308, 1751, v. 2, pp. 627–9.

327 G.F. Venel (1723–1775) W.A. Smeaton, DSB, v. 13, pp. 602–4; Goupil, ref. 140, pp. 125–32.

328 *Encyclopédie méthodique: Mathématique*, Paris, 1784, v. 1.

329 *Encyclopédie méthodique: Dictionnaire de physique*, Paris, 1793, v. 1.

330 See e.g. d'Alembert, ref. 236 and A. Baumé [(1728–1804) E. McDonald, DSB, v. 1, p. 527], *Chymie expérimentale et raisonnée*, 3 vols., Paris, 1773, v. 1, pp. 23–30.

331 Guyton de Morveau, ref. 220, v. 1, pp. 466–90. He discusses his experiments in this field in a letter to Richard Kirwan of 30 December 1786; see E. Grison,

M. Sadoun-Goupil and P. Bret, *A scientific correspondence during the chemical revolution: Louis-Bernard Guyton de Morveau and Richard Kirwan, 1782–1802*, Berkeley, CA, 1994, pp. 157–8.

332 A.F. de Fourcroy (1755–1809) W.A. Smeaton, DSB, v. 5, pp. 89–93; Fourcroy's criticism is cited by Thomas Beddoes in his translation of Bergman's *Elective affinities*, ref. 222, p. 321.

333 Guyton de Morveau, ref. 220, *Chymie*, Paris, 1792, v. 2, pp. 448–51.

334 A.F. de Fourcroy, *Encyclopédie méthodique chimie*, 1805, v. 4, pp. 37–40.

335 Desmarest, ref. 135, v. 2, p. 137.

336 D'Alembert, ref. 308, art. 'Capillaire', 1751, v. 2, pp. 627–9.

337 T.L. Hankins, 'Eighteenth-century attempts to resolve the *vis viva* controversy', *Isis* **56** (1965) 281–97; Hankins, ref. 111, chap. 9, pp. 195–213.

338 W.L. Scott, *The conflict between atomism and conservation theory: 1644–1840*, London, 1970.

339 R. Taton, *L'oeuvre scientifique de Monge*, Paris, 1951, pp. 339–40.

340 For estimates of the increase in the number of 'physicists' during the 18th century, see Hahn, ref. 128, and Heilbron, ref. 17.

341 J.L. Heilbron, *Weighing imponderables and other quantitative science around 1800*, Suppl. to v. 24 of *Hist. Stud. Phys. Sci.* Berkeley, CA, 1993, chap. 1.

342 W. Cullen (1710–1790) W.P. D. Wightman, DSB, v. 3, pp. 494–5; A.L. Donovan, *Philosophical chemistry in the Scottish Enlightenment*, Edinburgh, 1975.

343 P. Shaw (1694–1764) M.B. Hall, DSB, v. 12, pp. 365–6; F.W. Gibbs, 'Peter Shaw and the revival of chemistry', *Ann. Sci.* **7** (1951) 211–37; P. Shaw, *Three essays in artificial philosophy, or universal chemistry*, London, 1731; see the first part of the first essay, 'Of philosophical chemistry', esp. p. 13.

344 R.A.F. de Réaumur, *Art de faire éclore et d'élever en toute saison des oiseaux domestiques de toutes espèces*, Paris, 1749, v. 2, p. 328.

345 L. Euler, *Lettres à une princesse d'Allemagne sur divers sujets de physique et de philosophie*, 3 vols., St Petersburg, 1768–1772, Letter of 25 November 1760, no. 79 in v. 1, pp. 312–14; English trans., 2 vols., London, 1795, v. 1, pp. 346–50. The French version is reprinted in *Opera omnia*, ref. 70, 3rd Series, vs. 11 and 12, see v. 11, pp. 171–3. An earlier and fuller account of Euler's views is in a paper he gave to the Berlin Academy on 18 June 1744, summarised as 'Sur la nature des moindres parties de la matière' in *Hist. Acad. Roy. Sci. Berlin* (1745) 28–32, publ. 1746, and given in full in his *Opuscula varii argumenti*, Berlin, v. 1, 1746, pp. 287–300; both are in *Opera omnia*, ref. 70, 3rd Series, v. 1, pp. 3–15. C. Wilson, 'Euler on action-at-a-distance and fundamental equations in continuum mechanics', in P. M. Harman and A.E. Shapiro, eds., *The investigation of difficult things. Essays on Newton and the history of the exact sciences*, Cambridge, 1992, pp. 399–420.

346 J. Leslie (1766–1832) R.G. Olson, DSB, v. 8, pp. 261–2.

347 T. Young (1773–1829) E.W. Morse, DSB, v. 14, pp. 562–72. (Morse does not discuss Young's work on capillarity.); G. Peacock, *Life of Thomas Young*, London, 1855, chap. 7.

348 J. Leslie, 'On capillary action', *Phil. Mag.* **14** (1802) 193–205. Leslie gives no source for Laplace's conclusion but it was probably the first edition of Laplace's *Exposition du système du Monde*, 2 vols., Paris, 1796, v. 2, pp. 194–5. A later and fuller account is in Laplace's *Traité de mécanique céleste*, v. 4, Paris, 1805, Book 10, § 22, pp. 325–6; trans. by N. Bowditch, Boston, MA, 1839, v. 4, [9035], p. 645. For subsequent discussions of the speed of propagation of gravity before the general theory of relativity, see J.D. North, *The measure of the universe: a history of modern cosmology*, Oxford, 1965, pp. 43–51. There is still no direct evidence for this speed

but the theoretical arguments now seem to be compelling; all interactions mediated by massless carriers, which 'gravitons' seem to be, propagate with the speed of light, *c*. The question was raised recently by D. Keeports, 'Why *c* for gravitational waves?', *Amer. Jour. Phys.* **64** (1997) 1097, and received five answers, *ibid.* **65** (1998) 589–92.

349 T. Young, 'An essay on the cohesion of fluids', *Phil. Trans. Roy. Soc.* **95** (1805) 65–87; reprinted in Young, ref. 198, v. 2, pp. 649–60, with criticisms of Laplace's work, pp. 660–70; and in *Miscellaneous works of the late Thomas Young, M.D., F.R.S., etc.*, ed. G. Peacock, London, 1855, v. 1, pp. 418–53.

350 James Ivory, who was generally a severe critic of Young's work, gives him credit for this; see his art. 'Fluids, elevation of', in *Supplement to the fourth, fifth and sixth editions of Encyclopaedia Britannica*, 6 vols., London, 1815–1824, v. 4. pp. 309–23, see p. 319. Articles signed with the initials 'c.c.' are by Ivory, as is clear from a comment in the Preface to v. 1, p. xv.

351 L. Euler, 'Recherches sur la courbure des surfaces', *Mém. Acad. Roy. Sci. Berlin* (1760) 119–43, publ. 1767, in *Opera omnia*, ref. 70, 1st Series, v. 28, pp. 1–22.

352 For a defence of Young's priority in obtaining this result, see P. R. Pujado, C. Huh and L.E. Scriven, 'On the attribution of an equation of capillarity to Young and Laplace', *Jour. Coll. Interface Sci.* **38** (1972) 662–3.

353 *Miscellaneous works*, ref. 349, v. 1, p. 420.

354 Bikerman, ref. 272.

355 [H. Gurney], *Memoir of the life of Thomas Young, M.D., F.R.S.*, London, 1831, p. 35.

356 Young, ref. 198, v. 1, pp. 605–17.

357 F.O. [i.e. T. Young], art. 'Cohesion' in *Supplement to . . . Encyclopaedia Britannica*, ref. 350, v. 3, pp. 211–22. This was written in 1816 and published in February 1818, see the reprinting of it in *Miscellaneous works*, ref. 349, v. 1, pp. 454–83. According to his first biographer, Young wrote 63 entries for this *Supplement*, all of which he signed with consecutive pairs of initials chosen from the phrase 'Fortunam ex aliis', see Gurney, ref. 355, p. 30.

358 [T. Young], *Elementary illustrations of the Celestial Mechanics of Laplace*, Part 1 [all published], London, 1821, App. A, pp. 329–37; reprinted in *Miscellaneous works*, ref. 349, v. 1, pp. 485–90.

359 Young, ref. 198, v. 2, pp. 46–51; *Miscellaneous works*, ref. 349, v. 2, pp. 129–40; Todhunter and Pearson, ref. 186, v. 1, pp. 80–6.

3

Laplace

3.1 Laplace in 1805

In the field of capillarity it is usual to consider together the work of Young and Laplace, and it is true that they both obtained some of the same important results within a year of each other. Their aims and methods were, however, quite different. In reading Young we are reading 18th century natural philosophy; in reading Laplace we are reading 19th century theoretical physics [1]. This 'sea-change' in the early years of the new century is as dramatic as that of the 'scientific revolution' of the 17th century, and was due to the efforts of the great French school of mathematical physics of that time [2]. This is not the place to discuss the origin of this second revolution but to concentrate only on how it led to a revival of the subject of cohesion and to a second period of advance. The man responsible was Laplace [3].

The prevailing opinion in France at the end of the 18th century was that of Buffon and his followers; the cohesive forces were probably gravitational in origin and so followed the inverse-square law at large distances but departed from that law at short distances where the shapes of the particles affected the interaction. In 1796 Laplace discussed this view in the first edition of his *Exposition du système du monde*, noting, however, that the particles of matter would have to be of an inconceivably high density and extremely widely spaced if matter was to have its observed degree of cohesion and its known density [4]. In 1816, Laplace's protégé, J.B. Biot [5] was still supporting a gravitational origin with the specific rider that the influence of shape changed inverse square to inverse cubic at short distances [6]. Antoine Libes, less able mathematically than Laplace or Biot, argued in 1813 for inverse square at all distances [7]. Laplace said nothing further on the subject in the second edition of his book in 1799, but much more in the third and later editions from 1808 onwards [8]. His interest in cohesion had by then been aroused by two problems, the first of which was his friend Berthollet's wish to interpret chemistry in terms of Newtonian attractions [9].

One of Berthollet's great contributions to chemistry was his realisation that the course of chemical reactions depends as much on the amounts of substances involved as on their 'affinities'. This realisation led him to the concept of *la masse chimique*, and it was the ground of his criticism of Torbern Bergman in a tract conceived during his days in Egypt with Napoleon's expedition [10]. The importance of mass in this context may have disposed him to relate chemical reactions to gravitation. He takes up this theme at the opening of his *Essai de statique chimique* of 1803 [11]:

The powers that produce chemical phenomena are all derived from the mutual attraction of the particles of bodies, to which one gives the name affinity to distinguish it from astronomical attraction. It is probable that both are one and the same property; but astronomical attraction exerts itself only between masses placed at a distance at which the shape of the particles, their separations and their particular affections have no influence.

He goes on to say that chemical attractions are so altered by such particular circumstances that we can say little about their form with any assurance. He would welcome a mathematical theory of chemistry but accepts that its time has not yet come. Laplace was equally pessimistic when Davy put the idea forward at their meeting in 1813 [12]. By 1810, however, Berthollet was, under the influence of Laplace's work, affirming publicly that "the attractive force that produces capillary phenomena is the true source of chemical affinities" [13]. This view was not inconsistent with his identification of gravitational and chemical forces and was one that he had been expressing informally in his lectures at the École Polytechnique as early as 1803; he repeated it in about 1812 in a manuscript that was intended to be the basis of a never-to-be-published second edition of his *Essai* of 1803 [14].

Laplace contributed two notes to his friend's *Essai*. Their content suggests that he had not, in 1803, thought deeply about forces other than astronomical. In the first, Note V, he postulates that the repulsive force of heat between the particles of a gas is independent of their separation. His argument is that if one doubles the density of a gas one doubles the number of particles in the layer next to the wall and so doubles the pressure without any need to suppose that the forces themselves change with distance. Later in the book, Note XVIII, he says that his previous Note had been written in haste, and he now adopts the view that the force is as the reciprocal of the distance and so as the cube root of the volume [15]. This was a return to Newton's hypothesis. He adds that the force is also "proportional to the temperature". The scale is not specified although elsewhere he accepts that Gay-Lussac's work in 1802 implies that the air thermometer is the true measure of temperature; a ratio of 1.375 for the air pressure at 100 °C to that at 0 °C leads to a zero of the scale at -266.7 °C [16]. There were, however, many views among the supporters of the caloric theory on how this zero should be fixed and it was some years before Laplace firmly committed himself to this conclusion [17].

Berthollet and Laplace used the word *molécule* to denote the small particles in a fluid but its use did not imply an acceptance of the modern (or Dalton–Avogadro) view of molecules and their constituent atoms; it has been translated throughout this chapter by the less committing word 'particle'. Dalton himself complained about the imprecision in the use of such words as 'particle' and 'integrant part' or 'integrant particle'. He seemed content with the notion that such entities are the smallest that can be identified with the substance in question, e.g. water, with any further division into 'constituent particles', or his 'atoms', leading to entities of a different kind, e.g. hydrogen and oxygen [18]. The modern meaning of the word 'molecule' came into use only later in the century.

The second source of Laplace's interest in cohesion, and so in capillarity, was, as for Newton and Clairaut, an acceptance of a corpuscular theory of light and so a need to understand how light is refracted (that is, attracted) by matter, and in particular by the Earth's atmosphere. This was a matter of importance to astronomers, and Laplace first turns to it in 1805 in Book 10 of his *Mécanique céleste*, which concluded the fourth and, for the time being, final volume of this treatise. This book is something of a miscellany in which he collects together various topics that have arisen earlier in the work but which have not yet been dealt with. One of these was 'Des réfractions astronomiques' [19], and in it he introduces ϕ, the short-ranged but unknown force between a particle of light and one of air. The integral of ϕ with respect to the separation, r, and its higher moments or the integrals of ϕr^n, where $n > 0$, arise naturally in his treatment of this problem. The mathematical methods and the functions involved are those that he used shortly afterwards in his better-known and, as we can now see, better-judged treatment of capillarity. Thus by 1805 Laplace had settled on a Newtonian view of the attractive forces – they were short-ranged but of unknown functional form. He also brought to his thoughts on the structure of matter and its interactions the usual beliefs of the time in imponderable fluids, and notably in caloric which he held to be the agent of repulsion that stopped matter collapsing by keeping its particles apart. The corpuscular theory of light and a belief in imponderable fluids were aspects of Laplace's physics that were to be found wanting in the first part of the 19th century, and a younger generation of physicists, although raised in his methods, was soon to outgrow them [20]. This 'new physics' did not invalidate his work on capillarity but it was to overshadow it and to turn it once again into an unfashionable area of science.

Laplace held also to the static picture of gases and liquids that was the 'standard model' of the time; his particles did not move, at least when he was discussing the effect of the attractive forces between them on their cohesion. Daniel Bernoulli had put forward a kinetic theory of gases in 1738, but the idea was not a fruitful one at that time and it had generally been ignored [21]. The difficulties with static models of gases and liquids was not apparent until later in the century and played no part in what is usually seen as the downfall of Laplacian physics in the 1820s and 1830s.

3.2 Capillarity

Young read his paper on capillarity to the Royal Society on 20 December 1804. A year later, on 23 December 1805, Laplace read before the First Class of the Institut de France, the 'revolutionary' successor of the Académie Royale, his paper on the theory of capillary action. A summary of it appeared the next month in the *Journal de Physique*, the successor to Rozier's *Observations* [22], and a full account was published as a supplement to Book 10 (in the fourth volume) of his *Mécanique céleste* [23]. This was quickly followed by a second supplement whose aim, as stated in its opening sentence, was to perfect the theory already given and to extend its application [24]. In these works he carried out successfully what Clairaut had attempted, namely a derivation of the laws of capillarity from a supposed force of attraction between the particles. His success depended on his specific rejection of Clairaut's assumption that the range of the forces was comparable with the radius of the tube. He follows what he thought to be Hauksbee's deduction that the range was negligible in comparison with this distance [25].

There is no reason to suppose that Laplace knew of Young's paper, notwithstanding Young's later ill-chosen insinuations [26]. Nor apparently did he know of Leslie's paper although it was written at Versailles on 9 October 1802, when Leslie was in France during the brief Peace of Amiens. Communication between Britain and France was slow after the resumption of war in 1803. Laplace had, however, read Young by the time of his second Supplement of 1807, and mentions him and Segner briefly in his closing words. He must surely have known of Monge's paper of 1787 but he ignores it, perhaps because Monge, like Young, did not seek an explicit connection between the attractive forces and the capillary effects, or perhaps because of his personal dislike of Monge; Clairaut is the only one he acknowledges as having addressed this problem [27]. In the Introduction to his first paper [23] he makes a reference to an earlier and presumably unpublished attack on the problem, and then describes his present approach:

A long while ago, I endeavoured in vain to determine the laws of attraction that would represent these phenomena [i.e. those of capillarity]; but some late researches have rendered it evident that the whole may be represented by the same laws, which satisfy the phenomena of refraction; that is, by laws in which the attraction is sensible only at insensible distances; and from this principle we can deduce a complete theory of capillary action. [28]

He writes that "the attraction of a capillary tube has no other influence upon the elevation or depression of the fluid which it contains, than that of determining the inclination of the first tangent planes of the interior fluid surface, situated very near to the sides of the tube ..." [29]. This is a key point of both his and Young's work. Neither justifies it in detail; with Laplace it was a self-evident assertion that each solid–fluid pair would have a fixed angle of contact; with Young it was a consequence of his assumption of the three surface tensions, gas–liquid, gas–solid and liquid–solid. Laplace adds another assertion, also found much later to be substantially correct: "it is natural to suppose that the capillary attraction, like the force of gravity, is transmitted through other bodies" [30]. He implicitly assumes that the particles of matter are so small that he can sum their interactions by the mathematical operation of integration, an assumption that Poisson and others were later to question (see below). His picture of the cohesive forces has now left the gravitational model behind; he requires only that the forces are "sensible only at insensible distances", a phrase that he was to use often. Their origin and their form are unknown and, as he is to show, need not be known, although their dependence on the separation of the particles must be rapid enough for his integrals of the force and of some of its higher moments to converge. In his second Supplement he observes that for a force that falls off exponentially all the moments are finite, but this is only an example, not at the time a serious proposal for a force of this form [31]. A.T. Petit also invoked an exponential form in a thesis of 1811 in which he generalised some of Laplace's results [32]. Fourteen years later Laplace discussed an inverse-square law damped by an exponential as a possible modification of Newton's law of gravitation when the attraction took place through intervening layers of matter, but found that "the attraction of the particle placed at the centre of the Earth, acting at a point on its surface, is not diminished by a millionth part by the interposition of terrestial layers" [33].

With these preliminaries in place, Laplace tackles the problem of capillarity by first calculating the attractive force between a spherical liquid drop and a thin vertical 'canal' of liquid outside it and perpendicular to its surface (Fig. 3.1). M is the centre of the drop and at Q there is a volume element $u^2 du \sin\theta d\theta d\omega$, where u is the distance MQ, θ is the angle PMQ, and ω is the azimuthal angle between the plane of MPQ and a fixed vertical plane that contains MP. Let $\phi(f)$ be the (positive) force of attraction of a particle at Q for one at P in the column, where f is the separation of P and Q. Let PM be represented by r, then

$$f^2 = u^2 + r^2 - 2ur\cos\theta. \tag{3.1}$$

The vertical force on P of the particles in the volume element at Q is

$$\rho u^2 du \sin\theta d\theta d\omega \cos\alpha \ \varphi(f), \tag{3.2}$$

where α is the angle MPQ and ρ is the number density of the particles in the drop, that is, the number of particles per unit of volume. Laplace tacitly takes this to be unity, ignoring the niceties of dimensional correctness, and so omits it; let us do the

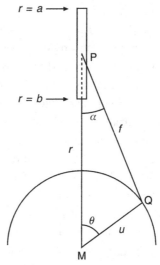

Fig. 3.1 Laplace's calculation of the force of attraction of a sphere for a thin 'canal' of material outside it.

same. (If the water is incompressible, then at first sight there is little to be gained by including this factor.) Now

$$(df/dr) = (r - u\cos\theta)/f = \cos\alpha, \tag{3.3}$$

so the force can be written

$$u^2 du \sin\theta\, d\theta\, d\omega (df/dr)\varphi(f), \tag{3.4}$$

which is the derivative with respect to r of

$$u^2 du \sin\theta\, d\theta\, d\omega [C - \Pi(f)], \tag{3.5}$$

where

$$\Pi(f) = \int_f^\infty \varphi(f')\, df'. \tag{3.6}$$

The function (3.5) is the potential at P due to the element of volume at Q, although Laplace does not use this name. The constant C is an arbitrary baseline or zero for the potential. We now integrate (3.5) over the angles ω from 0 to 2π, and θ from 0 to π. By differentiation of eqn 3.1 we have

$$f\, df = ur\sin\theta\, d\theta, \tag{3.7}$$

so the potential at P from a spherical shell of radius u and thickness du is

$$4\pi u^2 du\, C - \frac{2\pi u du}{r} \int_{r-u}^{r+u} f\Pi(f)\, df. \tag{3.8}$$

He introduces another symbol for this integral:

$$\psi(f) = \int_f^\infty f' \Pi(f') \, df'. \tag{3.9}$$

If $\varphi(f)$ has a small or 'insensible' range then so, he assumes, do its higher moments, $\Pi(f)$ and $\psi(f)$. The first term in eqn 3.8 is independent of r and so contributes nothing to the force that is obtained by differentiation with respect to r; we omit it henceforth. The remaining term in eqn 3.8 is

$$2\pi (u/r)du \, [\psi(r+u) - \psi(r-u)]. \tag{3.10}$$

The force on the whole column from a to b is therefore obtained by differentiating eqn 3.10 with respect to r to get the force, and then integrating it again to get the effect of the whole column. (We should now insert another factor, ρ, for the number density within the column but, again following Laplace, we omit it.) The result of this double operation is a force of

$$2\pi (u/a)du \, [\psi(a+u) - \psi(a-u)] - 2\pi (u/b)du \, [\psi(b+u) - \psi(b-u)]. \tag{3.11}$$

Now a, b, and $(a-b)$ are all large with respect to the range of the force, $\varphi(f)$, and so the terms with $\psi(a+u)$, $\psi(a-u)$, and $\psi(b+u)$ are negligible. We are left with the positive force of attraction of the shell of thickness du on the column from a to b of

$$2\pi (u/b)du \, \psi(b-u) \tag{3.12}$$

which is itself appreciable only when u is almost as large as b. The final integration over u, from 0 to b, gives the attractive force between the whole of the drop and the essentially infinitely long column touching it. This force is

$$\frac{2\pi}{b} \int_0^b u\psi(b-u) \, du. \tag{3.13}$$

We substitute $u = b - z$ in the integrand and write eqn 3.13 as two terms:

$$2\pi \int_0^b \psi(z) \, dz - \frac{2\pi}{b} \int_0^b z\psi(z) \, dz. \tag{3.14}$$

The integrands are negligible except when z is small so the upper limits can be replaced by ∞. Thus the force of attraction between a drop of radius b and the thin column of unit area touching it can be written as $K - (H/b)$, where

$$K = 2\pi \int_0^\infty \psi(z) \, dz \quad \text{and} \quad H = 2\pi \int_0^\infty z\psi(z) \, dz. \tag{3.15}$$

Laplace now uses an argument based on the symmetry of two touching spheres with respect to the tangent plane between them to repeat the derivation for a column within the drop, and so shows that the 'action' of the sphere on the column, per unit area, is

$$K + (H/b), \tag{3.16}$$

a quantity that was later called the 'internal pressure' within the drop. He generalises this result to obtain the internal pressure within a portion of liquid bounded by a surface with two principal radii of curvature, b_1 and b_2; namely [34],

$$K + \tfrac{1}{2}H[(1/b_1) + (1/b_2)]. \tag{3.17}$$

The second term in this expression is the excess pressure just inside a curved surface over that inside a plane surface, for which b_1 and b_2 are both infinite. It is therefore the same as the result that Young had obtained, and expressed in words, if we identify $\tfrac{1}{2}H$ with Young's surface tension. This was an identification that Laplace could not make in his first paper since he did not then know of Young's work, and in his later papers Laplace retained the symbol H but avoided the phrase 'surface tension'.

Laplace has now two tasks; first, to show that this expression for the pressure inside a curved surface leads to a satisfactory explanation of the known capillary phenomena, and, second, to give his interpretation of the two terms K and H. The first task had already been carried out in outline by Young on the basis of this expression and the constancy of the angle of contact of a given liquid–solid pair. Laplace carries it out again with great thoroughness. He shows that the rise in sufficiently narrow capillaries is inversely proportional to their diameters, that the rise between close parallel plates is the same as that in a tube of a radius equal to their separation, he gives a detailed explanation of Newton's 'oil of oranges' experiment, remarking that his advance on the work of that "great mathematician ... shows the advantages of an accurate mathematical theory" [35], he explains the forces between floating objects that are or are not wetted by the liquid [36], and he calculates the force needed to lift a solid disc from the surface of a liquid. This last calculation was for 'Dr Taylor's experiment', and Gay-Lussac [37], then a young protégé of Berthollet, contributed some new experiments on this topic. Laplace obtains also the general form of the differential equation that describes the shape of the meniscus in a tube under the combined effects of capillary attraction and gravity, but notes that this cannot be solved analytically except in special cases, such as for a tube so narrow that the meniscus forms part of the surface of a sphere. A few years later he was to use this impressive set of results to justify his *credo*:

One of the greatest advantages of mathematical theories, and one that best establishes their correctness, is their bringing together phenomena that seem to be disparate, and in determining their mutual relations, not in a vague or conjectural way, but by rigorous calculations. Thus the law of gravity relates the flux and reflux of the tides to the laws of the elliptical movement of the planets. It is the same here, the theory set out above relates the adhesion of discs to the surface of liquids to the rise of the same liquids in capillary tubes. [38]

For experimental work he relies in his first Supplement on measurements of the capillary rise of water carried out, at his request, by the Abbé R.-J. Haüy, assisted by J.-L. Trémery and (although Laplace does not mention him) the Italian, M. Tondi [39]. They found that, for three tubes, the product of the diameter and the height to which the water rose was about 13.5 mm^2. This was equivalent to Hauksbee's results, as quoted by Newton, for the rise between parallel plates, but it had been known since van Musschenbroek's experiments that it was only about half the rise in thoroughly clean tubes. For his second Supplement, Laplace called on Gay-Lussac for some new experiments that had "the correctness of astronomical observations" [40], and which showed a rise of twice that of Haüy and his colleagues – a change on which Young did not fail to comment. Gay-Lussac introduced the method, often used today, of determining the diameter and uniformity of the bore by measuring the length of a thread of a known weight of mercury. His results were corrected for the small departure of the water meniscus from a hemi-spherical shape and led to a surface tension, in modern units, of 74.2 mN m^{-1} at 8.5 °C [41], in excellent agreement with the value accepted today of 74.7 mN m^{-1}.

A point of some practical importance was the calculation of the depression of mercury in a barometer tube of known diameter. Laplace had designed the barometer used by Biot and Gay-Lussac to measure the heights in their balloon ascents in August and November 1804 [42]. Both Young and Laplace now had the mathematical and physical kit needed to calculate the depression, namely a knowledge (or presumption) of the constancy of the angle of contact of mercury and glass, and the relation, eqn 3.17, between the curvatures of the surface and the pressure difference across it. They had a reasonable knowledge of all the physical quantities involved: the density of mercury, the acceleration due to gravity, the angle of contact, and the surface tension or $\frac{1}{2}H$. There were, moreover, some measurements of the depression in tubes of different diameters made many years earlier by Lord Charles Cavendish and published in a paper by his more famous son Henry in 1776 [43]. In Fig. 3.2 these results are shown together with the curves calculated by Young in 1805 [44] and by Laplace in 1810 and 1826 [45]. Some years later Young revised his calculations and obtained results closer to those of Laplace [46]. Their curves have roughly the same shape as that found experimentally by Cavendish and are even closer to the modern results of Gould [47]. These calculations represent a great advance in the theory of capillarity over anything that had been accomplished in the previous century, and the credit for them certainly belongs to Young. Until 1804 there was no convincing explanation even of the proportionality of the capillary rise or fall, Δh, to the diameter of a narrow tube, d.

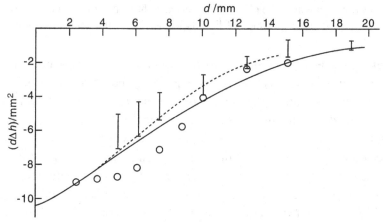

Fig. 3.2 The product of the depression of mercury, Δh, and the diameter of the capillary tube, d, as a function of d. The circles are the experimental points of Lord Charles Cavendish (before 1776) [43] and the vertical bars are the results of F.A. Gould (1923) [47]. The dashed line is that calculated by Young in 1805 [44] and the full line that by Laplace in 1810 [45].

The limiting constancy of the product $d\Delta h$ is now represented by the simple fact that the calculated curves have a finite non-zero intercept at $d = 0$, and the whole of the course of the curve and of the experimental results is the application of the new theory to tubes of appreciable diameter. The intercept of $d\Delta h$ at $d = 0$ is related to the surface tension, σ, and the angle of contact, θ, by

$$(d\Delta h)_{d=0} = 4\sigma \cos\theta/\rho g, \qquad (3.18)$$

where ρ is the density of the liquid and g is the acceleration due to gravity. The angle of contact for mercury is about 145° and so an intercept of -10 mm^2 implies a surface tension of 410 mN m^{-1}. The best modern value for clean mercury is 500 mN m^{-1}, but 410 is probably a fair value for the slightly oxidised mercury found in most barometer tubes.

Laplace's second task is the interpretation of his results for the cohesion of liquids, that is, of the magnitudes of the integrals K and H of eqn 3.15. He notes first that K is much larger than H/b, "because the differential [i.e. integrand] of the expression of H/b is equal to the differential of the expression K multiplied by z/b; and since the factor $\psi(z)$, in these differentials, is sensible only when the value of z/b is insensible, the integral H/b must be considerably less than the integral K." [48] He does not, at this point, attempt an estimate of the length H/K, but returns to this point at the end of the second Supplement where he writes, "It is almost impossible to determine, by experiment, the intensity of the attractive force of the particles of bodies [i.e. K]; we only know that it is incomparably greater than the capillary action." [49] He then attempts a theoretical estimate, based on his belief that the particles of light are deflected by molecular forces. The conclusion, that the ratio of K to the force of gravity is a distance greater than 10 000 times the distance of the Earth from the Sun, is so extreme that he at once dismisses it, contenting himself only with repeating that K is clearly very large [50]. He does not deduce explicitly that the ratio H/K is a measure of the mean length, $\langle z \rangle$, over which the force $\varphi(z)$ is active but it seems to be implicit in his discussion. Young, as we shall see, was to have a better physical grasp of the magnitude of K and so of the range of the forces.

The large size of K and the small size of H led to the natural interpretation that the former is the quantity responsible for the cohesion of solids and liquids, and, by extension, for their chemical attractions, while the latter describes a modification of this cohesion that is responsible for the much weaker capillary forces and the delicate phenomena that they cause. All this is set out in the 'General Considerations' that conclude the second Supplement. Quoting Berthollet's results, Laplace extends his argument into chemistry. He is now confident that the phenomena of cohesion and capillarity "and all those which chemistry presents, correspond to one and the same law [of attraction], of which there can be no doubt." [51] He gives what was by then the standard explanation of the elasticity of solids in terms of small displacements of the particles from their positions of equilibrium, and attributes the viscosity of liquids to the restraining influence of the attractive forces on their free flow, an influence that can be reduced by the repulsive force of heat. He rightly regards the viscosity as a hindrance in observing capillary phenomena, not a stickiness that causes it, as had often been thought previously. In one of several summaries of his work on cohesion that he wrote towards the end of his career he regretted that he had been able to make no progress in understanding the flow of liquids at a molecular level [52]; the position is little better today.

At the end of the second Supplement he mentions Segner's and Young's work, but emphasises the point that whereas they had merely postulated the existence of a surface tension, he had correctly deduced its existence as a consequence of a short-ranged force of attraction between the particles, and, moreover, he had obtained an explicit relation between the force $\varphi(r)$, where r is the separation of a pair of particles, and the tension $\frac{1}{2}H$. It was a difference of aim and achievement of which he was right to be proud. He did not like to use such words as 'surface tension' or 'membrane' to describe the source of capillary effects; he was content with the integral H. Benjamin Thompson, by then Count Rumford and living in Paris [53], was of the older school. He confessed that he could not understand Laplace's mathematics but, on 16 June 1806 and 9 March 1807, he read at the Institut two parts of a memoir in which he pointed out how the concept of a membrane at the surface of water explained many problems of the flotation of small bodies more dense than water [53]. The discussion between him and Laplace must have been a dialogue of the deaf [54].

The distinction between the two points of view was noticed but not fully appreciated by Young, who thought that Laplace's extensive derivations involved "the plainest truths of mechanics in the intricacies of algebraic formulas" [55]. Elsewhere he wrote anonymously that they were a mere "ostentatious parade of deep investigation . . . more influenced on some occasions, by the desire of commanding admiration, than of communicating knowledge." He continued:

The point, on which Mr. Laplace seems to rest the most material part of his Claim to originality, is the deduction of all phenomena of capillary action from The simple consideration of molecular attraction. To us it does not appear that The fundamental principle, from which he sets out, is at all a necessary Consequence of the established properties of matter. [56]

Young had, as we have seen, heterodox views on how the forces depended on the separation of the particles, but he neither thought it necessary, nor probably had he the skill, to relate the tension to an integral over these forces. Laplace's achievement was, however, something of a pyrrhic victory in that a knowledge of the integral, H, tells us nothing of the integrand, that is, of the forces themselves. Knowing both H and K provides more information but Laplace felt unable to estimate K with any confidence.

One inconsistency in Laplace's treatment was noted some years later by Poisson, who, after Laplace's death in 1827, became publicly more critical of the details of his mentor's work [57]. Particles at the surface of a liquid are subject to forces from one side only and so cannot be at equilibrium if the density is uniform up to a sharp surface at which it drops abruptly almost to zero. Laplace had mentioned this point in his discussion but apparently did not think it important [58]. In the *Nouvelle théorie de l'action capillaire* of 1831 [59], Poisson said that if equilibrium was to be maintained then the density must fall from its value in the bulk liquid to almost zero in the gas over a distance comparable with the range of the attractive forces. Gay-Lussac had found that the density of a powder was the same as that of the bulk solid, so the range of the forces was 'insensible', but not necessarily negligible [60]. If, therefore, as Laplace had supposed, the density changes abruptly, then the range of the forces is zero, the integral H becomes zero, and the surface tension vanishes. Poisson dresses the argument in more elaborate mathematical form, but this simple point is its basis. He also introduces correctly the factor ρ^2 in front of the integrals H and K in Laplace's original derivation, where ρ is the density of the particles in the bulk liquid. Bowditch, Laplace's translator, was convinced by Poisson's argument against a sharp interface and, in his footnotes to the *Mécanique céleste*, he repeats Laplace's derivation but now with the factor $[\rho(z)]^2$ inside the integrals H and K, this being a natural way of incorporating both of Poisson's amendments [61]. He claims that this change leaves Laplace's results unaltered in form but merely changes the numerical values of H and K, and so the quantitative relation between the attractive forces and the capillary rise. James Challis, the professor of astronomy at Cambridge, was asked to review the subject of capillarity for the fourth meeting of the British Association at Edinburgh in 1834 [62], and so was led to think about Poisson's objection. He concluded that the thickness of the surface layer was comparable with the size of the molecular cores which, he believed, was small compared with the range of the attractive forces. He was, in fact, wrong in both assumptions, but these were points not finally settled until many years later. In a report to the fifth meeting of the British Association, in Dublin in 1835, William Whewell also dismissed Poisson's objection, but now on essentially the same grounds as Bowditch [63].

Poisson's criticisms were taken more seriously on the Continent. Arago [64] wrote in his obituary of Poisson:

One asks oneself how it is possible that Laplace can go so far as to represent quantitatively the phenomena of capillary rise, while neglecting in his calculation the true, unique cause of these phenomena. I declare that this is a great mathematical scandal which should be resolved by those who have the time and talent needed to decide between those two great men, Laplace and Poisson. [65]

H.F. Link, in Berlin, came to a similar conclusion: "The results of these [Poisson's] investigations cannot be happy for physics. A mathematician of the first rank, Laplace, overlooks those important conditions, which, one can now see, put his formulae in opposition to all experience." [66] But by the 1830s the subject had dropped out of the mainstream of physics, and when the subject was taken up again sixty years later there were better ways of resolving the problem. Even a substantial paper by Gauss, in which he dealt more directly than Laplace had done with the question of the constancy of the angle of contact of liquid and solid, failed to arouse real interest (see below) [67].

A second difficulty with Laplace's results was his neglect of any discussion of the short-ranged repulsive forces. He says explicitly that his integrals H and K are to be taken from zero to infinity, and properly observes that if they are not to diverge at the upper limit then a restriction is needed on the range of the attractive force or, at least, on the way it becomes 'insensible' at large distances. He says, however, nothing about the behaviour of the integrands at the lower limit. He knew, of course, that he could not take an integral of a function of the form $-ar^{-n}$ down to $r = 0$, and so he must have supposed some form of repulsion to have intervened, but he says nothing about it beyond a general attribution to a supposed caloric fluid. Young, whose own views on the repulsive forces were provocatively unconventional, reproached him for this neglect [68]. Laplace replied some years later:

In Nature, the particles of bodies are acted on by two opposing forces: their mutual attraction and the repulsive force of heat. When liquids are placed in a vacuum, the two forces are found to be almost in equilibrium; if they follow the same law of change with distance the integral that expresses the capillary effect will become insensible; but if their laws of change are different, and if, as is necessary for the stability of the equilibrium, the repulsive force of heat decreases more rapidly than the attractive force, then the integral expression of the capillary effects [i.e. H] is sensible, even in the case where the integral expression of the chemical effects [i.e. K] has become zero, and the capillary phenomena take place in a vacuum just as in air, in conformity with experiment. The theory that I have given of these phenomena includes the action of the two forces of which I have just spoken, in taking for the integral expression of the capillary effect the difference of the two integrals relating to the molecular attraction and the repulsive force of heat, which disposes of the objection of the learned physicist Mr Young, who has criticised this theory for its neglect of the latter force. [69]

His assumption of a repulsive force of shorter range than the attractive is consistent with the picture of Boscovich and of many other writers; it is one that we accept now almost without thought. He still does not deal, however, with the mathematical problem of the divergence of the integrals H and K at the lower limit of zero separation, if the repulsive force has there become infinite in order for the particles to have size. This was a question that he never faced squarely; indeed, since it was never put to him by Young or any other critic, it may be that he did not see it as a problem but was content with the notion of a repulsion arising from the caloric attached to each particle.

In his last writings on the forces betwen the particles of matter, in the fifth and final volume of the *Mécanique céleste*, published in parts between 1823 and 1825, he sets out his conclusions as follows:

Each particle in a body is subject to the action of three forces; 1st, the attraction of the surrounding particles; 2nd, the attraction of the caloric of the same particles, plus their attraction for its caloric; 3rd, the repulsion of its caloric by the caloric of these particles. The first two forces tend to bring the particles together; the third to separate them. The three states, *solid, liquid* and *gaseous*, depend on the relative efficacy of these forces. In the solid state the first force is the greatest, the influence of the shape of the particles is very considerable and they are joined in the direction of their greatest attraction. The increase in caloric diminishes this effect by expanding the body; and when the increase becomes such that the effect is very small, or zero; the second force predominates and the body assumes the liquid state. The interior particles can then move relative to each other; but the attraction of each particle by the particles that surround it and by their caloric, retains the ensemble in the same space, with the exception of the particles at the surface, which the caloric removes in the form of vapours, until the pressure of these vapours stops the action. Then, on a further increase of caloric, the third force overcomes the other two; all the particles of the liquid, in the interior as well as on the surface, separate from each other; the liquid acquires suddenly a very considerable volume and force of expansion; it will dissipate itself into vapour unless it is forcibly restrained by the walls of the vessel or tube that contains it. This is the state of highly compressed gas to which M. Cagnard-Latour has reduced water, alcohol, ether, etc. In this state the first two forces are still effective, but the density of the fluid does not follow

Mariotte's law. One can see that for this to be satisfied, and also the laws of MM. Dalton and Gay-Lussac, it is necessary that the fluid be reduced to the aeriform state in which the third force alone is effective. [70]

This passage is of interest from several points of view. It shows, firstly, his continuing belief in heat as the agent of repulsion in all three states and so is at one with his caloric theory of gases which had an internal consistency that enabled it to hold its own until well into the 19th century [17]. He was, for example, able to use this theory in his well-known resolution of the problem of calculating the speed of sound in air by appealing to the difference between what we now call adiabatic and isothermal compression. Secondly, we see that the passage does not resolve the problem of the integrals over the repulsive forces; indeed, it seems to compound it. He believes that all three forces are short-ranged, and he had said earlier that the repulsive forces are shorter than the others, but he requires also that the caloric–caloric repulsion is the dominant force in the gas when the particles, although clearly full of caloric, are much more widely separated than in the liquid. It may be possible to produce a quantitatively satisfactory picture that resolves this paradox, but he does not attempt it. Thirdly, he has recognised the importance of the rather crude experiments of Baron Cagniard de la Tour [71] which first showed the existence of what we now call the gas–liquid critical point (see Section 4.1). Finally, in this passage he repeats his belief that the attractive forces in a solid are specific and localised. We still accept that molecular shape has a great effect on the temperature of melting. The forces are more general and diffuse in a liquid where they arise from the particle–caloric attraction. Earlier he had put this thought into different words:

Then each particle [in a liquid], in all positions, suffers the same attractive forces and the same repulsive force of heat; it yields to the slightest pressure, and the liquid enjoys a perfect fluidity. [72]

This belief that, in a liquid, each particle swims in a smooth force-field of attraction arising from all (or many) of the other particles, is an important one that was first formulated explicitly by Laplace. It was to become of increasing importance as the 19th century advanced and even now is often used as the first approximation in treating a new problem. Modern statistical mechanics knows it by the name of the 'mean-field approximation', and we shall refer to it often.

We have seen that Laplace thought that his integral K was exceedingly large but, since it played no role in capillarity, he did not try to make a realistic estimate of it. Young, rushing in where Laplace feared to tread, did make an estimate of the value of what he called "the corpuscular attraction", saying, "... there is reason to suppose the corpuscular forces of a section of a square inch of water to be equivalent to the weight of a column about 750 000 feet high, at least if we allow the cohesion to be independent of the density." [46] In modern units this makes the attractive

force, expressed as a pressure, equal to 25 kbar. The corpuscular attraction, or
Laplace's K, has no precise equivalent in modern theory, but the property closest to
it is the change of internal energy, U, with the volume, V, that is, $(\partial U/\partial V)_T$, which
is about 1–5 kbar for most liquids. Dupré (see Section 4.1) later used the latent heat
of evaporation per unit volume which is of similar value. Young's estimate is therefore a reasonable one although somewhat high. Unfortunately he does not tell us how
he arrives at this figure, but the most likely route is from Canton's measurement
of the compressibility of water, which he mentions briefly in the same article,
and from his belief that compressibility is related to tensile strength [73]. Later in
the article he uses the word 'elasticity' for the same property, and in another article
he gives "850 000 feet" as the modulus of elasticity of ice [74]. He could now have
identified his "corpuscular attraction" with Laplace's K, and his surface tension
with $\frac{1}{2}H$, and so obtained a mean range of the attractive force, $\langle z \rangle$, from the ratio
H/K, but to have done this would have been an admission of the usefulness of
his rival's "algebraic formulas". He therefore arrives at essentially the same result
by a parallel but more obscure argument, at the end of which he deduces that "the
contractile force is one-third of the whole cohesive force of a stratum of particles, equal in thickness to the interval, to which the primitive equable cohesion
extends." [46] (The adjective 'equable' refers to his assumption that the cohesive
force is constant at all separations within its range.) His estimate of the range is
therefore (3 × surface tension ÷ corpuscular attraction) or "about the 250 millionth
of an inch". If we take, in modern units, a surface tension of water of 70 mN m^{-1},
which he knew accurately, and his estimate of the corpuscular attraction which is
23×10^8 N m^{-2}, then this range is 10^{-10} m, or 1 Å. A modern estimate of the
range would be about 5 Å, so Young's physical intuition had guided him to what
we would see as a reasonable estimate. This remarkable result is the first quantitative estimate of any aspect of interparticle cohesion that we can recognise as having
been derived by a physically sound method of reasoning. The tentative efforts of
two greater men, Newton and Laplace, had been guided by their commitment to
particular theories, notably a corpuscular theory of light, that resulted in numerical
values that we can now see are wrong. Unfortunately Young published this work as
a pseudonymous article in a supplementary volume of *Encyclopaedia Britannica*,
so it neither brought him any credit nor did it have any discernible effect on the
development of the field.

He then went on to draw a natural but false conclusion. He supposed that the
stationary particles in saturated water vapour were at a separation at which the
attractive forces were just strong enough to overcome the repulsive, and so cause
the vapour to condense to a liquid. He estimates that at 60 °F (15.6 °C) the reduction in volume in going from vapour to liquid is a factor of 60 000, which implies
a reduction in the mean separation of the cube root of this, or a factor of 39.

He deduces, therefore, that the range of the attractive force is about 40 times the diameters of the particles, so that any one particle in a liquid is under the influence of many others, an argument that can be used to justify the assumption of a mean-field approximation. His ratio of 40 would have been very different if he had chosen a different temperature; thus at the normal boiling point it would have been 12. He was worried by this apparent dependence of the range on temperature since he knew that the vapour pressure of a liquid changed more rapidly with temperature than did its surface tension or elasticity, but decided that

. . . on the whole it appears tolerably safe to conclude, that, whatever errors may have affected the determination, the diameter or distance between two particles of water is between the two thousand and the ten thousand millionth of an inch [i.e. 0.1 to 0.02 Å]. [46]

A more realistic estimate of the upper limit of particle size could have been obtained from the many experiments on thin films and, in particular, from Benjamin Franklin's famous experiment of 1773 of the stilling of water waves by pouring a little oil on the surface [75]. He found that a teaspoonful ($2\,cm^3$?) spread rapidly over half an acre ($2000\,m^2$) of the two-acre pond on Clapham Common near London. He attributed the rapidity of the spreading to a repulsion between the particles of oil but made no comment on the implication of the thinness of the layer, about 10 Å, which we now know is about the length of a typical molecule of a vegetable oil. The thinness to which gold leaf could be beaten had often been cited as a measure of the smallness of the particles, so this line of argument was probably known to him, but clearly an estimate of an upper limit to their size was not his aim and, in view of his comments on the mutual repulsion of the oil particles, he may not have thought that his layer was continuous and compact, that is, there may have been no lateral contact between the particles [75]; if so, he would have been correct but it was not until the end of the next century that this question was resolved (see Section 4.5). Young saw it somewhat differently:

The attractive power of water being greater than that of oils, a small portion of oil thrown on water is caused to spread on it with great rapidity by means of the force of cohesion; for it does not appear that want of chemical affinity, between the substances concerned, diminishes their cohesive power. . . . [76]

James Ivory [77] was, perhaps, one of the first British mathematicians to master the new French mathematics and, in particular the *Mécanique céleste* of Laplace. He wrote on capillarity in the same Supplement to the *Encyclopaedia Britannica* that had carried Young's work, but he almost ignored Young's contributions, crediting him only with the observation that the angle of contact is constant. Instead all is ascribed to Leslie, his fellow student at St Andrews and at Edinburgh, and to Laplace. His conclusion reads:

... but if the truth is to be told, it may be affirmed that; reckoning back from the present time to the speculations of the Florentine academicians, the formula of Laplace, and the remark of Professor Leslie relating to the lateral force, are the only approaches that have been made to a sound physical account of the phenomena. [78]

It was a biased verdict but one made understandable by the obscurity of Young's writing and reinforced by Ivory's distrust of many of his contemporaries, including Young.

In Italy a young physicist at Pavia, Giuseppe Belli [79], took up the subject of molecular attraction in 1814, apparently under the influence of Laplace's papers [80]. He starts from the fact that the force of attraction between two metal plates is independent of their thickness, a fact that he quotes from Haüy's textbook [81]. He then calculates the force between two plates on the assumption that the interparticle force follows an inverse integral power of the separation. The observed independence of thickness requires that the power be greater than 4. If it were 4 exactly then his exposition becomes "defective". It is easy to show that this borderline case leads to a logarithmic dependence on thickness, but he does not do this [82]. A force of power -4 corresponds to an interparticle potential of -3. The fact that potentials are inadmissible unless they decay more rapidy than the inverse power of the dimensionality of the space of the system is now a central feature of classical statistical mechanics. It is implicit in Newton's calculations in the *Principia* (see Section 2.1) but Belli seems to have been the first to discuss the point clearly. Eighteenth century calculations of the force between particles and spheres, and Belli's extension of them to that between two spheres, raise other difficulties, which Belli does not escape [83].

He then moves, in proper Laplacian manner, to consider the two phenomena of the refraction of light and of capillarity. For the second he maintains that the lower limit of the inverse power of the force must be 5 not 4, presumably because of the extra factor of separation in the integral H, but his argument is hard to follow because of the faulty labelling of his diagrams.

He refutes the proposition that the attractive forces are gravitational, modified at short distances by the non-spherical shapes of the particles, by making explicit calculations of the gravitational force between non-spherical bodies. He considers the force on a particle at the bottom of a drop of liquid suspended below a horizontal plate. If the gravitational attraction of the Earth were to be balanced by the opposing gravitational attraction of the drop then, he maintains, the density of the drop would have to be 12×10^9 times that of the Earth. Laplace had raised a similar point earlier [4].

We can recognise some valid theoretical points in this paper, and some that are now less convincing, but a first publication by a hitherto unknown 22-year-old physicist from Pavia attracted little or no attention at the time. With, however, the involvement of men such as Young and Laplace the subjects of cohesion and capillarity had recovered from Leslie's jibe that their pursuit had been left "to the culture of a secondary order of men." Laplace's policy of reducing physics to the study of the attractions between particles (of matter and of light) that were mediated and supplemented by the actions of imponderable fluids, was one followed by French physicists during the early years of the century. He was able to set the agenda not only by reason of his intellectual domination but also by the patronage he could exercise in the filling of salaried posts and in the choice of subjects in which the First Class of the Institut would award prizes and allocate funds. Thus Biot, a protégé of Laplace, and his younger colleague Arago undertook for the Institut a substantial experimental and theoretical study of the refraction of light by gases in which they tried to estimate the strength of the forces between the particles of matter and those of light. They believed that this study would prove to be a practicable route to the measurement of the forces responsible for Berthollet's chemical affinities [84].

Biot and Arago were part of the young team, many trained at the École Polytechnique, that Laplace and Berthollet gathered around them at Arcueil, to the south of Paris, where they had neighbouring houses and where they built a laboratory [85]. The dominance of Laplace's view of physics in the decade from 1805 to 1815 was exerted largely through this circle. It was, for example, in the Memoirs of the Society of Arcueil that Étienne Malus published his discovery of the polarisation of light by reflection, of which he gave a corpuscular explanation in terms of repulsive forces [86].

In 1808 the Emperor Napoleon, himself a member of the First Class of the Institut, called for 'An historical report on the progress of the mathematical and physical sciences since 1789'. A deputation led by the President, Bougainville, waited on him in February. Delambre [87] gave the report on the mathematical sciences and Cuvier [88] that on the physical. Delambre confined himself, in the main, to a factual summary of the achievements of the last twenty years in what we should now call applied mathematics and experimental physics. He praised Laplace's work and made note of that on capillarity. Cuvier, after an excessively flowery introduction, came to the heart of the Laplacian programme:

The prodigious number of facts which extends from the simple aggregation of the particles of a salt to the structure of organic bodies and to the most complex functions of their life, seems, however, to be attributed most directly to the general phenomenon of molecular attraction, and we could not choose a more convenient thread to guide us through this maze. [89]

He then starts his report with two subjects, the theory of crystals and the theory of affinities, "two sciences entirely new and born in the period that we have to review". With crystals he was on sure ground; it was essentially a new science.

Crystallography, the structure and symmetry of crystals, was a subject then flying from the Laplacian nest, but the theory of the elasticity of solids was to prove to be the one field where Laplacian physics was to remain fruitful and where it survived, although much criticised, when the rest of his scheme fell under the assaults of the Young Turks from 1815 onwards. With the subject of affinities, Cuvier was less fortunate in his prognosis. He says that it had had a primitive origin but he claimed that it been revolutionised by Berthollet. In fact Berthollet's treatment was to mark the end of the Newtonian chemical tradition that had started with Freind a hundred years earlier. It had not been, for a long time, a useful tool even in the hands of those who had nominally adhered to it. They had been more concerned to establish chemistry as an autonomous science and, once Berthollet's short-lived influence had waned, this was to be the way forward in the 19th century.

3.3 Burying Laplacian physics

Both the weaknesses of Laplace's programme and the loss of his powers of patronage became increasingly apparent after the restoration of the monarchy in 1815 [20]. The corpuscular theory of light was the first casualty, to be followed by a slow loss of faith in the reality of the caloric fluid. In chemistry, Dalton's atomic theory and the electrochemistry of Davy and Berzelius soon proved to be more fertile guides to research than Berthollet's affinities. New branches of physics arose that did not fit into Laplace's programme, notably the magnetic forces of electric currents which did not conform to the picture of central forces between particles. The first mathematicians and physicists to bring forward mechanical and optical views that did not fit his picture were those outside his circle and his influence: Fresnel [90], Fourier [91], Sophie Germain [92] and later Navier [93]. They were joined eventually by those from his entourage: Biot, Arago and Petit; only Poisson kept the faith, even when he was querying some of the mathematical methods.

The abandoning of Laplace's views in these new branches of physics led ultimately to the generation of the field theories that were such a prominent feature of the second half of the 19th century, but the change was gradual. Both Fresnel and Cauchy [94] envisaged a molecular aether and a German school backed for a time an electromagnetic theory that rested on forces between moving particles [95]. None of this new physics had anything to contribute to the problem of cohesion in liquids where we can see, with hindsight, that Laplace's ideas were broadly correct. A modern physicist recognises his treatment of capillarity as a simple mean-field approximation for a system with pairwise additive intermolecular forces; it is the legitimate ancestor of much current work in the field [96]. Nevertheless the sheer volume and exuberance of the physics of light, electromagnetism and, later, heat, inevitably buried Laplace's achievement with his failures.

Even a field in which we should see some scope for discussing the role of the intermolecular forces, the thermal conductivity of solids, developed in a way that not so much contradicted Laplace as ignored him. Fourier's mature views are set out in his *Théorie analytique de la chaleur* of 1822 [97]. As early as 1807 he had presented to the Institut a phenomenological treatment of heat conduction in a solid. This was criticised by Lagrange and Laplace for mathematical faults in the derivation and solution of his differential equations, but a revised version won the prize in the competition set by the First Class in 1810. Poisson publicly reviewed Fourier's papers of 1807 and 1815, and provided his own alternative derivation on strict Laplacian lines for the transmission of heat from particle to particle [98]. The submission of Fourier's prize essay overlapped with another series of competitions on the elasticity of plates which is more germane to our field (see below). Here again the prize went eventually to a non-Laplacian essay, and here again Poisson provided a Laplacian counter-effort.

Fourier is not against all corpuscular explanations – at one point he gives a standard account of the displacement of the particles in a solid from their positions of equilibrium by external forces [99] – but he is clear that 'heat' cannot be reduced to 'mechanics' [100]. He does not commit himself to the nature of heat, and does not need to, since, as he emphasises, his equations are valid independently of any such assumption [101]. He acknowledges, however, that heat

... is the origin of all elasticity; it is the repulsive force which preserves the form of solid masses and the volume of liquids. In solid masses, neighbouring particles [*molécules*] would yield to their mutual attraction, if its effect was not destroyed by the heat which separates them. [99]

Nevertheless, when it comes to developing his treatment such Laplacian notions are discarded. His *molécules*, it is clear, then become merely locations at which the temperature is recorded, or infinitely small elements of volume ($dxdydz$), "la molécule rectangulaire" [102]. This was a usage that Laplace himself had adopted in his early work, writing in 1796 that "the volume of any molecule remains constant, if the fluid is incompressible, and depends only on pressure, following a fixed law, if the fluid is elastic and compressible." [103]

Fourier opens his book with a 'Preliminary Discourse' of which the first sentence is: "Primary causes are unknown to us; but are subject to simple and constant laws, which may be discovered by observation, the study of them being the object of natural philosophy." [104] This sentence naturally aroused the admiration of Auguste Comte [105] who was to make similar declarations about the limited aims of the natural sciences. Such positivism was foreign to the Laplacian programme but it was to become the dominant mode of thought in France and, to a lesser degree, in other countries also [106].

In the early 18th century those who tried to interpret cohesion in terms of forces between the intimate particles of matter had to contend with the criticism of the Cartesians and Leibnizians that they had not produced a plausible mechanism by which such forces could act. When the parallel criticism of the gravitational force collapsed in the face of its irrefutable success in accounting for the observations of the astronomers, then the objection to molecular attraction at a distance was muted or tacitly abandoned. But now, when Laplace had carried the Newtonian programme forward with a satisfactory resolution of all the capillary problems that had so intrigued the natural philosophers of the 18th century, the counter-attack came from the opposite direction; such interpretations were unnecessarily specific in their mechanisms and should be abandoned in favour of phenomenological descriptions that avoided all appeal to molecular attraction or other microscopic mechanisms. The force remained but the particles were to be abandoned. A quest for descriptions that avoided particulate mechanisms was not wholly new; such ideas had been put forward during the 18th century by both physicists and philosophers (in the modern sense of these terms) [107]. Thus 'pressure' was an unspecified surface force for Euler and for Lagrange, while for Laplace it was the bulk consequence of molecular and caloric interactions, only to become for most physicists a macroscopic stress again in the 19th century [108]. 'Heat conduction' went through a similar cycle. What was new from about 1820 onwards was that a macroscopic and often positivist description (using that word in a broad sense) became the dominant mode of thought.

So the Laplacians lost the battle, or left the field, in the areas of electricity and magnetism, of light, and, later, of heat and thermodynamics; but what of the subject of the properties of matter? They could make little more progress with the properties of gases and liquids since they were restrained by a static molecular picture of matter and a corpuscular theory of heat. Solids are, however, a state of matter in which heat, and so the motion of the molecules, plays only a secondary role, and here they did not abandon the field. Throughout the century a battle was fought between the molecular and macroscopic interpretations of the elasticity of solids. This was a field of great practical importance to the civil and mechanical engineers of the time and these practical men were decidedly non-molecular in their prejudices. Indeed the vigour of the engineering profession and of its works probably had as decisive an influence on the abandoning of Laplacian ideas as any metaphysical preferences of the positivists. This emphasis on practical affairs was strong in mechanics and thermodynamics [109]. Carnot [110], Navier and Clapeyron [111] were all engineers and Joule [112], a Daltonian chemist by training, came from a practical background. His early physical work was largely free from molecular speculations, as was that of William Thomson [113], a devotee of Fourier's work. Cauchy, another engineer by training, alternated between molecular and non-molecular treatments of the elasticity of solids. Let us turn therefore to this field and see how the battle was fought, but with first a brief account of what was known or believed about the crystal structure of solids.

3.4 Crystals

The properties of solids had played a less important role than those of liquids in the study of cohesion in the 18th century. There were two distinct lines of study that were to coalesce much later but which were separate at the start of the 19th century. The first was that of speculation on the shapes and arrangements of the constituent particles of well-defined crystals [114] and the second was that of the elasticity of solids [115]. The first was rooted in mineralogy and so ultimately in chemistry, and the second arose from the concerns of the engineers. We need to know a little about the first before tackling the second.

In the 17th century Robert Hooke and Christiaan Huygens had had realistic ideas about how crystals of well-defined geometrical shapes could be assembled by packing together arrays of spheres or ellipsoids. They did not require that the entities that they chose filled all space; contact between them was sufficient. Freind had summarised this approach in his *Chymical lectures* of 1712:

And since the force of attraction is stronger in one side of the same particle than another, there will constantly be a greater concretion of salts upon those sides, which attract most strongly. From hence it may easily be demonstrated, that the figure [i.e. shape] of the least particles, is entirely different from that which appears in the crystal. But we must leave this to the mathematicians lest we shou'd seem to encroach upon their province. [116]

The opposite view, namely that the particles occupy all space and so must have shapes that are related to those of the crystals, was also held in the 18th century. In 1777 Guyton de Morveau wrote that:

Every regular solid body produced by crystallisation can be composed only of particles that have a form compatible [*une forme génératrice*] with that which results from their union: it is impossible that any number of cubes whatever can have the appearance of a sphere, since we suppose the need for the most perfect contact between all the elements: this principle, as we have said, can one day serve to determine the shape of the constituent particles of all crystalline solids. [117]

Such ideas were developed more fully by Haüy, who drew on the observation of Romé de l'Isle [118] that the angles of a crystal of a given material are constant even if the overall habit of the crystal is not. Haüy recognised that the individual chemical elements could not be the building blocks of such geometrically perfect forms; he believed that assemblies of the elements formed what he first called in 1784 the *molécules constituantes* [119]. In his more fully developed *Traité de minéralogie* of 1801 [120] he changed his notation and distinguished between the *molécules élémentaires* (e.g. one part of soda and one of muriatic acid, in common salt) and the *molécules integrantes* (also Laplace's term) formed from these, whose geometric faces were parallel to the natural joints revealed by cleaving the crystal (simple cubes for common salt). These played a role similar to that of the unit cell in modern crystallography. The differing overall shapes of crystals of the same substance he attributed to the removal of parts of layers of these units. The ratio of the length of tread to riser in the resulting staircase was a small integer, a 'rationality of intercepts' that later came to be called the law of rational indices. One can see in its implications a parallel for solids with Gay-Lussac's law of combining volumes for gases. Haüy's *molécules integrantes* generally filled all space like Guyton de Morveau's units, but he found occasionally that they could only be packed together so that they touched on edges thus leaving some unfilled space [121]. Once such exceptions were admitted then the argument for the precise geometric shapes of these units became less compelling.

In 1813 W.H. Wollaston gave the Bakerian lecture to the Royal Society [122] and chose as his subject the formation of crystal structures by the packing of spheres. He was obviously embarrassed when told that the scheme was not original, as he had thought, but had been put forward over a hundred years earlier by Hooke. He nevertheless went ahead with his lecture, with acknowledgements to Hooke, and so laid the foundation for many later 19th century schemes of the same kind [123]. L.A. Seeber, the professor of physics at Freiburg, added to this picture the observation that the thermal expansion of crystals could not be explained by the static packing of inert spheres but required that there be attractive and repulsive forces between the units [124].

In the changing climate of opinion after 1815 it was not surprising that such Laplacian views of crystal structure were challenged, nor that the opposition came again from outside Paris. C.S. Weiss [125], the professor of mineralogy at Berlin, had worked with Haüy and had translated his work into German, with some critical comments. He rejected its atomistic basis, being concerned rather to establish the geometrical side of crystallography on abstract principles of symmetry [126]. A similar path was followed with greater rigour by the better-known Friedrich Mohs of Freiburg [127] in his textbook of mineralogy of 1822–1824 [128], which was translated into English by his former assistant Wilhelm Karl Haidinger [129]. Mohs saw minerals as part of natural history and his classification was based on considerations of symmetry, geometry, colour and other physical attributes; he regarded the chemical composition as of secondary importance and, like Weiss, did not discuss molecular building blocks. This macroscopic view was to be the way forward for crystallography in the 19th century. Speculations on the atomic structures of crystals were to be unfruitful, with the restricted exception of the principle of isomorphism. It was to be a hundred years before x-ray diffraction was to give the crystallographers a tool with which to determine the molecular facts. Arguments based on symmetry are always powerful in the physical sciences and were soon to make their presence felt in the hitherto unrelated field of the elasticity of solids.

3.5 Elasticity of plates

In February 1808 Cuvier had defended the Laplacian programme in general and crystallography in particular before the Emperor Napoleon. A few months later a different aspect of the properties of solids came to the fore. The German physicist and musician, E.F.F. Chladni [130] visited Paris and demonstrated before the First Class of the Institut and before Napoleon the great variety of the vibratory states of glass plates. He held these at two or more points around their edges and set them vibrating by stroking the edge with a violin bow. The nodes of the vibrations were made visible by the lines on the surface along which a powder sprinkled on the plates came to rest. These nodal lines formed a great number of patterns although each was repeatable if the points of clamping and the frequency of the exciting vibrations were reproduced accurately. (Ørsted was making similar experiments in Copenhagen at this time, for which he offered an electrical explanation [131].)

Here was a problem for the mathematical physicists; what equations governed the modes of vibration of circular plates, and could they be solved? Hitherto, problems of elasticity and the strength of materials had been the province of the practical men, and although Euler, d'Alembert and others had contributed some theoretical results these had been mainly for stretched cords, beams and other one-dimensional problems. At the direct request of Napoleon, and almost certainly at the prompting of Laplace, the Institut offered a prize outside its usual series for a disquisition on the theory of the elasticity and vibration of plates and a comparison with Chladni's results [132]. Laplace probably saw here a chance for his young protégé, the 27-year-old Siméon-Denis Poisson, to show his abilities. The preamble to the announcement of the prize notes that Poisson had recently read before the First Class, of which he was not yet a member, a paper on the vibration of sound in tubes. Laplace soon made his own views clear in a long note he attached to a memoir on the passage of light through a transparent medium [133]. The memoir was read before the First Class on 30 January 1808, so presumably the note was added after Chladni's visit to Paris. In it he wrote:

To determine the equilibrium and motion of an elastic sheet that is naturally rectilinear, and is bent into any curve whatever, one has to suppose that at each point, its spring [*ressort*] is in inverse ratio to the radius of curvature. But this rule is only secondary, and derives from the attractive and repulsive action of the particles, which is a function of their separation. To put this derivation forward, one must conceive that each particle of an elastic body is in equilibrium in its natural state, subject to the attractive and repulsive forces it experiences from the other particles, the repulsive forces being due to heat or other causes.

Laplace was naturally one of the judges for the prize and intending competitors could not have had a clearer hint of how he thought the problem should be tackled. Another judge, Lagrange, was, however, not committed to this molecular approach.

In the event Poisson did not compete for the prize and the only entry received by the closing date of 1 October 1811 was from Sophie Germain, a 35-year-old lady who had learnt her mathematics by private study and by correspondence, first with Gauss on number theory and then with Legendre [134] on elasticity, notwithstanding the fact that Legendre was also one of the judges. She based her treatment on the earlier work of Euler on the bending of rods and on the *Méchanique analitique* of 1788 of Lagrange [135]. She assumed, by a simple generalisation of Euler's result for a thin rod (repeated by Laplace as his "secondary rule") that the restoring force on a surface, initially planar and now bent, is proportional to the sum of the reciprocals of the two principal radii of curvature. She did not defend this generalisation and the sixth-order differential equation that she obtained did not follow from it. The most noteworthy feature of her entry was, however, that she never mentioned Laplacian particles, a natural consequence of her lack of an entrée into his school, and her choice of Euler and Lagrange as models to follow.

On 4 December 1811, Legendre wrote to her warning her that she would not receive the prize and telling her that Lagrange had derived from her (unproved) assumption a fourth-order differential equation for the deflection z as a function of the planar coordinates, x and y, and the time t. He showed that this equation reduced to Euler's one-dimensional result for a thin rod if $dz/dy = 0$, which hers did not [136]. Lagrange's equation, with appropriate boundary conditions, is accepted today as the correct description of the motion of the central portion of a vibrating plate [137].

The competition was set again and new entries were required by 1 October 1813. By then Lagrange had died and Poisson, who had joined the First Class in 1812 on the early death of Malus, became one of the judges. Again Sophie Germain was the only competitor. She knew now the equation she was aiming for and she duly arrived at it, but her analysis was still faulty and her starting point still without the primary justification that Laplace and Poisson would have liked. She did make useful progress in solving Lagrange's equation under appropriate conditions and her entry, although not awarded the prize, received an honourable mention.

The competition was set for a third time with a closing date of 1 October 1815. By then Poisson had taken up the subject (but not within the competition) and had naturally treated the problem of elasticity as one of the change in the forces between neighbouring particles [138]. His analysis of the bending of a surface without thickness, in which all the particles are initially in the same plane, was based on the assumption that the bending reduced the interparticle separations and so increased the repulsive forces. Its principal aim was to arrive at Lagrange's equation, which, although not yet in the public domain, he knew was of more than passing interest since Germain had used it to reproduce some of Chladni's results.

Germain was not deterred by Poisson's intervention; in her third entry she again simply argued that, for small deformations, the elastic force had to be proportional to the difference of shape of the deformed and the undeformed surfaces. She extended her discussioh to (and made some experiments on) surfaces whose natural or undeformed state was already curved. The judges were still not satisfied with her derivation of the differential equation but decided that her comparisons with Chladni's results and her own new work on curved surfaces justified the award of the prize. The Institut never published her essay but it is close in content to her own publication, written a few years later, after some advice from Fourier who had returned to Paris from the provinces in 1815 [139].

She later extended her treatment to surfaces of varying thickness and now, for the first time, mentioned "the particles that comprise the thickness of a solid", but by restricting the discussion to solids "of which the thickness is very small" she was able to resolve the problem into one of thin sheets and so to continue to discuss it in terms of changes of curvature [140]. Her last contribution to the field was a short paper in the *Annales de Chimie et de Physique* which she submitted in 1828 in an attempt to intervene in an argument that was developing between Poisson and Navier; they ignored her comments [141]. She still maintained that the only incontestable fact about the forces of elasticity is the tendency of bodies endowed with such forces "to re-establish the form that an external effect has caused them to lose." She is not convinced that we need interparticle forces but if they are introduced they cannot be repulsive only, as Poisson had apparently implied in his 1814 memoir and had just repeated in the *Annales* [142]. Both attractive and repulsive forces are needed; they balance in the natural state, and if the particles are pushed together the repulsive increase more strongly than the attractive.

Her unwillingness to invoke molecular hypotheses and intermolecular forces arose from her choice of mentors, Euler, Legendre, and, later, Fourier. It was an unwillingness common to many 19th century 'elasticians', some of whom shared her broadly positivist views but were little influenced by her example. These views were most apparent in an essay published by her nephew in 1833 after her early death [143], and it was these views rather than her mathematics that led to the publication of her works in 1879 [144]. Some French elasticians did not abandon the Laplacian approach and, throughout the century, there was a conflict between those who were content with the macroscopic concepts that came to be called (in English) stress and strain, and those who hankered after a deeper interpretation in terms of interparticle forces [145]. The division paralleled the later one between those who were satisfied with the apparent certainties of classical thermodynamics and those who sought a deeper interpretation of its laws in the molecular mechanics of the kinetic theory.

The first substantial attack on the problem of the elasticity of solids, as distinct from that of rods and plates [146], came from Navier, Cauchy and Poisson. The last was always a Laplacian but the other two kept a foot in both camps.

3.6 Elasticity of solids

Navier was the first to tackle the problem of the elasticity of solid bodies [147]. He had joined the staff of the École des Ponts et Chaussées in 1819, at the age of 34, as instructor in mechanics, and at once entered the field. On 23 November 1819 he read at the Academy, of which he was not yet a member, a paper on the elasticity of bent rods, and on 14 August 1820, a paper on the bending of loaded plates. These were later published, the second in abstract [148], and this one was also circulated among his confrères on lithographed sheets [149]. Both papers acknowledge that the cause of elasticity is the interparticle forces, and in the second he introduces what was to become his basic hypothesis, namely that the net force between a pair of particles [150] vanishes in the natural state of the body and is proportional to their change of separation in the strained state. His general attack is, however, essentially macroscopic, particularly in the earlier lithographic version of his memoir on plates [151].

These papers were preliminaries to his attack on the general problem of solids of arbitrary shape, which was the subject of the memoir that he read to the Academy on 14 May 1821, a memoir which is sometimes regarded as the birth of the modern theory of elasticity [152]. This also appeared in abstract [153] but publication of the full texts of all three memoirs was held up by the reviewing panels appointed by the Academy. These included Poisson and Fourier but not Cauchy, as Navier believed. He complained in vain at this delay but it was only in 1827, three years after his own election to the Academy, that the most important, that on solids, appeared in print [154]. Parallel papers on fluids, read on 18 March and 16 December 1822 appeared at once in abstract [155], but again full publication was delayed until 1827 [156].

In this work on solids Navier's approach was molecular, a choice that led to opposite reactions from his two rivals in the field, Poisson and Cauchy. The former, who regarded himself as the authority on the Laplacian style of physics, was led to make a similar attempt at a theory of the elasticity of solids. Cauchy, on the contrary, produced papers that aimed to free the theory from an explicitly molecular basis. He was the quicker off the mark. Inspired by Navier's memoir of May 1821, which he had heard read at the Academy, by parallel work by Fresnel on the propagation of light through an aether treated as an elastic solid, by his own ideas on the mechanics of fluids, and by the need, as he saw it, to put the teaching of his engineering classes on a sound mathematical basis, he developed a theory of elasticity that did not rest on a molecular hypothesis. He reported on this work at a meeting of the Academy on 30 September 1822 and prepared an abstract of it the next year [157]; the full text appeared in revised form in 1827 to 1829 in his *Exercices de mathématiques* [158].

Poisson did not enter the fray until 1 October 1827 when he read a short paper at the Academy on 'corps sonores' [159]. He was followed in the discussion that day [160] by Cagniard de la Tour who said that he was making experiments in this field, and by Cauchy who sketched his own non-molecular theory and deposited a sealed packet that contained an outline of three of his papers that were to appear in his *Exercices* [161]. Poisson followed his short paper with another in which he introduced what we now call 'Poisson's ratio', that is, a measure of the change in the diameter of a rod on stretching it [162]. Then came his book-length memoir of 14 April and 24 November 1828 [142] in which he tried to trump Navier and outflank Cauchy.

Navier felt badly used by both Cauchy and Poisson. He believed that Cauchy had held up his memoir of 14 April 1821 in order to publish his own work [163], and he thought that Poisson had not given him credit for his work of 1820 to 1821. The conflict with Poisson led to a long exchange of notes in the *Annales de Chimie* that started with Navier's letter of 28 July 1828 and ran until early the next year when Arago, the editor, put an end to it [164]. Saint-Venant later defended Navier; he thought that Poisson's and Arago's criticisms were either "without foundation or exaggerated" [165].

This vast body of work from Navier, Poisson and Cauchy cannot be described here in full, nor is that necessary [147]. What is attempted is an elucidation of the assumptions about the origin of elasticity that each made at different stages of his thinking and a short explanation of how these assumptions led to different expressions for the elastic behaviour.

Poisson's position is the easiest to summarise for he never deviated from the Laplacian assumption of short-ranged forces between pairs of particles. He usually made no assumption about the form of these forces but occasionally gave hypothetical examples. He could sometimes be cavalier about whether the forces were attractive or repulsive or a difference betwen the two. This was a point on which he was criticised by both Germain and Navier, but it is clear that he properly regarded both as necessary to achieve equilibrium in a dense fluid or a solid, and his carelessness about which he used was only a matter of convenience in discussing the particular problem he had in hand. His most explicit discussion of the forces is in his memoir read at the Academy on 12 October 1829, which was published in abstract

in the *Annales de Chimie* and in full in the *Journal de l'École Polytechnique* [166].
Essentially the same ideas, but not so fully articulated, are in his great memoir of
April 1828. He starts by saying that the force between the particles is "attractive
or repulsive: it depends on the nature of the particles and their quantity of heat".
(The word 'or' is ambiguous and it was such phraseology that offended Germain
and Navier.) He then introduces an important idea. We have seen that Laplace,
in his second Supplement of 1807, had made an assumption that we should now
call a mean-field approximation [167]; Poisson now specifies the condition needed
for such an approximation to be valid, but does not mention Laplace's earlier use
of it:

Bodies are formed of disjoint particles [*molécules*], that is, of portions of ponderable matter
that are of insensible size, separated by empty spaces or pores whose dimensions are also
imperceptible to our senses. The particles are so small and approach each other so closely
that a portion of a body that contains an extremely large number can also be supposed to
be extremely small, and the size of its volume to be insensible.

Later he writes:

In all cases we shall suppose that the sphere of activity at each point in a body, although
its radius be insensible, contains nevertheless an extremely large number of particles. This
hypothesis, the only one that I have made in my new Memoir, will, without doubt, be
admitted by physicists as being in conformity with nature. [168]

The supposition that the size of the particles is much smaller than the range of the
attractive forces, and that both are 'insensible' was, perhaps, in Laplace's mind
as early as 1796 [169] but he did not repeat it explicitly in his statement of the
mean-field approximation of 1807. It was derived by Young in 1816, as we have
seen, on the basis of what we now know to be an unsound argument, but Poisson
could not have known of its publication in 1818. Poisson repeated this supposition
in 1831 in his *Nouvelle théorie* [59], and from there it made its way into the English
literature via Challis's review of 1834 [62]. It was plausible at the time but was to
give rise to trouble later in the century when it was realised that the range of the
attractive forces did not greatly exceed the diameters of the molecules.

  Poisson then suggests an explicit form for the forces. Two particles of mass m
and m' containing amounts of caloric c and c' respectively, and separated by a
distance r exert on each other a force R, where

$$R = cc'\gamma - mm'\alpha - mc'\beta - m'c\beta', \qquad (3.19)$$

or

$$R = Fr - fr, \qquad (3.20)$$

where $F(r)$ and $f(r)$, as we should now write them, are the repulsive (+) and
attractive (−) forces. The first arises from the repulsion of the two portions of caloric,

and the latter from the mutual attraction of the masses and from the attraction of matter for caloric. This all followed Laplace's ideas. The function of r represented by γ is universal, but α, β and β' are specific to each species of matter. He does not imply that the term $(-mm'\alpha)$ is the gravitational force but is merely using the masses as measures of the amount of matter. He notes that the force R may not lie along the line of the "centres of gravity" since the force fields may not be spherically symmetrical. For crystalline, and therefore generally non-isotropic solids, he adds "secondary forces" that are not central and are responsible for holding the particles in a regular array [170]. Such forces were responsible also for "chemical decompositions".

Poisson had, in 1828, given an example of a possible form for the attractive or repulsive force as a function of the separation, r, of the particles namely

$$ab^{-(r/n\alpha)^m},$$

where b is greater than unity and may conveniently be set equal to e, the base of natural logarithms, since he calls this function an exponential. Here α is the mean separation of the particles and m and n are large numbers. Such a force remains finite, and equal to a, when the separation r becomes zero. A finite limit does not accord with the concept of 'impenetrability' which, in its simplest form, requires the force to become infinite at some separation r_0, greater than zero. Such a concept, however, played little part in the Laplacian scheme; it was probably thought of as an unnecessary piece of 17th or 18th century metaphysical baggage that should be ignored.

His most disturbing criticism of all earlier work, including his own, was to challenge the replacement of sums of the actions of his "disjoint particles" by integrals over their positions. It might be thought that his new hypothesis that the range of the interactions was long compared with the sizes and so with the mean spacing of the particles was one that led naturally to the replacement of sums by integrals, but that was not how he saw it. He repeated this criticism in the final appendix of his *Nouvelle théorie* of 1831. He obtained expressions for what it is convenient to call the stress [171] by expanding in a Taylor series a function of the interparticle force, $r^{-1}f(r)$, about a neighbouring position, r'. His leading term for the stress was proportional to a coefficient, K, and the first-order term to a second coefficient, k, where

$$K = \frac{2\pi}{3\alpha^6} \int_0^\infty r^3 f(r)\,dr, \quad \text{and} \quad k = \frac{2\pi}{15\alpha^6} \int_{r=0}^{r=\infty} r^5 d[r^{-1}f(r)] \quad (3.21)$$

An integration by parts gives $K + k = 0$, if $r^4 f(r)$ is zero at $r = 0$ and at $r = \infty$. (Poisson wrote that $f(r)$ must be zero at both limits but was corrected by Navier.) If there is to be no stress in an unstrained solid then $K = 0$, and hence $k = 0$ also, which implies the absurd result that there is no stress in a strained solid. He was challenged on this point by Navier who observed, correctly, that there are many possible forces for which the limit of $r^4 f(r)$ is not zero at $r = 0$. Poisson's own exponential functions satisfy this criterion but a truly impenetrable particle, with $f(r)$ positive infinite within some hard core, does not. They never resolved this point between them and Poisson himself was not consistent in his avoidance of integrals [172]. It is now accepted that it is legitimate to replace sums by integrals within a mean-field approximation.

Navier made use of interparticle forces in most of his work although he was close to Fourier and to positivist circles. He accepted that the forces were of short range [173] but went beyond Poisson in saying that since there is no force on a particle in a body in its natural state then the force on it in a slightly strained state is proportional to the distance the particle has been moved. At first sight this statement seems to be no more than (in modern terms) the statement that a particle in an unstrained body at equilibrium is at the minimum of a parabolic potential well, so that the force is analytic and initially linear in the displacement, given that the displacement is measured with respect to the local environment of each particle. The statement [174] was criticised by Poisson who objected that it went beyond the simple purity of the Laplacian hypothesis, and who may have seen a flaw in the unclear way that Navier expressed it. In a static classical mechanical treatment a particle in a solid is at a potential minimum of the total field from all surrounding particles, but it is not at the minimum of its pairwise interaction with its nearest neighbours; rather it is repelled by them and this repulsion is balanced by the attraction of the more distant particles, and so the nearest-neighbour forces are not proportional to the displacements in a strained body. Navier did not at first make this distinction and Poisson did not explicitly adduce it, but it is brought out more clearly by Arago in the note with which he closed the discussion in the pages of the *Annales de Chimie*. In his reply in another journal [164], Navier says that he has an open mind on the question of the equilibrium arising from pairs of particles or from the whole assembly. (William Thomson later fell into the same error as Navier on this point, although only in an informal discussion of molecular packing [175].)

There is a close analogy between the equations that govern the elastic displacements in solids and those that describe the viscous flow of liquids. Navier studied both phenomena and in adapting his interparticle forces to his work on liquids he introduced a term that depended also on the speed of separation of the particles [155, 156]. This hypothesis played no part, however, in his treatment of elasticity.

Neither Poisson's nor Navier's method of attack on the problem is satisfactory because of the lack of generality of their concept of stress. It led them both to the view that one constant is sufficient to describe the elasticity of an isotropic body composed of particles that interact only in pairs by means of forces that act along the lines joining each pair; this was Poisson's k and an equivalent constant that Navier denoted by ε. This conclusion was to prove, throughout the 19th century, a much-debated point between those who might be called the neo-Laplacians, who accepted it, and a less molecularly-minded group who were to insist on two independent constants for an isotropic solid and more for crystals of cubic or lower symmetry. Cauchy set the scene by first eschewing the molecular approach to obtain stress–strain relations for a continuum elastic solid in a form that is today accepted as satisfactory and, indeed, necessary for a proper treatment of the problem. He then introduced a system of central interparticle forces and showed how this led to a reduction in the number of independent coefficients of elasticity. The argument then centred on the conditions needed for Cauchy's reduction to be valid. His work, even more than that of Navier in 1821, marks the start of a mathematical theory of the elasticity of solids.

His first improvement on Navier's work came in the 1823 abstract of his early work on continua [157]. He criticises Navier's assumption that the forces acting on a portion of solid act perpendicularly to its surface. This is true for a fluid at equilibrium but in a solid the force can act "perpendicularly or obliquely to the surface". He says that Fresnel had told him of a parallel generalisation for the forces acting on a solid optical aether in a body that exhibits double refraction. Navier replied at once to say that his assumption was both legitimate and necessary [163].

Cauchy's generalisation of the concept of pressure or stress requires that it be expressed by what we now call a second-rank tensor. His approach became clear in the definitive article that he wrote in 1828 [158(e)]. This opens with the uncompromising statement:

In research on the equations that express the conditions of equilibrium or the laws of internal motion of solid or fluid bodies, one can consider these bodies either as continuous masses the density of which changes from one point to another by insensible degrees, or as a system of distinct material points, separated from each other by very small distances.... It is from the [first] point of view that we shall here now consider solid bodies.

Let us therefore see, in modern notation, what Cauchy, and after him Lamé, Green and others, achieved with this continuum approach.

In a right-handed system of orthogonal axes, x_1, x_2 and x_3, the stress, or force per unit area, on a small flat area in the $x_2 x_3$ plane is a normal stress of σ_{11} in the x_1 direction and two transverse stresses, σ_{12} and σ_{13}, in the x_2 and x_3 directions respectively. If the turning moment of the forces acting on a small prism with sides parallel to the axes is to be zero then the stress tensor must be symmetric, $\sigma_{ij} = \sigma_{ji}$, so that it has in general six components: $\sigma_{11}, \sigma_{22}, \sigma_{33}, \sigma_{12}, \sigma_{13}$ and σ_{23}. In a condensed

notation introduced by Voigt [176] in 1910 and now generally used [177], these
are denoted by the subscripts 1, 2, 3, 6, 5 and 4, respectively. Similarly, the strain
in the body can be described also by a second-rank tensor. If the displacement of a
portion of a body at point x is defined by a displacement field $t(x)$, then the tensor
with components $\partial t_i / \partial x_j$ describes the relative displacements. The symmetric part
of this tensor, with elements

$$2\varepsilon_{ij} = (\partial t_i / \partial x_j) + (\partial t_j / \partial x_i) \tag{3.22}$$

is called the strain tensor, and the anti-symmetric part describes rotational displace-
ments. The six symmetric terms ε_{11}, ε_{22}, ε_{33}, ε_{12}, ε_{13} and ε_{23} are again abbreviated
in Voigt's notation to ε_1, ε_2, ε_3, $\frac{1}{2}\varepsilon_6$, $\frac{1}{2}\varepsilon_5$ and $\frac{1}{2}\varepsilon_4$. These six elements are, however,
not independent since they are derivatives of a single vector field $t(x)$. This differ-
ence between the stress and the strain tensors will become relevant much later in
the story (Section 5.5). The relation between stress and strain was taken by Cauchy
to be a generalisation of Hooke's law, that is,

$$\sigma_m = \sum c_{mn} \varepsilon_n, \quad \text{or more briefly,} \quad \sigma_m = c_{mn} \sigma_n, \tag{3.23}$$

where in the second equation we have used Einstein's summation convention; the
sum is taken over each index that is repeated on the right-hand side, that is, over n
in this case. The elastic constants c_{mn} are 36 in number but if the work of straining
a body is to be a perfect differential of the elements of strain then there is again a
symmetry condition, $c_{mn} = c_{nm}$, so that there are, in general, only 21 independent
elastic coefficients.

  If the material is isotropic, as Poisson, Navier, and Cauchy assumed, then the
work of deformation, w, which is second order in the strain,

$$dw = \sigma_m \, d\varepsilon_m = c_{mn} \varepsilon_n \, d\varepsilon_m, \tag{3.24}$$

must be independent of the orientation of the axes. The tensor formed from $\varepsilon_1 \ldots \varepsilon_6$
has then only two quadratic invariants, the square of the dilation Δ and a quantity
sometimes denoted by Θ where

$$\Delta = \varepsilon_1 + \varepsilon_2 + \varepsilon_3, \quad \Theta = \varepsilon_1\varepsilon_2 + \varepsilon_2\varepsilon_3 + \varepsilon_3\varepsilon_1 - \tfrac{1}{4}\big(\varepsilon_4^2 + \varepsilon_5^2 + \varepsilon_6^2\big). \tag{3.25}$$

The work of deformation is a function of these quantities and can be written

$$w = \big(\tfrac{1}{2}\lambda + \mu\big)\Delta^2 - 2\mu\Theta. \tag{3.26}$$

The coefficients λ and μ are the two independent constants of elasticity of an
isotropic medium in the notation introduced by Lamé [178] in his *Leçons* of 1852
and now widely used [179]. The reduction of 21 to 2 elastic constants can be

expressed in terms of the coefficients c_{mn} as follows:

$$c_{11} = c_{22} = c_{33} = \lambda + 2\mu,$$
$$c_{12} = c_{13} = c_{23} = \lambda,$$
$$c_{44} = c_{55} = c_{66} = \mu, \tag{3.27}$$

with all other constants equal to zero. The stress–strain relation for an isotropic solid can be written in the double-suffix notation as

$$\sigma_{ij} = \lambda(\varepsilon_{11} + \varepsilon_{22} + \varepsilon_{33})\delta_{ij} + 2\mu\varepsilon_{ij}, \tag{3.28}$$

where δ_{ij} is Kronecker's delta which is equal to unity if $i = j$, and is zero otherwise. Cauchy's symbols K and k, which differ from those defined by Poisson with these symbols, are given in terms of Lamé's symbols by $K = \lambda$ and $k = 2\mu$.

A two-constant theory of the elasticity of an isotropic solid was achieved by Cauchy in 1828 [158(e)], and later by George Green and others [180]. Cauchy could recover Navier's one-constant theory if he put one of his constants equal to twice the other; $k = 2K$, or $\lambda = \mu$. Bodies of lower symmetry have more independent elastic constants; thus a cubic crystal has three, conventionally chosen to be c_{11}, c_{12}, and c_{44}. In the isotropic case these are linked by the equation $c_{44} = \frac{1}{2}(c_{11} - c_{12})$.

The inverse of eqn 3.23 expresses the strains in terms of the stresses and the elastic moduli or compliance constants s_{mn};

$$\varepsilon_m = s_{mn}\,\sigma_n, \quad \text{where } s_{mn} = s_{nm} \quad \text{and} \quad s_{lm}c_{mn} = \delta_{ln}, \tag{3.29}$$

so that if the elastic constants are known the compliance constants can be calculated, and vice versa. In the isotropic case we have now $s_{44} = 2(s_{11} - s_{12})$ and, in terms of Lamé's constants,

$$s_{11} = (\lambda + \mu)/\mu(3\lambda + 2\mu), \quad s_{12} = -\lambda/2\mu(3\lambda + 2\mu), \quad s_{44} = 1/\mu. \tag{3.30}$$

If a wire or other body of uniform cross-section is stretched then we have $\sigma_1 > 0$ and all other $\sigma_i = 0$. We have then the strains,

$$\varepsilon_1 = s_{11}\,\sigma_1, \quad \varepsilon_2 = s_{12}\,\sigma_1, \quad \varepsilon_3 = s_{13}\,\sigma_1, \quad \varepsilon_4 = \varepsilon_5 = \varepsilon_6 = 0, \tag{3.31}$$

and so for (the modern definition of) Young's modulus for an isotropic solid [181],

$$E = \sigma_1/\varepsilon_1 = 1/s_{11} = \mu(3\lambda + 2\mu)/(\lambda + \mu), \tag{3.32}$$

and for Poisson's ratio, the ratio of the lateral contraction to the extension,

$$\nu = -\varepsilon_2/\varepsilon_1 = -\varepsilon_3/\varepsilon_1 = \lambda/2(\lambda + \mu). \tag{3.33}$$

The compressibility of an isotropic solid is

$$\kappa = 3\varepsilon_1/\sigma_1 = 3(s_{11} + 2s_{12}) = 3/(3\lambda + 2\mu), \tag{3.34}$$

since the suffixes 1, 2, and 3 are equivalent for uniform compression. The modulus of elasticity that corresponds to pure shear is μ.

We again recover Poisson's one-constant theory if we put $\lambda = \mu$, so that Poisson's ratio becomes equal to $\frac{1}{4}$, as he deduced in 1827 [162]. In general, however, the constants are not simply related to each other but they are constrained in their magnitudes by the need for the work of deformation to be positive. This condition requires that

$$\mu > 0, \quad \text{and} \quad (3\lambda + 2\mu) > 0, \quad \text{or} \quad E > 0, \quad \kappa > 0, \quad \text{and} \quad \tfrac{1}{2} > \nu > -1. \tag{3.35}$$

The limit of $\lambda \to \infty$, $\kappa \to 0$, $\nu = \frac{1}{2}$, and $E = 3\mu$ is that of an incompressible solid. In practice ν is positive (except for some unusual composite materials) and generally lies between Poisson's value of $\frac{1}{4}$ and its upper limit of $\frac{1}{2}$.

The elasticians of the time made much of the parallelism between a deformed isotropic elastic solid and a flowing liquid. This is most clearly expressed, in modern symbols, by writing the stress tensor in a liquid, σ_{ij}, in terms of a velocity-gradient tensor, υ_{ij},

$$\sigma_{ij} = -p\delta_{ij} + \pi_{ij}, \quad \text{where } \pi_{ij} = \eta'(\upsilon_{11} + \upsilon_{22} + \upsilon_{33})\delta_{ij} + 2\eta\upsilon_{ij}, \tag{3.36}$$

and where p is the static pressure and the second equation is the analogue of eqn 3.28. The two coefficients η and η' are coefficients of viscosity and are the analogues of μ and λ. The first is the coefficient of shear viscosity, and that of bulk viscosity is conventionally defined as $(\eta' + 2\eta/3)$. The viscosity of liquids was, however, and still is, too difficult a subject for it to throw any light on the intermolecular forces.

Cauchy followed his paper on the elasticity of a continuous medium [158(e)] with others [158(f), (g)] in which, without explanation or apology, he reverted to a molecular approach. One outcome was that the assumption of pairwise additive central interparticle forces did indeed lead to a reduction in the number of independent elastic constants – in general from 21 to 15, through what are now called the 'Cauchy relations' [182]:

$$c_{12} = c_{66}, \quad c_{13} = c_{55}, \quad c_{14} = c_{56},$$
$$c_{23} = c_{44}, \quad c_{25} = c_{46}, \quad c_{36} = c_{45}. \tag{3.37}$$

(Voigt called them the 'Poisson relations' [183], and, later, the 'relations of Poisson and Cauchy' [184].) Considerations of symmetry can reduce the number 21 to a much smaller figure. Thus in a cubic crystal

$$c_{11} = c_{22} = c_{33}, \quad c_{12} = c_{23} = c_{13}, \quad c_{44} = c_{55} = c_{66}, \tag{3.38}$$

and all the other constants are zero. Thus there are in general 3 independent constants in a cubic crystal, but in one with pairwise additive interparticle forces the Cauchy relations provide a further reduction to 2, through the condition $c_{12} = c_{44}$. As we have seen, in the isotropic case the reduction is from 2 to 1 through the condition $\lambda = \mu$.

Cauchy's other work on the molecular model was not so successful. He obtained two sums over the interparticle forces which he called G and R. If the range of the force was large compared with the interparticle spacing then these sums could be reduced to integrals and from these he deduced that $G = -R$. To agree with Navier's results for an isotropic solid the ratio G/R had to approach zero. He never did resolve this problem, although he further generalised the continuum approach before leaving the field for a time when he went into voluntary exile from France after the revolution of 1830. Saint-Venant later analysed Cauchy's confusion on this point [185].

Two more engineers entered the field in 1828. Gabriel Lamé and Émile Clapeyron were graduates of the École Poytechnique who in 1820 had gone to St Petersburg where they had worked on practical problems of iron bridges and similar structures. Now, in a memoir in which they describe themselves as 'Colonels de Génie au service de Russie', they joined in the attack on the problem of elasticity, about which Lamé later wrote: "We think that this problem, unfortunately very difficult and not yet fully solved, is the most important that can be tackled by those engineers who concern themselves with the physical sciences." [186] Their memoir [187] contains little that is wholly new and it is not clear what they knew of the work of Navier and Cauchy; there are no references. It is important, however, for it marks Lamé's entrance into the field; Clapeyron was to concern himself in the 1830s more with steam engines and, after his 'discovery' of Carnot's work, with what came to be called thermodynamics.

Their memoir is in two main parts, the first of which is essentially a repetition of Navier's work with the minor exceptions that they require the particles to be equally spaced and that the force of attraction is proportional to the sum of the masses of the interacting particles and not to their product, as was usual. They make no comment on or use of this innovation which may have been a slip of the pen, or it may have followed the usage of a mathematically similar paper by Libes in 1802 [7]. The second part of their memoir is closer to Cauchy's continuum treatment in that they introduce the six components of the stress tensor. They clearly preferred the continuum model to the molecular and, in his *Leçons* of 1852, Lamé, having used the molecular hypothesis earlier in the book, came to an outspoken conclusion. The book ends by him asking whether

... all questions concerning molecular physics have been retarded, rather than advanced, by the extension – at least premature if not false – of the laws of celestial mechanics. Mathematicians, preoccupied by the immense work needed to complete Newton's discoveries, and accustomed to finding a mathematical explanation of all celestial phenomena in the principle of universal gravity, have ended by persuading themselves that attractions, or ponderable matter alone, should be able to offer similar explanations of most terrestial phenomena. They have taken it as a point of departure for their researches into different branches of physics, from capillarity to elasticity. It is no doubt probable that the progress of general physics will one day lead to a principle, analogous to that of universal gravity (which would be only a corollary of it), which would serve as the basis of a rational theory and include at the same time both celestial and terrestial mechanics. But to presuppose [the existence of] this unknown principle, or to try to deduce it wholly from one of its parts, is to hold back, perhaps for a long time, the epoch of its discovery. [188]

Thus battle was joined. The continuum theory led to a plethora of elastic constants – there were 21 in general, 3 for a cubic crystal, and 2 for an isotropic solid. If the material were deemed to be formed of particles acting on each other with short-ranged central forces then the number was reduced – to 15 in general, 2 for a cubic crystal, and 1 for an isotropic solid. The equations needed to effect this reduction were the Cauchy relations. But was the reduction justified? The ideas of Laplace, although at the time virtually confined to this specialised branch of physics, were not without their supporters throughout the 19th century. This party was called by Pearson the supporters of the 'rari-constant' theory, and they were opposed by those who supported the 'multi-constant' theory [189]. In the first camp he put Poisson, Navier, Cauchy (with reservations), Rudolf Clausius [190], F.E. Neumann [191] and Barré de Saint-Venant, and in the second, Lamé, G.G. Stokes [192], William Thomson and J.C. Maxwell. Even those in the continuum camp often regarded the use of the multi-constant theory as something forced upon them if they were to describe adequately the physics of real solids, and did not think that the use of this theory precluded them from using molecular language and methods elsewhere in their papers. There was a similar situation in the fields of thermodynamics and hydrodynamics. Classical thermodynamics was a powerful theoretical tool in the middle and second half of the 19th century which had initially no molecular foundations. With the development of the kinetic theory of gases the question arose of how to give a molecular foundation to thermodynamics by invoking the advances made in kinetic theory. Some wished to maintain the macroscopic 'purity' of the classical theory, others sought for the deeper understanding of its results that seemed to flow from a molecular interpretation. Similarly, in hydrodynamics it was perceived that the subject demanded a continuum treatment, but it was hard to see what caused the viscosity of a liquid, for example, without supposing a molecular constitution of matter. Josef Stefan in Vienna was one who struggled long with this problem without resolving it [193].

The criticisms of the multi-constant party were threefold; first, that the hypothesis of forces between pairs of particles was unproved, second, that the analysis of the rari-constant party was faulty, and third, that the experimental evidence was against them. William Thomson and P.G. Tait managed to encapsulate all three criticisms into one sentence when they wrote: "Under Properties of Matter, we shall see that an untenable theory (Boscovich's), falsely worked out by mathematicians, has led to relations among the coefficients of elasticity which experiment has proved to be false." [194] The first criticism need not detain us; few of those who freely used interparticle forces would have denied that the reality of these was a hypothesis that was open to challenge, however strong their conviction that it was correct. The second and third are more serious.

An early criticism of the analysis came from Stokes in 1845. He did not hold with Poisson's distinction between the effects of near and distant neighbours of the molecule whose displacement was under consideration [195]. He had apparently not read Cauchy's work at this time. Technical criticism came also from Thomson who told Stokes in 1856 that he could devise a mechanical system of particles which, he said, conformed to the molecular hypothesis but did not satisfy the Cauchy relations [196]. In his Baltimore Lectures of 1884 he belatedly made good that promise with a model of particles linked by wires and cranks [197], but, as Pearson remarked, his model lacked all conviction [198]. It may have been inspired by Maxwell's first mechanical model for his electromagnetic theory. Lamé [199] and Samuel Haughton in Dublin [200] both thought that it was the improper use of integrals in place of sums that was responsible for the reduction in the number of constants. The rari-constant theory not surprisingly attracted the contempt of Duhem who attacked both the hypothesis and the analysis in 1903 [201].

A technical defence of Cauchy's molecular analysis was given by Clausius in 1849 [202]. Rather than abandon central forces between the particles he assumed that experiments that contradicted the rari-constant theory were affected by inelastic (or 'after-effect') displacements of the particles. He also emphasised the importance of Cauchy's definition of the word 'homogeneous' [203]. This point proved to be the crux of the matter. Cauchy had defined the homogeneous state of a body as one in which, in modern terms, each particle is at a centre of symmetry or inversion point of the whole lattice, but his definition was not generally understood and continued to give trouble. Saint-Venant tried to put the matter straight in 1860 when he wrote:

We know the distinction established by M. Cauchy between an *isotropic* body and one that is simply *homogeneous*. It is isotropic if the same molecular displacements lead everywhere and in all directions to the same elastic responses. It is merely homogeneous if its matter shows the same elasticity at all points in corresponding directions [*directions homologues*] but not in all directions around the same point. Thus regular crystalline materials are homogeneous without being isotropic. [204]

He held, all his life, to a belief that he thought almost self-evident, that theory should start from the assumption that the energy of an assembly of particles was the sum of their kinetic energies of translation and configurational potential energy (our terms) and that the latter was itself a function only of the interparticle separations [205].

This is satisfactory, as far as it goes, but it does not get to the root of Cauchy's restriction on the molecular constitution that is needed to achieve what he meant by homogeneity. The importance of Cauchy's restriction in any derivation of the rari-constant theory was not obvious at a time when ideas of crystal symmetry were little developed, but Thomson, independently of Clausius, came to realise what the problem was. He had, as we have seen, originally dismissed the rari-constant theory with contempt – "a theory which never had any good foundation" [206] – but he eventually modified his opposition and asked instead if there were conditions under which it might be expected to hold. He considered a simple molecular model, an array of close-packed spherical particles in which each has 12 nearest neighbours [207]. William Barlow, in his first crystallographic paper of 1883, had pointed out that there were two different regular ways of packing spheres at the maximum density (when they occupy the fraction $(\pi/3\sqrt{2}) = 0.740$ of the space) [208]. One of these, the cubic close-packed structure, has a centre of symmetry, but the other, the hexagonal close-packed, does not, although, as Thomson observed, it can be regarded as two interpenetrating lattices each of which is centro-symmetric. Thomson did not use the words 'centre of symmetry', but he showed that only the first structure was homogeneous in the sense of that word used by Cauchy and those who followed him. Nevertheless he was only able to obtain a rari-constant theory for this structure by assuming that the central forces decreased with distance in a particular way.

The theoretical problem was not settled until the 20th century. In 1906 A.E.H. Love at Oxford gave a modern version of Cauchy's derivation which has occasionally been cited as the authoritative source [209]. Most writers, however, ascribe the first full and satisfactory treatment of the problem to Max Born in his monograph of 1915, *Dynamik der Kristallgitter* [210]; it was an ascription that he himself accepted [211]. This was the first book in the field after the x-ray experiments of von Laue and the Braggs had shown beyond doubt that crystals were composed of repeating atomic units. Born showed in general what Thomson had shown for a particular case, namely that crystal lattices can be regarded as formed of a number of simpler interpenetrating lattices. These can have centres of symmetry when the overall lattice does not. This book did not end the argument which rumbled on until the middle of the 20th century [212]. There then appeared the best and most accessible treatment in the chapter that K. Huang wrote for Born and Huang's *Dynamical theory of crystal lattices* [213]. Some physicists now speak of the 'Cauchy–Born relations' [214].

The result of a hundred years of debate is that it is now established that the Cauchy relations hold for a system of particles to which classical (not quantum) mechanics apply, which owes its cohesion to pairwise additive central forces, which adopts a stable structure in which each particle is at a centre of symmetry of the whole lattice, and which initially is in a state free from strain. These are conditions with which Laplace or Poisson would surely have felt quite comfortable and which were, on the whole, implicitly adopted by the rari-constant party. This party can be seen, at least in this context, as Laplace's 19th century heirs when they studied elasticity in the hope that it might throw more light on the intermolecular forces than had Laplace's treatment of capillarity. This aim was summed up by Pearson in 1893 when he wrote that the theory was "tending to introduce us by means of the elastic constants into the molecular laboratory of nature – indeed this is the transcendent merit of rari-constancy, if it were only once satisfactorily established!" [215]

But do real solids satisfy these conditions or, to put the question the other way round, do the elastic constants of real solids satisfy the Cauchy relations? For some years the experimental evidence was slight. Poisson had, in 1827, relied on the single experiment of Cagniard de la Tour to back his theoretical estimate of $\frac{1}{4}$ for the ratio of the lateral contraction to the extension of an isotropic cylinder subject to a unidirectional stress, that is of 'Poisson's ratio' [162]. This ratio was the first parameter chosen to test the rari-constant theory. For an isotropic solid for which Cauchy's relations hold it is $\frac{1}{4}$, but if they do not hold it can be as large as $\frac{1}{2}$. It is, however, a difficult property to measure and it was not clear which bodies were isotropic. The difficulty of obtaining such bodies was first underlined by Félix Savart's careful analysis, in 1829, of the modes of oscillation of rock crystal (quartz) taken in different different crystallographic directions. It was soon clear that the elastic constants were not the same in all directions [216].

The single experiment cited by Poisson did not carry much conviction with dispassionate observers. A more systematic attack on the problem of Poisson's ratio was mounted by Guillaume Wertheim in the 1840s. He was German-born (and baptised Wilhelm) but moved to Paris in 1841 at the age of 26, where he became a naturalised French citizen [217]. His first work in this field appeared in 1842 when he accepted the rari-constant theory and was led by it to some vague speculations on the relation between mechanical properties and interparticle forces [218]. A series of further papers led to his memoir of 1848 on Poisson's ratio for a range of metals and alloys [219]. He showed that the ratio is significantly larger than $\frac{1}{4}$ and often close to $\frac{1}{3}$, but instead of concluding that his results showed that his materials required a two-constant theory, he suggested that the one-constant theory be retained but with $\lambda = 2\mu$, which leads to a Poisson's ratio of $\frac{1}{3}$. This conclusion satisfied neither party; it was not acceptable as a one-constant theory since it had no theoretical basis (although Cauchy saw no objection to it [220]), and it did not at first sight support those who were arguing for a two-constant theory. His results raised doubts about the isotropy and/or homogeneity of his materials. The experiments were accepted, and indeed still are [221], but his deductions from them were criticised by Clausius [202] and by Saint-Venant [222].

Later results confirmed the message; for most, but not all, materials the ratio is larger than $\frac{1}{4}$. In the 1880s, E.-H. Amagat [223] made a careful set of measurements as an adjunct to his work on the compressibility of gases and liquids. For tubes of glass and 'crystal' (fused quartz) he found, after choosing "the most regular parts possible", mean values of the ratio of 0.245 and 0.250 respectively [224]. For most metals he found values of 0.3 to 0.4, as had Wertheim, but for lead [225] the ratio was 0.425–0.428 and for rubber [226] it was almost 0.5 [227]. He argued that the approach of the ratio to its upper limit of $\frac{1}{2}$ in lead and rubber was evidence for their more liquid-like character; that is, he proposed that this limit could be reached not only for an incompressible solid, for which $\lambda \to \infty$, but also by a material that cannot resist shear, for which $\mu \to 0$.

The early experiments of Woldemar Voigt confirmed the rari-constancy of annealed glass [228], and he went on to make more extensive measurements of the several elastic constants of well-defined crystals with the aim of testing Cauchy's relations directly [183]. He followed his mentor Franz Neumann, under whose supervision he had written his thesis at Königsberg, in making experiments that took explicit account of the symmetries of the crystals; most of his predecessors had worked with glassy or polycrystalline materials. For the cubic crystals he found that the elastic constants c_{12} and c_{44} were equal for sodium chloride (*Steinsalz*), for which $c_{12}/c_{44} = 1.02$, but not for calcium fluoride (*Flusspath*), for which this ratio was 1.32. He deduced that since "Poisson's relation $c_{12} = c_{44}$ is not fulfilled for fluorspar, the material must consist of strongly polar molecules", that is, ones for which the intermolecular forces are not central. His many other experiments led to similar conclusions. Thus by the end of the 19th century there was ample evidence that most materials did not satisfy the Cauchy relations, nor have a Poisson's ratio of $\frac{1}{4}$, but that a few carefully chosen materials did conform to the rari-constant rules. The more practical elasticians and engineers concluded correctly that the rari-constant theory was of little use to them, and that remains the position to this day. It can even be briskly dismissed as "an error", or even as "absurd" [229]. Some of the more theoretically inclined elasticians even added their voices to the opposition to the idea of interacting point atoms, an opposition that had some considerable following at the end of the century. Thus Love wrote in 1906:

The hypothesis of material points and central forces does not now hold the field. This change in the tendency of physical speculation is due to many causes, among which the disagreement of the rari-constant theory with the results of experiment holds a rather subordinate position. . . . It is now recognized that the theory of atoms must be a part of a theory of the aether, and that the confidence that was felt in the hypothesis of central forces between material particles was premature. [230]

Others were less pessimistic and, as we have seen, explored instead the conditions under which Cauchy's relations might be expected to hold, and the types of materials that could be shown to conform to them. This led in the 20th century to a brief revival of interest in the elastic constants and the light that they could throw on the intermolecular forces, but that discussion belongs to a later chapter.

The study of cohesion as a fundamental part of physics has, in this chapter, been left in the 1820s while we have pursued the clash between the advocates of the rari- and multi-constant theories. Today the study of the elastic properties of materials is both a specialised branch of applied mathematics and a practical subject of importance to mechanical and civil engineers, but it is not an important component of courses of physics [231]. In the 19th century it occupied an unusual position. It was important enough to attract serious work from many of the leading physicists of the time, such as Clausius, Franz Neumann, Voigt, Lamé, Regnault, Amagat, Stokes, William Thomson and Maxwell, to name but three each from Germany, France and Britain. Some of this importance arose from the parallels that they saw between the elastic properties of solids and of the aether as a medium for the propagation of light waves, and some from the needs of the great engineering enterprises of the time. Pearson, writing in 1886, said of the decade 1840–1850: "Not in one country alone, but throughout the length and breadth of Europe we find men foremost in three of the great divisions of science (theoretical, physical and technical) labouring to extend our knowledge of elasticity and of subjects akin to it." [232] In spite of this importance it remained, nevertheless, a curiously detached branch of science. Of those physicists listed above, Clausius and Maxwell were the founders in the 1860s of the kinetic theory of gases, and Thomson followed that subject closely, yet none made any effort to integrate their work in the two fields, although the kinetic theory made no sense without molecules and forces between them. Part of the problem was a reluctance to believe that the nature of matter, particulate or otherwise, was the same in all three phases, solid, liquid and gas. As we shall see, Clausius firmly believed this but the others were not so sure. Even today, when we accept that the same molecular entities are present in the three states of, say, argon, we use rather different theoretical methods in solids for translating the effects of the forces between these entities into the observed physical properties. One reason is the greater importance of quantal effects in solids, but the difference is not confined to this problem. Even in the 19th century physicists apparently saw little advantage in trying to integrate the study of solids with that of liquids and gases.

It is interesting to compare the different form of the debates in the 18th and the 19th centuries between those who believed in particles with forces between them that apparently acted at a distance, and those who refused to countenance such ideas. In the late 17th and in the 18th centuries the second party included some noteworthy figures – Huygens, Leibniz, Euler and, at times, some members of the Bernoulli family – but their opposition never cohered into an alternative doctrine. In the 19th century the opposition was less single-minded since many physicists adopted both hypotheses at different times or for tackling different problems, but those who insisted that the elastic properties of solids could not be explained by central forces between particles had a good case which was cogently argued and which was justified by the behaviour of most materials.

Laplace's fundamental notion of interparticle forces "sensible only at insensible distances" fuelled the debate between the elasticians. His ideas were not lost in what is sometimes called the fall of Laplacian physics, but were buried in this specialised branch of the subject. They remained central to the ideas of Poisson, Cauchy, Saint-Venant and Clausius. They returned, at the hands of van der Waals and others, to the mainstream of physics later in the century, when they had been fruitfully united with a kinetic view of matter.

Notes and references

1 Cantor observes similiarly that Young's optics is closer to that of Euler than to that of Fresnel, his near contemporary; see G.N. Cantor, *Optics after Newton. Theories of light in Britain and Ireland, 1704–1840*, Manchester, 1983, p. 15. Garber does not call the work of the French school 'theoretical physics', and believes that that discipline arose first later in the century in Germany and Britain; E. Garber, *The language of physics: The calculus and the development of theoretical physics in Europe, 1750–1914*, Boston, MA, 1999.

2 See e.g. S.F. Cannon, *Science in culture: the early Victorian period*, New York, 1978, 'The invention of physics', chap. 4, pp. 111–36; M. Crosland and C. Smith, 'The transmisssion of physics from France to Britain: 1800–1840', *Hist. Stud. Phys. Sci.* **9** (1978) 1–61; A. Cunningham and P. Williams, 'De-centring the 'big picture': *The origins of modern science* and the modern origins of science', *Brit. Jour. Hist. Sci.* **26** (1993) 407–32.

3 P.-S. de Laplace (1749–1827). There is no adequate biography of Laplace but he received an unusually long entry in DSB, v. 15, pp. 273–403, by C.C. Gillispie and others. This has been revised and re-issued as C.C. Gillispie, *Pierre Simon de Laplace, 1749–1827*, Princeton, NJ, 1997. The short section on his work on cohesion, pp. 358–60 of DSB and pp. 203–8 of Gillispie, 1997, is by R. Fox. There are another two pages on this subject in H. Andoyer, *L'oeuvre scientifique de Laplace*, Paris, 1922. See also the lecture, R. Hahn, *Laplace as a Newtonian scientist*, Los Angeles, CA, 1967.

4 P.-S. Laplace, *Exposition du système du monde*, 2 vols., Paris, 1796, v. 2, pp. 196–8.

5 J.-B. Biot (1774–1862) M.P. Crosland, DSB, v. 2, pp. 133–40.

6 J.-B. Biot, *Traité de physique expérimentale et mathématique*, 4 vols., Paris, 1816, v. 1, chap. 12, 'Sur les forces qui constituent les corps dans les divers états de solides, de liquides et de gaz', pp. 247–63, see p. 252.

7 A. Libes, *Traité complet et élémentaire de physique*, 2nd edn, 3 vols., Paris, 1813, v. 1, p. 374; v. 2, pp. 1–20; 'Théorie de l'attraction moléculaire ou de l'affinité chimique ramenée à la loi de la gravitation', *Jour. Physique* **54** (1802) 391–8, 443–9.

8 Laplace, ref. 4, 'De l'attraction moléculaire', 2nd edn, 1798, pp. 286–7; 3rd edn, 1808, pp. 296–321. The last edition, the 6th of 1835, is the one reprinted as v. 6 of his *Oeuvres complètes*, [hereafter OC], 14 vols., Paris, 1878–1912, pp. 349–92.

9 C.-L. Berthollet (1748–1822) S.C. Kapoor, DSB, v. 2, pp. 73–82; M. Sadoun-Goupil, *Le chimiste Claude-Louis Berthollet, 1748–1822. Sa vie – son oeuvre*, Paris, 1977. The

relationship between Berthollet and Laplace is a central theme of M.P. Crosland, *The Society of Arcueil: a view of French science at the time of Napoleon I*, London, 1967, see chap. 5.

10 C.-L. Berthollet, *Recherches sur les lois de l'affinité*, Paris, 1801; English translation by M. Farrell, *Researches into the laws of chemical affinity*, London, 1804.

11 C.-L. Berthollet, *Essai de statique chimique*, 2 vols., Paris, 1803, v. 1, pp. 1–2; Sadoun-Goupil, ref. 9, pp. 162–85.

12 J. Davy, *Memoirs of the life of Sir Humphry Davy, Bart.*, 2 vols., London, 1836, v. 1, p. 470. The passage is quoted by T.H. Levere, *Affinity and matter: Elements of chemical philosophy, 1800–1865*, Oxford, 1971, p. 54, and by M. Goupil, *Du flou au clair? Histoire de l'affinité chimique: de Cardan à Prigogine*, Paris, 1991, p. 212. Laplace was more hopeful by 1820.

13 From Berthollet's 'Introduction' to the French translation (1810) of Thomas Thomson's *System of chemistry* of 1809, quoted by Sadoun-Goupil, ref. 9, p. 213; M. Sadoun-Goupil, 'Introduction' to C.-L. Berthollet, *Revue de l'Essai de statique chimique*, Paris, 1980, pp. 1–52, see p. 19. This *Revue* opens with a new chapter, 'De l'attraction moléculaire', which was closely based on Laplace's work.

14 Berthollet, ref. 13, 1980.

15 Berthollet, ref. 11, v. 1, pp. 245–7 and 522–3. The ascription of the first Note to Laplace is made on p. 165. Both Notes are in OC, ref. 8, v. 14, pp. 329–32.

16 P.-S. Laplace, *Traité de mécanique céleste*, 4 vols., Paris, 1798–1805, v. 4, pp. xx–xxiii and 270. A fifth volume was published in parts in 1823–1825, with a posthumous supplement in 1827; OC, ref. 8, vols. 1–5. An English translation of the first four volumes, with extensive notes, was made by Nathaniel Bowditch, *Mécanique céleste by the Marquis de la Place*, 4 vols., Boston, MA, 1829–1839. [N. Bowditch (1773–1838) N. Reingold, DSB, v. 2, pp. 368–9, and the memoir on pp. 1–168 of v. 4 of his translation.] References here are to the original French edition by volume and page number, and to Bowditch's translation by his marginal numbering of paragraphs or sentences. Quotations are generally in the English of Bowditch's translation.

17 Fox dates this commitment to 1821, see R. Fox, *The caloric theory of gases from Lavoisier to Regnault*, Oxford, 1971, p. 168, and, for further discussion, H. Chang, 'Spirit, air, and quicksilver: The search for the "real" scale of temperature', *Hist. Stud. Phys. Biol. Sci.* **31** (2001) 249–84.

18 J. Dalton (1766–1844) A. Thackray, DSB, v. 3, pp. 537–47; J. Dalton, 'Inquiries concerning the signification of the word Particle, as used by modern chemical writers, as well as concerning some other terms and phrases', *(Nicholson's) Jour. Nat. Phil. Chem. Arts* **28** (1811) 81–8. See also L.A.Whitt, 'Atoms or affinities? The ambivalent reception of Daltonian theory', *Stud. Hist. Phil. Sci.* **21** (1990) 57–88.

19 Laplace, ref. 16, v. 4, pp. 231–76; Bowditch, ref. 16, [8137–541].

20 R. Fox, 'The rise and fall of Laplacian physics', *Hist. Stud. Phys. Sci.* **4** (1974) 89–136; J.L. Heilbron, *Weighing imponderables and other science around 1800*, Suppl. to v. 24, Part 1, *Hist. Stud. Phys. Sci.*, Berkeley, CA, 1993.

21 D. Bernoulli, *Hydrodynamica, sive, De viribus et motibus fluidorum commentarii*, Strasbourg, 1738, pp. 200ff.; English translation by T. Carmody and H. Kobus, *Hydrodynamics by Daniel Bernoulli*, New York, 1968, pp. 226ff. This section is reprinted in an English translation by S.G. Brush, *Kinetic theory*, 3 vols., Oxford, 1965–1972, v. 1, pp. 57–65.

22 P.-S. Laplace, 'Sur la théorie des tubes capillaires', *Jour. Physique* **62** (1806) 120–8; OC, ref. 8, v. 14, pp. 217–27. J. Dhombres, 'La théorie de la capillarité selon Laplace: mathématisation superficielle ou étendue?', *Rev. d'Hist. Sci.* **42** (1989) 43–77.

Dhombres lists 13 publications by Laplace on capillarity and related phenomena published between 1806 and 1826; one of these, however, that of 1807, on Laplace's 'Second Supplement' (see below), is by Biot.

23 P.-S. Laplace, 'Supplément au dixième livre du Traité de mécanique céleste. Sur l'action capillaire'. This Supplement of 1806, which is paginated separately, is usually bound into the 4th volume which is dated 1805; OC, ref. 8, v. 4, pp. 349–417.

24 P.-S. Laplace, 'Supplément à la théorie de l'action capillaire', 1807. This is also usually bound into v. 4 of the *Mécanique céleste*; OC, ref. 8, v. 4, pp. 419–98. A less technical account of some of the work in the second Supplement appeared in three papers in the *Journal de Physique*; 'Sur l'attraction et la répulsion apparente des petits corps qui nagent à la surface des fluides', **63** (1806) 248–52; 'Extrait d'un mémoire de l'adhésion des corps à la surface des fluides', *ibid.* 413–18; and 'Sur l'action capillaire', *ibid.* 474–84; OC, ref. 8, v. 14, pp. 228–32, 247–53, and 233–46.

25 Hauksbee's experimental reputation was high among the French Newtonians. His is the name that is mentioned most frequently (after that of Newton himself) in v. 3 of A. Libes, *Histoire philosophique des progrès de la physique*, 4 vols., Paris, 1810–1813.

26 T. Young, *A course of lectures on natural philosophy and the mechanical arts*, 2 vols., London, 1807, v. 1, p. 794 and Fig. 530-1. See also the reprint of his 1805 paper in v. 2, pp. 649–60 to which he made minor corrections and added ten pages of translation of Laplace, with a critical commentary.

27 See J.J. Bikerman, 'Capillarity before Laplace: Clairaut, Segner, Monge, Young', *Arch. Hist. Exact Sci.* **18** (1977–1978) 102–22.

28 Laplace, ref. 23, p. 2; Bowditch, ref. 16, [9178–9].

29 Laplace, ref. 23, p. 5; Bowditch, ref. 16, [9201].

30 Laplace, ref. 23, p. 3; Bowditch, ref. 16, [9182].

31 Laplace, ref. 24, 1807, p. 5; Bowditch, ref. 16, [9790].

32 A.T. Petit (1791–1820) R.Fox, DSB, v. 10, pp. 545–6; A.T. Petit, 'Théorie mathématique de l'action capillaire', *Jour. École Polytech.* 16me cahier, **9** (1813) 1–40. Petit's thesis is discussed by I. Grattan-Guinness, *Convolutions in French mathematics, 1800–1840*, 3 vols., Basel, 1990, v. 2, pp. 447–9.

33 Laplace, ref. 16, 1825, v. 5, Book 16, chap. 4; OC, ref. 8, v. 5, pp. 445–60, see p. 451.

34 Laplace, ref. 23, p. 18; Bowditch, ref. 16 [9301]. Van der Waals repeated Laplace's derivation in his thesis of 1873, see Section 4.3. For other modern derivations, see Dhombres, ref. 22, Grattan-Guinness, ref. 32, v. 2, pp. 442–7, and Heilbron, ref. 20, pp. 158–61.

35 Laplace, ref. 23, p. 7; Bowditch, ref. 16, [9209].

36 For a modern account of his work on this topic, see J.J. Bikerman, 'Theories of capillary attraction', *Centaurus* **19** (1975) 182–206.

37 J.L. Gay-Lussac (1778–1850) M.P. Crosland, DSB, v. 5, pp. 317–27.

38 Laplace, ref. 4, 3rd edn, 1808, p. 309.

39 R.-J. Haüy (1743–1822) R. Hooykaas, DSB, v. 6, pp. 178–83; A. Lacroix, 'La vie et l'oeuvre de l'abbé René-Just Haüy', *Bull. Soc. Française de Minérologie* **67** (1944) 15–226. Jean-Louis Trémery (1773–1851), "ingénieur en chef des Mines", assisted Haüy in his crystallographic work (Lacroix, p. 143). The mineralogist Matteo Tondi (1762–1835) worked in Paris for most of the period from 1799 to 1813 (Lacroix, pp. 72–4; *Enciclopedia Italiana*, Rome, v. 33, 1937, p. 1027). Their part in the capillarity experiments is acknowledged in the second edition of 1806 of Haüy's *Traité élémentaire de physique*, 2 vols., Paris, v. 1, pp. 209–47, 'Tubes capillaires', see p. 224. Bikerman, ref. 36, is wrong in suggesting that the 'M. Haüy' who supplied Laplace with experimental results is not the Abbé R.-J. Haüy.

40 Laplace, ref. 24, 1807, p. 52; Bowditch, ref. 16, [10302].

41 Gay-Lussac's results and calculations based on them are in Biot, ref. 6, v. 1, chap. 22, 'Des phénomènes capillaires', pp. 437–65.

42 See Sadoun-Goupil, ref. 9, p. 75.

43 For both Lord Charles Cavendish (1704–1783) and Henry Cavendish (1731–1810), see the double biography by C. Jungnickel and R. McCormmach, *Cavendish*, Amer. Phil. Soc., Philadelphia, PA, 1996; and for Henry, see R. McCormmach, DSB, v. 3, pp. 155–9. The barometric results are in H. Cavendish, 'An account of the meteorological instruments used at the Royal Society's House', *Phil. Trans. Roy. Soc.* **66** (1776) 375–401.

44 T. Young, 'An essay on the cohesion of fluids', *Phil. Trans. Roy. Soc.* **95** (1805) 65–87, reprinted in ref. 26 and in *Miscellaneous works of the late Thomas Young, M.D., F.R.S.*, ed. G. Peacock, London, 1855, v. 1, pp. 418–53.

45 P.-S. Laplace, 'Sur la dépression du mercure dans un tube de baromètre, due à sa capillarité', in the *Connaissance des temps pour l'an 1812*, 1810, but quoted here from OC, ref. 8, v. 13, pp. 71–7. These calculations were "revised" in 1826 with the help of his assistant Alexis Bouvard (1767–1843) [A.F.O'D. Alexander, DSB, v. 2, pp. 359–60], 'Mémoire sur un moyen de détruire les effets de la capillarité dans les baromètres', published in the *Connaissance des temps pour l'an 1829*, 1826, and reprinted in OC, ref. 8, v. 13, pp. 331–41; they are little changed.

46 F.O. [i.e. T. Young], art. 'Cohesion', in *Supplement to the fourth, fifth, and sixth editions* of Encyclopaedia Britannica, 6 vols., London, 1815–1824, v. 3, pp. 211–22; reprinted in *Miscellaneous works*, ref. 44, v. 1, pp. 454–83.

47 F.A. Gould, 'Manometers and barometers', in R. Glazebrook, ed., *A dictionary of applied physics*, London, 1923, v. 3, pp. 140–92, see p. 160.

48 Laplace, ref. 23, pp. 13–14; Bowditch ref. 16, [9257]. The same sentence occurs in the 3rd edn, 1808, of ref. 4, p. 316.

49 Laplace, ref. 24, p. 72; Bowditch, ref. 16, [10488].

50 Laplace, ref. 24, p. 74; Bowditch, ref. 16, [10498–9].

51 Laplace, ref. 24, p. 71; Bowditch, ref. 16, [10475].

52 P.-S. Laplace, 'Considérations sur la théorie des phénomènes capillaires', *Jour. Physique* **89** (1819) 292–6; OC, ref. 8, v. 14, pp. 259–64.

53 B. Thompson, Count Rumford (1753–1814) S.C. Brown, DSB, v. 13, pp. 350–2; *Benjamin Thompson, Count Rumford*, Cambridge, MA, 1979. The first part only of his memoir was printed by the Institut, of which he was a foreign member; Rumford, 'Expériences et observations sur l'adhésion des molécules de l'eau entre elles', *Mém. Classe Sci. Math. Phys. Inst. France* **7** (1806) 97–108. Both parts were printed in the Geneva journal, *Bibliothèque Britannique, Science et Arts* **33** (1806) 3–16; **34** (1807) 301–13; **35** (1808) 3–16, and are in English in Count Rumford, *Collected works*, 5 vols., Cambridge, MA, 1969, see v. 2, pp. 478–87. The editor of the *Bibliothèque Britannique* commented on the coincidence of Young, Laplace and Rumford all tackling the same problem at the same time, *ibid.* **33** (1806) 97–9, and he printed abstracts of the papers of the first two; Laplace, 99–115 (abstract by Biot); Young, 193–209; Laplace, 283–90; **34** (1807) 23–33.

54 Brown, ref. 53, 1979, pp. 281–4.

55 Young, ref. 26, v. 2, p. 670; *Miscellaneous works*, ref. 44, v. 1, p. 453.

56 [T. Young] Review of 'Théorie de l'action capillaire; par M. Laplace. Supplément au dixième livre du Traité de Mécanique Céleste, pp. 65, 4to, Paris, 1806. Supplément, pp. 80, 1807', *Quart. Rev.* **1** (1809) 107–12, see p. 109.

57 S.-D. Poisson (1781–1840) P. Costabel, DSB, v. 15, pp. 480–90; M. Métivier, P. Costabel, and P. Dugac, ed., *Siméon-Denis Poisson et la science de son temps*,

Paliseau, 1981. This book contains a list of Poisson's works, with notes, pp. 209–65. See also D.H. Arnold, 'The Mécanique Physique of Siméon Denis Poisson: The evolution and isolation in France of his approach to physical theory (1800–1840)', in *Arch. Hist. Exact Sci.* **28** (1983) '1. Physics in France after the Revolution', 243–66; '2. The Laplacian program', 267–87; '3. Poisson: mathematician or physicist?', 289–97; '4. Disquiet with respect to Fourier's treatment of heat', 299–320; '5. Fresnel and the circular screen', 321–42; '6. Elasticity: The crystallization of Poisson's views on the nature of matter', 343–67; *ibid.* **29** (1983) '7. Mécanique Physique', 37–51; '8. Applications of the Mécanique Physique', 53–72; '9. 'Poisson's closing synthesis: Traité de Physique Mathématique', 73–94; *ibid.* **29** (1984) '10. Some perspective on Poisson's contributions to the emergence of mathematical physics', 287–307; Grattan-Guinness, ref. 32, v. 2; E. Garber, 'Siméon-Denis Poisson: Mathematics versus physics in early nineteenth-century France', in *Beyond history of science. Essays in honor of Robert E. Schofield*, ed. E. Garber, Bethlehem, PA, 1990, pp. 156–76.

58 Laplace, ref. 24, pp. 74–5; Bowditch, ref. 16, [10502′ ff.]. Laplace also noted that the composition of the surface layer in a mixture, such as that of alcohol and water, would differ from that in the bulk liquid.

59 S.-D. Poisson, *Nouvelle théorie de l'action capillaire*, Paris, 1831. This book was the first volume of what was intended to be a comprehensive treatise on physics. An abstract, with the same title, had appeared in *Ann. Chim. Phys.* **46** (1831) 61–70. H.F. Link (1767 or 1769–1851), successively Professor of Chemistry and then Botany at Berlin [Pogg., v. 1, col. 1469–70], gave a long summary of the book in *Ann. Physik* **25** (1832) 270–87; **27** (1833) 193–234; with an 'Answer' from [G.F.] Parrot of St Petersburg on 234–8 and Link's reply on 238–9. (In the first of these articles his name is given as H.S. Linck and the confusion is only partially removed by a footnote in the second: "Auch heisse ich nicht H.S. Link".) For a modern summary of Poisson's work, see Arnold, ref. 57, part 8, and A. Rüger, 'Die Molekularhypothese in der Theorie der Kapillarerscheinungen (1805–1873)', *Centaurus* **28** (1985) 244–76. Poisson had produced a second edition of Clairaut's *Théorie de la figure de la Terre* in 1808 but his only editorial comment on the chapter on capillarity was a reference to Laplace's recent work.

60 Poisson, ref. 59, p. 6.

61 Bowditch, ref. 16, [9841ff.]. He lists in v. 4, p. xxxvi all the places where he has reworked Laplace's treatment to take account of Poisson's criticisms.

62 J. Challis (1803–1882) O.J. Eggen, DSB, v. 3, pp. 186–7; J. Challis, 'Report on the theory of capillary attraction', *Rep. Brit. Assoc.* **4** (1834) 253–94; 'On capillary attraction, and the molecular forces of fluids', *Phil. Mag.* **8** (1836) 89–96. This article contains a small correction to the B.A. review.

63 W. Whewell, 'Report on the recent progress and present condition of the mathematical theories of electricity, magnetism, and heat', *Rep. Brit. Assoc.* **5** (1835) 1–34; see also, for a further refutation, [J.A.] Quet, *Recueil de rapports sur les progrés des lettres et les sciences en France: De l'électricité, du magnétisme et de la capillarité*, Paris, 1867, pp. 245–74.

64 D.F.J. Arago (1786–1853) R. Hahn, DSB, v. 1, pp. 200–3.

65 This éloge was read before the Academy on 16 December 1850, and was printed in the *Oeuvres complètes de François Arago*, Paris, v. 2, 1854, pp. 593–689. It is followed by Arago's funeral oration, pp. 690–8.

66 Link, ref. 59, (1833) p. 230.

67 C.F. Gauss (1777–1855) K.O. May, DSB, v. 5, pp. 298–315; W.K. Bühler, *Gauss: a biographical study*, Berlin, 1981; C.F. Gauss, 'Principia generalia theoriae figurae

fluidorum in statu aequilibrii', *Comm. Soc. Reg. Sci. Göttingen* **7** (1830) 39–88, translated into German as 'Allgemeine Grundlagen einer Theorie der Gestalt von Flüssigkeiten im Zustand des Gleichgewichts', in Ostwald's *Klassiker der exacten Wissenschaften*, Leipzig, 1903, No. 135. See also Rüger, ref. 59. Mossotti's contribution to this field also added little to what was known; O.F. Mossotti, 'On the action of the molecular forces in producing capillary phenomena', *(Taylor's) Scientific Memoirs* **3** (1843) 564–77.

68 Young, ref. 26, v. 2, pp. 661–2.

69 Laplace, ref. 52, p. 293, OC, ref. 8, v. 14, p. 261.

70 P.-S. Laplace, ref. 16, v. 5, Book 12 (1823), 'De l'attraction et de la répulsion des sphères, et des les lois de l'équilibre et du mouvement des fluides élastiques', pp. 87–144, see pp. 92–3; OC, v. 5, pp. 99–160, see pp. 104–5. There is a précis of Book 12 in I. Todhunter and K. Pearson, *A history of the theory of elasticity and of the strength of materials*, 2 vols., London, 1886, 1893, v. 1, pp. 161–6.

71 C. Cagniard de la Tour (1777–1859) J. Payen, DSB, v. 3, pp. 8–10.

72 Laplace, ref. 24, pp. 67–71, see pp. 68–9; Bowditch, ref. 16 [10461–87], see [10463]. Bowditch writes 'attractive force' but Laplace has 'forces', which seems to express better the essence of a mean-field approximation.

73 For a modern discussion of these points, see G.D. Scott and I.G. MacDonald, 'Young's estimate of the size of molecules', *Amer. Jour. Phys.* **33** (1965) 163–4; E.A. Mason, 'Estimate of molecular sizes and Avogadro's number from surface tension', *ibid.* **34** (1966) 1193; A.P. French, 'Earliest estimates of molecular size', *ibid.* **35** (1967) 162–3.

74 O.R. [i.e. T. Young], art. 'Carpentry', in *Supplement . . . to Encyclopaedia Britannica*, ref. 46, 1817, v. 2, pp. 621–46; reprinted in part in *Miscellaneous works*, ref. 44, v. 2, pp. 248–61.

75 [B. Franklin], 'Extract of a letter to Doctor Brownrigg from Doctor Franklin', *Phil. Trans. Roy. Soc.* **64** (1774) 447–60. A history of early studies of the stilling of water waves by a layer of oil was written by A. van Beek, 'Mémoire concernant la propriété des huiles de calmer les flots, et de rendre la surface de l'eau parfaitement transparente', *Ann. Chim. Phys.* **4** (1842) 257–89. See also C.H. Giles, 'Franklin's teaspoonful of oil', *Chem. Industry* (1969) 1616–24, and, with S.D. Forrester, 'Wave damping: the Scottish contribution', *ibid.* (1970) 80–7.

76 Young, ref. 26, v. 1, p. 625.

77 J. Ivory (1765–1842) M.E. Baron, DSB, v. 7, p. 37; A.D.D. Craik, 'James Ivory, F.R.S.: 'The most unlucky person that ever existed'', *Notes Rec. Roy. Soc.* **54** (2000) 223–47.

78 [J. Ivory] art. 'Fluids, elevation of', in *Supplement . . . to Encyclopaedia Britannica*, ref. 46, 1820, v. 4, pp. 309–23, see p. 320.

79 G. Belli (1791–1860) Pogg., v. 1, col. 140–1, 1535–6.

80 G. Belli, 'Osservazioni sull' attrazione molecolare', *Gior. Fis. Chim. ec., di Brugnatelli* **7** (1814) 110–26, 169–202. There is a summary of this paper in Todhunter and Pearson, ref. 70, v. 1, pp. 93–6.

81 Belli, ref. 80, p. 175. For Haüy's book, see ref. 39.

82 Belli, ref. 80, p. 187.

83 See e.g. J.S. Rowlinson, 'Attracting spheres: some early attempts to study interparticle forces', *Physica A* **244** (1997) 329–33.

84 J.B. Biot and F. Arago, 'Mémoire sur les affinités des corps pour la lumière, et particulièrement sur les forces réfringentes des différens gaz', *Mém. Classe Sci. Math. Phys. Inst. France* **7**, 2me partie (1806) 301–87.

85 Crosland, ref. 9.

86 E. Malus (1775–1812) K.M. Pedersen, DSB, v. 9, pp. 72–4; E. Malus, 'Sur une propriété de la lumière réfléchie', *Mém. Phys. Chim. Soc. d'Arcueil* **2** (1809) 143–58; 'Sur une propriété des forces répulsives qui agissent sur la lumière', *ibid.* 254–67.

87 J.-B. J. Delambre (1749–1832) I.B. Cohen, DSB, v. 4, pp. 14–18.

88 G. Cuvier (1769–1832) F. Bourdier, DSB, v. 3, pp. 521–8.

89 'Présentation à son Majesté Impériale et Royale en son Conseil d'État', *Hist. Classe Sci. Math. Phys. Inst. France* **8** (1808) 169–229, see 204. Extended and revised versions of both reports were also published separately; the first as *Rapport historique sur les progrès des sciences mathématiques depuis 1789, et sur leur état actuel . . .*, Paris, 1810, and the second as *Rapport . . . des sciences naturelles . . .*, Paris, 1810, with a second edition in 1828.

90 A.J. Fresnel (1788–1827) R.H. Silliman, DSB, v. 5, pp. 165–71.

91 J.B.J. Fourier (1768–1830) J. Ravetz and I. Grattan-Guinness, DSB, v. 5, pp. 93–9; J. Herivel, *Joseph Fourier: The man and the physicist*, Oxford, 1975, esp. 'Epilogue', pp. 209–41; Grattan-Guinness, ref. 32, v. 2, chap. 9, pp. 584–632.

92 S. Germain (1776–1831) E.E. Kramer, DSB, v. 5, pp. 375–6; L.L. Bucciarelli and N. Dworsky, *Sophie Germain: An essay in the history of the theory of elasticity*, Dordrecht, 1980; A. Dahan Dalmedico, 'Mécanique et théorie des surfaces; les travaux de Sophie Germain', *Hist. Math.* **14** (1987) 347–65; 'Étude des méthodes et des "styles" de mathématisation: la science de l'élasticité', chap. V.2, pp. 349–442 of *Sciences à l'époque de la Révolution française: recherches historiques*, ed. R. Rashed, Paris, 1988.

93 C.-L.-M.-H. Navier (1785–1836) R.M. McKeon, DSB, v. 10, pp. 2–5.

94 A.-L. Cauchy (1789–1857) H. Freudenthal, DSB, v. 3, pp. 131–48; B. Belhoste, *Augustin-Louis Cauchy, a biography*, New York, 1991; A. Dahan Dalmedico, *Mathématisations: Augustin-Louis Cauchy et l'école française*, Argenteuil and Paris, 1992, Part 4, 'L'élasticité des solides', pp. 215–98.

95 A.E. Woodruff, 'Action at a distance in nineteenth century electrodynamics', *Isis* **53** (1962) 439–59; G.N. Cantor and M.J.S. Hodge, *Conceptions of ether; studies in the history of ether theories, 1740–1900*, Cambridge, 1981.

96 J.S. Rowlinson and B. Widom, *Molecular theory of capillarity*, Oxford, 1982.

97 J. Fourier, *Théorie analytique de la chaleur*, Paris, 1822; English translation, with a list of Fourier's papers, by A. Freeman, *Analytical theory of heat*, Cambridge, 1878.

98 This controversy is discussed by Arnold, ref. 57, part 4, and by Herivel, ref. 91, pp. 153–9. Poisson's review of 1808, signed only with the letter P, is reprinted by G. Darboux in *Oeuvres de Fourier*, 2 vols., Paris, 1888, 1890, v. 2, pp. 215–21.

99 Fourier, ref. 97, pp. 37–9; English trans., pp. 39–40.

100 Fourier, ref. 97, pp. 13–14; English trans., p. 23.

101 Fourier, ref. 97, pp. 597–8; English trans., p. 464.

102 Fourier, ref. 97, pp. 84, 89–90; English trans. pp. 78, 84.

103 Laplace, ref. 4, v. 1, p. 309; Bucciarelli and Dworsky, ref. 92, p. 132, note 5.

104 Fourier, ref. 97, p. i; English trans., p. 1. See also G. Bachelard, *Étude sur l'évolution d'un problème de physique: la propagation thermique dans les solides*, Paris, 1927, esp. chap. 4, pp. 55–72 on Comte and Fourier.

105 I.A.M.F.X. Comte (1798–1857) L. Laudan, DSB, v. 3, pp. 375–80.

106 See, for example, Biot's summary of contemporary views on the nature of caloric in ref. 6, v. 1, pp. 19–23.

107 See e.g. R. Harré, 'Knowledge', chap. 1, pp. 11–54, and S. Schaffer, 'Natural philosophy', chap. 2, pp. 55–91, of G.S. Rousseau and R. Porter, ed., *The ferment of*

knowledge: Studies in the historiography of eighteenth-century science, Cambridge, 1980.

108 P. Duhem, 'L'évolution de la mécanique', *Rev. gén. des sciences* (1903) 119–32.

109 D.S.L. Cardwell, *From Watt to Clausius: The rise of thermodynamics in the early industrial age*, London, 1971; C. Smith, *The science of energy. A cultural history of energy physics in Victorian Britain*, London, 1998.

110 N.L.S. Carnot (1796–1832) J.F. Challey, DSB, v. 3, pp. 79–84.

111 B.-P.-É. Clapeyron (1799–1864) M. Kerker, DSB, v. 3, pp. 286–7.

112 J.P. Joule (1818–1889) L. Rosenfeld, DSB, v. 7, pp. 180–2; D.S.L. Cardwell, *James Joule: a biography*, Manchester, 1989.

113 W. Thomson (1824–1907) J.Z. Buchwald, DSB, v. 13, pp. 374–88; C. Smith and M.N. Wise, *Energy and empire: a biographical study of Lord Kelvin*, Cambridge, 1989, esp. chap. 6, pp. 149–202, 'The language of mathematical physics'.

114 There are numerous histories of crystallography. Two of the most relevant to Section 3.4 are J.G. Burke, *Origin of the science of crystals*, Berkeley, CA, 1966, and M. Eckert, H. Schubert, G. Torkar, C. Blondel and P. Quédec, 'The roots of solid-state physics before quantum mechanics', chap. 1, pp. 3–87, of L. Hoddeson, E. Braun, J. Teichmann and S. Weart, ed., *Out of the crystal maze: Chapters from the history of solid-state physics*, Oxford, 1992. For metals in the 18th century, see C.S. Smith, 'The development of ideas on the structure of metals', in M. Clagett, ed., *Critical problems in the history of science*, Madison, WI, 1959, pp. 467–98.

115 Todhunter and Pearson, ref. 70.

116 J. Freind, *Chymical lectures: In which almost all the operations of chymistry are reduced to their true principles . . .*, London, 1712, p. 147.

117 L.B. Guyton de Morveau, H. Maret and J.-F. Durande, *Élémens de chymie théorique et pratique*, 3 vols., Dijon, 1777–1778, v. 1, pp. 73–8, 'De la crystallisation', see pp. 75–6. The book comprises lectures read at the Dijon Academy in 1774. Guyton's co-authors were two medical men, H. Maret (1726–1786) and J.-F. Durande (1732–1794), the Professor of Botany.

118 J.-B.L. Romé de l'Isle (1736–1790) R. Hooykaas, DSB, v. 11, pp. 520–4.

119 R.-J. Haüy, *Essai d'une théorie sur la structure des cristaux, appliquée à plusieurs genres de substances crystallisées*, Paris, 1784, see 'Article premier', pp. 47–56. Haüy's work on crystals, and that of some of his predecessors and successors, is described in detail in a series of articles by K.H. Wiederkehr in *Centaurus* **21** (1977) 27–43, 278–99; **22** (1978) 131–56, 177–86. See also Lacroix, ref. 39; and the articles that follow: C. Mauguin, 'La structure des cristaux d'après Haüy', *Bull. Soc. Française de Minérologie* 227–63; J. Orcel, 'Haüy et la notion d'espèce en minérologie', *ibid.* 265–337; S.H. Mauskopf, 'Crystals and compounds: Molecular structure and composition in nineteenth-century French science', *Trans. Amer. Phil. Soc.* **66** (1976) Part 3. Haüy's work is the subject of Issue no. 3 of *Rev. d'Hist. Sci.* **50** (1997) 241–356.

120 R.-J. Haüy, *Traité de minéralogie*, 5 vols., Paris, 1801, v. 1, 'Discours préliminaire', pp. i–lii, and pp. 1–109, 283ff.; Mauguin, ref. 119.

121 Haüy, ref. 120, pp. 464–79.

122 W.H. Wollaston (1766–1828) D.C. Goodman, DSB, v. 14, pp. 486–94; W.H. Wollaston, 'On the elementary particles of certain crystals', *Phil. Trans. Roy. Soc.* **103** (1813) 51–63. Some of Wollaston's models are now in the Science Museum, London.

123 For a review of later work in this style, see W. Barlow and H.A. Miers, 'The structure of crystals – Report of the Committee . . .', *Rep. Brit. Assoc.* **71** (1901) 297–337.

124 L.A. Seeber (1793–1855) Pogg., v. 2, col. 891; L.A. Seeber, 'Versuch einer Erklärung des innern Baues der fester Körper', *Ann. Physik* **76** (1824) 229–48, 349–72.

125 C.S. Weiss (1780–1856) W.T. Holser, DSB, v. 14, pp. 239–42.

126 C.S. Weiss, 'Ueber eine verbesserte Methode für die Bezeichnung der verschiedenen Flächen eines Crystallisations-systems; . . .', *Abhand. Phys. Klasse König-Preuss. Akad. Wiss.* (1816–1817) 286–336, and other papers in this journal from 1814 onwards. For Weiss and his successors, see E. Scholz, 'The rise of symmetry concepts in the atomistic and dynamistic schools of crystallography, 1815–1830', *Rev. d'Hist. Sci.* **42** (1989) 109–22.

127 F. Mohs (1773–1839) J.G. Burke, DSB, v. 9, pp. 447–9.

128 F. Mohs, *Treatise on mineralogy, or the natural history of the mineral kingdom*, 3 vols., Edinburgh, 1825. The original German edition was published in 1822–1824. His principles are set out in the 'Introduction' to a shorter and earlier book, *The characters of the classes, orders, genera, and species; or, the characteristics of the natural history system of mineralogy*, Edinburgh, 1820, pp. iii–xxvii.

129 W.K. Haidinger (1795–1871) J. Wevers, DSB, v. 6, pp. 18–20. He sets out the principles proposed by Mohs in W. Haidinger, 'On the determination of the species, in mineralogy, according to the principles of Professor Mohs', *Trans. Roy. Soc. Edin.* **10** (1824) 298–313. He was a Foreign Member of that Society and later of the Royal Society of London; see *Proc. Roy. Soc.* **20** (1871–1872) xxv–xxvii.

130 E.-F.-F. Chladni (1756–1827) S.C. Dostrovsky, DSB, v. 3, pp. 258–9.

131 H.C. Ørsted (1777–1851) L.P. Williams, DSB, v. 10, pp. 182–6. For these experiments, see 'Letter of M. Orsted, Professor of Philosophy at Copenhagen, to Professor Pictet of Geneva, upon sonorous vibrations', *Phil. Mag.* **24** (1806) 251–6. (The date on this letter of 26 May 1785 is clearly a misprint.) For his later experiments, see K. Jelved, A.D. Jackson and O. Knudsen, *Selected scientific works of Hans Christian Ørsted*, Princeton, NJ, 1998, 'On acoustic figures', pp. 261–2; 'Experiments on acoustic figures', 1808, pp. 264–81; and for his views on matter and the interactions in it, see his 'View of the chemical laws of nature obtained through recent discoveries', 1812, pp. 310–92.

132 E.-F.-F. Chladni, *Traité d'acoustique*, Paris, 1809. An appendix sets out the terms of the prize "for giving a mathematical theory of the vibrations of elastic surfaces, and for comparing it with experiment", pp. 353–7.

133 P.-S. Laplace, 'Mémoire sur le mouvement de la lumière dans les milieux diaphanes', *Mém. Classe Sci. Math. Phys. Inst. France* (1809) 300–42; OC, ref. 8, v. 12, pp. 267–98, see p. 288.

134 A.-M. Legendre (1752–1833) J. Itard, DSB, v. 8, pp. 135–43.

135 J.L. Lagrange, *Méchanique analitique*, Paris, 1788.

136 Bucciarelli and Dworsky, ref. 92, pp. 54–6.

137 Bucciarelli and Dworsky, ref. 92, p. 131, note 19.

138 S.-D. Poisson, 'Mémoire sur les surfaces élastiques', *Mém. Classe Sci. Math. Phys. Inst. France*, 2me partie (1812) 167–225. The memoir was read on 1 August 1814 and the volume was published in 1816. For the contrasting approaches of Germain and Poisson, see Grattan-Guinness, ref. 32, v. 2, pp. 461–70.

139 S. Germain, *Recherches sur la théorie des surfaces élastiques*, Paris, 1821. She acknowledges Fourier's advice in the 'Avertissement', pp. viii–ix. See also Bucciarelli and Dworsky, ref. 92, pp. 85–97, and Todhunter and Pearson, ref. 70, v. 1, pp. 147–60.

140 S. Germain, *Remarques sur la nature, les bornes et l'étendue de la question des surfaces élastiques, et l'équation générale de ces surfaces*, Paris, 1826, pp. 3–4.

141 S. Germain, 'Examen des principes qui peuvent conduire à la connaissance des lois de l'équilibre et du mouvement des solides élastiques', *Ann. Chim. Phys.* **38** (1828) 123–31.

142 S.-D. Poisson, 'Mémoire sur l'équilibre et le mouvement des corps élastiques', *Ann. Chim. Phys.* **37** (1828) 337–54. This was read before the Academy on 14 April and 24 November 1828 and published in full in *Mém. Acad. Roy. Sci.* **8** (1825) 357–570, 623–7, published in 1829.

143 S. Germain [ed. J. Lherbette], *Considérations générales sur l'état des sciences et des lettres aux différentes époques de leur culture*, Paris, 1833.

144 S. Germain, *Oeuvres philosophiques*, Paris, 1879.

145 For a discussion of what was meant at different times by pressure in a flowing fluid or a strained solid, see A. Dahan Dalmedico, 'La notion de pression: de la métaphysique aux diverses mathématisations', *Rev. d'Hist. Sci.* **42** (1989) 79–108.

146 The later history of the elasticity of plates and thin shells adds nothing to our story, see A.E.H. Love, *A treatise on the mathematical theory of elasticity*, 2 vols., Cambridge, 1892, 1893, 'Historical introduction' to v. 2, pp. 1–23, and chaps. 19–22, pp. 186–288.

147 There are numerous histories of elasticity, but they naturally treat the subject from the standpoint of the development of the general theory and so rarely go deeply into the problems of the interparticle forces. The early work of Saint-Venant is useful since he himself was a major contributor to the field. [A.J.C. Barré de Saint-Venant (1797–1886) J. Itard, DSB, v. 12, pp. 73–4; O. Darrigol,'God, waterwheels, and molecules: Saint-Venant's anticipation of energy conservation', *Hist. Stud. Phys. Biol. Sci.* **31** (2001) 285–353]. See his'Historique abrégé des recherches sur la résistance et sur l'élasticité des corps solides' in C.L.M.H. Navier, *Resumé des leçons données à l'École des Ponts et Chaussées sur l'application de la mécanique à l'établissement des constructions et des machines; Première section, De la résistance des corps solides*, 3rd edn, ed. A.J.C. Barré de Saint-Venant, Paris, 1864, pp. xc–cccxi. The work of Todhunter and Pearson, ref. 70, is valuable for the extent of its coverage, and that of Grattan-Guinness, ref. 32, for a full account of French mathematical work in the field. Eighteenth century work is not relevant to the subject in hand but is discussed by C. Truesdell, 'The creation and unfolding of the concept of stress' in his *Essays in the history of mechanics*, Berlin, 1968, pp. 184–238, and in his 'The rational mechanics of flexible or elastic bodies, 1638–1788', which is v. 11, part 2, of the 2nd Series of *Leonhardi Euleri omnia opera*, Zürich, 1960.

148 C.L.M.H. Navier, 'Sur la flexion des verges élastiques courbes', *Bull. Sci. Soc. Philomathique Paris* (1825) 98–100, 114–18; 'Extrait des recherches sur la flexion des plans élastiques', *ibid.* (1823) 92–102.

149 Saint-Venant, ref. 147, p. cxlvi.

150 Navier uses the usual word 'molécule', but his meaning is made clear by his qualification of it in other papers as 'points matérials, ou molécules' and as 'molécules matérielles'. Again, the less committing word 'particle' is used in quotations from his work.

151 See Bucciarelli and Dworsky, ref. 92, p. 141, notes 12, 13.

152 E.g. Todhunter and Pearson, ref. 70, v. 1, p. 1.

153 C.L.M.H. Navier, 'Sur les lois de l'équilibre et du mouvement des corps solides élastiques', *Bull. Sci. Soc. Philomathique Paris* (1823) 177–81.

154 C.L.M.H. Navier, 'Mémoire sur les lois de l'équilibre et du mouvement des corps solides élastiques', *Mém. Acad. Roy. Sci.* **7** (1824) 375–94, read 14 May 1821, and published in 1827.

155 C.L.M.H. Navier, 'Sur les lois des mouvements des fluides, en ayant égard à l'adhésion des molécules', *Ann. Chim. Phys.* **19** (1821) 244–60, 448; '. . . du mouvement . . .', *Bull. Sci. Soc. Philomathique Paris* (1825) 75–9.

156 C.L.M.H. Navier, 'Mémoire sur les lois du mouvement des fluides', *Mém. Acad. Roy. Sci.* **6** (1823) 389–440, read 18 March 1822 and published in 1827.

157 A.-L. Cauchy, 'Recherches sur l'équilibre et le mouvement intérieur des corps solides, ou fluides élastiques ou non élastiques', *Bull. Sci. Soc. Philomathique Paris* (1823) 9–13.

158 His work and the development of his thought can be followed through a series of articles: A.-L. Cauchy, *Exercises de mathématiques*, Paris, 2nd year (1827): (a) 'De la pression dans les fluides', 23–4; (b) 'De la pression ou tension dans un corps solide', 42–59; (c) 'Sur la condensation et la dilation des corps solides', 60–9; (d) 'Sur les relations qui existent, dans l'état d'équilibre d'un corps solide ou fluide, entre les pressions ou tensions et les forces accélératrices', 108–11; 3rd year (1828): (e) 'Sur les équations qui expriment les conditions d'équilibre, ou les lois du mouvement intérieur d'un corps solide, élastique, ou nonélastique', 160–87; (f) 'Sur l'équilibre et le mouvement d'un système de points matériels sollicités par des forces d'attraction ou de répulsion mutuelle', 188–212; (g) 'De la pression ou tension dans un système de points matériels', 213–36; (h) 'Sur quelques théorèmes relatifs à la condensation ou à la dilation des corps', 237–44; 4th year (1829): (i) 'Sur les équations différentielles d'équilibre ou de mouvement pour un système de points matériels sollicités par les forces d'attraction ou de répulsion mutuelle', 129–39. These articles are reprinted in vols. 7–9 of *Oeuvres complètes d'Augustin Cauchy*, 2nd series, Paris, 1889–1891.

159 S.-D. Poisson, 'Note sur les vibrations des corps sonores', *Ann. Chim. Phys.* **36** (1827) 86–93.

160 See *Ann. Chim. Phys.* **36** (1827) 278.

161 This packet was not opened until 1974 when it was published with an introduction by C. Truesdell, 'Rapport sur le pli cacheté, . . . dans la séance du 1er octobre 1827, par M. Cauchy, . . ., 'Sur l'équilibre et le mouvement intérieur d'un corps solide considéré comme un système de molécules distinctes les unes des autres'', *Compt. Rend. Acad. Sci.* **291** (1980) Suppl. 'Vie académique', 33–46. It is a sketch of his work in ref. 158(f)–(i).

162 S.-D. Poisson, 'Note sur l'extension des fils et des plaques élastiques', *Ann. Chim. Phys.* **36** (1827) 384–7. See also p. 451 of his great paper of 1828, ref. 142.

163 [A. Fresnel], 'Observations de M. Navier sur un mémoire de M. Cauchy', *Bull. Sci. Soc. Philomathique Paris* (1823) 36–7.

164 Navier's complaints are in *Ann. Chim. Phys.* **38** (1828) 304–14; **39** (1828) 145–51; and **40** (1829) 99–107. Poisson's replies are in **38** (1828) 435–40; and **39** (1828) 204–11. (In the Royal Society copy of this journal there are marginal notes in French, apparently contemporary, but scarcely legible, that suggest that Euler had had something useful to contribute on this subject.) Arago's closing 'Note du rédacteur' is in **40** (1829) 107–11. This was answered by Navier in a paper read at the Academy in May 1829 and published as 'Note relative à la question de l'équilibre et du mouvement des corps solides élastiques', (*Férussac's*) *Bull. Sci. Math.* **11** (1829) 243–53.

165 Saint-Venant, ref. 147, p. clxv.

166 S.-D. Poisson, 'Mémoire sur l'équilibre et le mouvement des corps solides élastiques et des fluides', *Ann. Chim. Phys.* **42** (1829) 145–71; 'Mémoire sur les équations générales de l'équilibre et du mouvement des corps solides élastiques, et fluides', *Jour. École Polytech.* 20me cahier, **13** (1831) 1–174.

167 Laplace, ref. 24 (1807), pp. 68–9; Bowditch, ref. 16 [10461–520].

168 Poisson, ref. 166 (1829), pp. 149, 153.

169 Laplace, refs. 4 and 8.

170 S.-D. Poisson, 'Mémoire sur l'équilibre des fluides', read at the Academy on 24 November 1828, the same day as the conclusion of his great memoir on elasticity; published in abstract in *Ann. Chim. Phys.* **39** (1828) 333–5 and in full in *Mém. Acad. Roy. Sci.* **9** (1826) 1–88, published 1830.

171 The words 'stress' and 'strain' were not used in English with their modern precise meanings until introduced in the middle of the 19th century by W.J.M. Rankine, but the concepts are implicit in the work of Poisson and Navier, where they are usually called 'forces' and 'displacements', and more explicit in Cauchy's work. For 'strain', see W.J.M. Rankine, 'Laws of elasticity of solid bodies', *Camb. Dubl. Math. Jour.* **6** (1851) 47–80, 172–81, 185–6; **7** (1852) 217–34, on 49, and for 'stress', 'On axes of elasticity and crystalline form', *Phil. Trans. Roy. Soc.* **146** (1856) 261–85. It is convenient to use both words in an anachronistic way to describe work from the 1820s onwards.

172 Navier, ref. 164, 1829; see also Arnold, ref. 57, part 6.

173 Saint-Venant, ref. 147, p. clix.

174 It is found first in the 1823 memoir on elastic plates, ref. 148.

175 W. Thomson, 'Molecular constitution of matter', *Proc. Roy. Soc. Edin.* **16** (1890) 693–724.

176 W. Voigt (1850–1919) S. Goldberg, DSB, v. 14, pp. 61–3.

177 W. Voigt, *Lehrbuch der Kristallphysik*, Leipzig, 1910.

178 G. Lamé (1795–1870) S.L. Greitzer, DSB, v. 7, pp. 601–2.

179 G. Lamé, *Leçons sur la théorie mathématique de l'élasticité des corps solides*, Paris, 1852, p. 50.

180 G. Green (1793–1841) P.J. Wallis, DSB, v. 15, pp. 199–201; D.M. Cannell, *George Green, mathematician and physicist, 1793–1841: The background to his life and work*, London, 1993; G. Green, 'On the laws of reflexion and refraction of light at the common surface of two non-crystallized media', *Trans. Camb. Phil. Soc.* **7** (1839) 1–24, 113–20, reprinted in *Mathematical papers of the late George Green*, ed. N.M. Ferrers, London, 1871, pp. 245–69. The paper was read before the Society on 11 December 1837.

181 Truesdell, ref. 147, 1968, ascribes the first use of this modulus to Euler.

182 The modern use of this phrase seems to be due to Love, in the second and later editions of his *Treatise*, ref. 146. The second edition is virtually a new book and contains in Note B, at the end, a modern version of Cauchy's work in ref. 158(g).

183 W. Voigt, 'Bestimmung der Elasticitätsconstanten von Beryll und Bergkrystall', *Ann. Physik* **31** (1887) 474–501, 701–24; '... von Topas und Baryt', *ibid.* **34** (1888) 981–1028; '... von Flussspath, Pyrit, Steinsalz, Sylvin', *ibid.* **35** (1888) 642–61.

184 W. Voigt, 'L'état actuel de nos connaissances sur l'élasticité des crystaux', *Rapports présentés au Congrès International de Physique*, Paris, 1900, v. 1, pp. 277–347.

185 Saint-Venant, ref. 147, pp. clxiii and 653–6.

186 G. Lamé, *Notice autobiographique*, Paris, [1839?], p. 14. This pamphlet was designed to support his case for election to the Academy and gives the background for many of his papers.

187 G. Lamé and E. Clapeyron, 'Mémoire sur l'équilibre intérieure des corps solides homogènes'. This was sent to the Academy in April 1828 and published in *Mém. div. Savans Acad. Roy. Sci.* **4** (1833) 463–562. It had already appeared in (*Crelle*) *Jour. reine angew. Math.* **7** (1831) 150–69, 337–52, 381–413, where it was preceded by the report made on it for the Academy by Navier and Poinsot, pp. 145–9. Fourier's

response, as Secretary, to this favourable report is printed by Lamé in ref. 186, pp. 14–15.

188 Lamé, ref. 179, p. 332.

189 Todhunter and Pearson, ref. 70, v. 1, pp. 496–505; K. Pearson (1857–1936) C. Eisenhart, DSB, v. 10, pp. 447–73.

190 R. Clausius (1822–1888) E.E. Daub, DSB, v. 3, pp. 303–11.

191 F.E. Neumann (1798–1895) J.G. Burke, DSB, v. 9, pp. 26–9. Neumann's allegiance to molecular interpretations was, at best, lukewarm, see K.M. Olesko, *Physics as a calling: discipline and practice in the Königsberg Seminar for Physics*, Ithaca, NY, 1991.

192 G.G. Stokes (1819–1903) E.M. Parkinson, DSB, v. 13, pp. 74–9.

193 J. Stefan (1835–1893) W. Böhm, DSB, v. 13, pp. 10–11; B. Pourprix and R. Locqueneux, 'Josef Stefan (1835–1893) et les phénomènes de transport dans les fluides: la jonction entre l'hydrodynamique continuiste et la théorie cinétique des gaz', *Arch. Int. d'Hist. Sci.* **38** (1988) 86–118.

194 W. Thomson and P.G. Tait, *Treatise on natural philosophy*, 2nd edn, Cambridge, 1883, v. 1, part 2, § 673, p. 214.

195 G.C. Stokes, 'On the theories of the internal friction of fluids in motion, and of the equilibrium and motion of elastic solids', *Trans. Camb. Phil. Soc.* **8** (1849) 287–319, reprinted in his *Mathematical and physical papers*, Cambridge, v. 1, pp. 75–129. The paper was read on 14 April 1845.

196 *The correspondence between Sir George Gabriel Stokes and Sir William Thomson, Baron Kelvin of Largs*, ed. D.B. Wilson, 2 vols., Cambridge, 1990, v. 1, Letter 145.

197 Lord Kelvin, *Baltimore lectures on molecular dynamics and the wave theory of light*, London, 1904, Lecture 11, pp. 122–34. A.S. Hathaway's original mimeographed reproduction of the lectures of 1884 has been printed by R. Kargon and P. Achinstein, *Kelvin's Baltimore lectures and modern theoretical physics*, Cambridge, MA, 1987, pp. 106–14. Kelvin had previously devised mechanical models that exhibited elasticity without any 'repulsion' between the units; see his Friday evening Discourse at the Royal Institution of 4 March 1881, 'Elasticity viewed as possibly a mode of motion', *Proc. Roy. Inst.* **9** (1882) 520–1, and his Address at the meeting of the British Association in Montreal two months before the Baltimore Lectures, 'Steps towards a kinetic theory of matter', *Rep. Brit. Assoc.* **54** (1884) 613–22. These are reprinted in his *Popular lectures and addresses*, London, v. 1, 1889, pp. 142–6 and 218–52.

198 Todhunter and Pearson, ref. 70, v. 2, part 2, pp. 364, 456–9.

199 Lamé, ref. 179, pp. 77–8.

200 S. Haughton (1821–1897) DNB, Suppl., 1909; D.J.C[unningham]., *Proc. Roy. Soc.* **62** (1897–1898) xxix–xxxvii. S. Haughton, 'On a classification of elastic media, and the laws of plane waves propagated through them', *Trans. Roy. Irish Acad.* **22** (1855) 97–138. This paper was read in January 1849, before the publication of Lamé's *Leçons*.

201 P.-M.-M. Duhem (1861–1916) D.G. Miller, DSB, v. 4, pp. 225–33. Duhem's criticism in his 'L'évolution de la mécanique' of 1903, ref. 108, is quoted at length, in English, in J.F. Bell, *The experimental foundations of solid mechanics*, which is volume VIa/1 of the *Handbuch der Physik*, ed. S. Flügge, Berlin, 1973, see pp. 249–50.

202 R. Clausius, 'Ueber die Veränderungen, welche in den bisher gebräuchlichen Formeln für das Gleichgewicht und die Bewegung elastischer fester Körper durch

neuere Beobachtungen nothwendig geworden sind', *Ann. Physik* **76** (1849) 46–67.

203 For Cauchy's definitions, see ref. 158(f), p. 198, and 158(g), pp. 230, 236.

204 A.J.C. Barré de Saint-Venant, 'Mémoires sur les divers genres d'homogénéité mécanique des corps solides élastiques, …', *Compt. Rend. Acad. Sci.* **50** (1860) 930–3. He repeated this definition a few years later, ref. 147, App. 2, p. 526.

205 The final form of Saint-Venant's views is in the long notes he attached to §§ 11 and 16 of his translation of the textbook of A. Clebsch, *Théorie de l'élasticité des corps solides*, Paris, 1883.

206 W.Th[omson]., art. 'Elasticity' in *Encyclopaedia Britannica,* 9th edn, London, 1877.

207 W. Thomson, ref. 175, and, as Lord Kelvin, 'On the elasticity of a crystal according to Boscovich', *Proc. Roy. Soc.* **54** (1893) 59–75, reprinted as App. I of ref. 197, 1904.

208 W. Barlow (1845–1934) W.T. Holser, DSB, v. 1, pp. 460–3; W. Barlow, 'Probable nature of the internal symmetry of crystals', *Nature* **29** (1883–1884) 186–8, 205–7; see also, L. Sohncke, 383–4, and Barlow, 404, on the same subject.

209 A.E.H. Love (1863–1940) K.E. Bullen, DSB, v. 8, pp. 516–17; Love, ref. 146, 2nd edn.

210 M. Born (1882–1970) A. Hermann, DSB, v. 15, pp. 39–44; M. Born, *Dynamik der Kristallgitter*, Leipzig, 1915; 'Über die elektrische Natur der Kohäsionskräfte fester Körper', *Ann. Physik* **61** (1920) 87–106.

211 M. Born, 'Reminiscences of my work on the dynamics of crystal lattices', pp. 1–7 of *Lattice dynamics*, Proceedings of the International Conference held at Copenhagen, August 5–9, 1963, ed. R.F. Wallis, Oxford, 1965, and 'Rückblick auf meine Arbeiten über Dynamik der Kristallgitter', pp. 78–93 of H. and M. Born, *Der Luxus des Gewissens*, Munich, 1969.

212 See, for example, P.S. Epstein, 'On the elastic properties of lattices', *Phys. Rev.* **70** (1946) 915–22; C. Zener, 'A defense of the Cauchy relations', *ibid.* **71** (1947) 323; I. Stakgold, 'The Cauchy relations in a molecular theory of elasticity', *Quart. Appl. Math.* **8** (1950) 169–86.

213 M. Born and K. Huang, *Dynamical theory of crystal lattices*, Oxford, 1954, chap. 3, 'Elasticity and stability', pp. 129–65.

214 G. Zanzotto, 'The Cauchy–Born hypotheses, nonlinear elasticity and mechanical twinning in crystals', *Acta Cryst.* **A52** (1996) 839–49, and sources quoted there.

215 Todhunter and Pearson, ref. 70, v. 2, part 1, p. 99.

216 F. Savart (1791–1841) S. Dostrovsky, DSB, v. 12, pp. 129–30; F. Savart, 'Recherches sur l'élasticité des corps qui cristallisent régulièrement', *Ann. Chim. Phys.* **40** (1829) 5–30, 113–37, and in *Mém. Acad. Sci. Roy.* **9** (1826) 405–53, published 1830; English trans. in *(Taylor's) Scientific Memoirs* **1** (1837) 139–52, 255–68.

217 G. Wertheim (1815–1861) Pogg., v. 2, col. 1302–3; His life and work are described by Bell, ref. 201, pp. 56–62, 218–59.

218 G. Wertheim, 'Recherches sur l'élasticité. Premier mémoire', *Ann. Chim. Phys.* **12** (1844) 385–454.

219 G. Wertheim, 'Mémoire sur l'équilibre des corps solides homogènes', *Ann. Chim. Phys.* **23** (1848) 52–95.

220 [A. Cauchy], 'Rapport sur divers mémoires de M. Wertheim', *Compt. Rend. Acad. Sci.* **32** (1851) 326–30.

221 Bell, ref. 201, pp. 257–9.

222 Saint-Venant, ref. 147, pp. ccxci–iii, and App. 5, pp. 656–9.

223 É.-H. Amagat (1841–1915) J. Payen, DSB, v. 1, pp. 128–9.

224 É.-H. Amagat, 'Recherches sur l'élasticité des solides et la compressibilité du mercure', *Jour. Physique* **8** (1889) 197–204, 359–68.

225 É.-H. Amagat, 'Recherches sur l'élasticité des solides', *Compt. Rend. Acad. Sci.* **108** (1889) 1199–202.

226 É.-H. Amagat, 'Sur la valeur du coefficient de Poisson relative au caoutchouc', *Compt. Rend. Acad. Sci.* **99** (1884) 130–3.

227 For a modern perspective, see W. Köster and H. Franz, 'Poisson's ratio for metals and alloys', *Metall. Rev.* **6** (1961) 1–55.

228 W. Voigt, 'Ueber das Verhältniss der Quercontraction zur Längsdilation bei Stäben von isotropem Glas', *Ann. Physik* **15** (1882) 497–513; 'Ueber die Beziehung zwischen den beiden Elasticitätsconstanten isotroper Körper', *ibid.* **38** (1889) 573–87.

229 Bucciarelli and Dworsky, ref. 92, pp. 66, 71.

230 Love, ref. 146, 2nd edn, 1906, 'Historical introduction', pp. 1–31, see pp. 14–15.

231 The volumes of the Springer *Handbuch der Physik* on this subject in the Radcliffe Science Library at Oxford are visibly the least worn and so presumably the least read. The Physics and Chemistry Library at Cornell chose not to buy these volumes.

232 Todhunter and Pearson, ref. 70, v. 1, p. 832.

4

Van der Waals

4.1 1820–1870

The half-century that followed the decline of Laplace's influence in the 1820s was an exciting if confusing time for both physicists and chemists. Laplace and his contemporaries had created many of the mathematical tools that would be needed by the rising generation of theoretical physicists but these tools were to be used in decidedly non-Laplacian ways in the flourishing fields of thermodynamics, optics, electricity and magnetism. The men who were responsible for these developments were mainly German and British; French influence declined rapidly from about 1830. An important early figure was Franz Neumann but it was the brilliant generation that followed who were to lead these fields – Stokes (b.1819), Helmholtz (1821) [1], Clausius (1822), William Thomson (1824), Kirchhoff (1824) [2], and Maxwell (1831) [3]. Some of the views that they were to articulate were held instinctively by Faraday [4], the modest but acknowledged leader of the experimental scientists. The physicists often maintained that every theory should ultimately be reducible to mechanics but they nevertheless created theoretical structures that did not lend themselves to such a reduction. The fertility of field theories led, in Britain at least, to a disparagement of theories based on action at a distance, but in Germany matters were less polarised. The influence of Kant's philosophy led Helmholtz in particular to retain this concept, and Clausius and Boltzmann were later to be equally happy with it, at least as a pragmatic basis for molecular modelling. An example of its use is the velocity- and acceleration-dependent forces between charged particles with which Weber tried to save electrodynamics from the embrace of field theory [5]. Clausius and Boltzmann tried to reduce the second law of thermodynamics to mechanics and although their efforts were unsuccessful Boltzmann's work became the starting point for the development of non-equilibrium statistical mechanics.

Outside the specialised field of the elasticity of solids there was little work from the major workers in the years up to 1857 that was relevant to the understanding of the cohesion of matter. We can see, in retrospect, both external and internal reasons for this neglect. The external competition from more fashionable fields was strong and, in the cases of thermodynamics and of electricity and magnetism, was reinforced by the need to solve the practical problems of the steam engine and of electrical telegraphy. The often positivist spirit of the times was against molecular speculation. John Herschel, in his Presidential Address to the British Association in 1845, said:

The time seems to be approaching when a merely mechanical view of nature will become impossible – when the notion of accounting for *all* the phaenomena of nature, and even of mere physics, by simple attractions and repulsions fixedly and unchangeably inherent in material centres (granting any conceivable system of Boscovichian alternations), will be deemed untenable. [6]

The internal problem was, as usual, the lack of understanding necessary to underpin the next advance. The biggest obstacle was the static view of matter of Laplace and his school, with the concomitant lack of understanding of 'heat', which often included a belief in a caloric mechanism of molecular repulsion. There was, moreover, the continuing uncertainty among both physicists and chemists about the reality of atoms and their relation, if any, to the particles or 'molécules' of Laplace's school. But obstacles that are clear in retrospect are not as clear at the time. The usual reaction of scientists when they see that a field is not making progress is not to question why, but to go and do something else; science is "the art of the soluble". In this case the major scientists went to other more profitable fields and those who were to lay the groundwork for the next advance were often men from a practical background who were looking at problems only remotely connected with cohesion. This Section is an all too brief summary of the relevant work from about 1820 to 1860 and an attempt to show how, by the decade of the 1860s, the field was again ripe for development.

The first moves towards tackling the difficulties that lay in the way of a theory of matter and its cohesion came from Leslie's 'secondary order of men', those outside the main stream of physicists. Newton had said that 'heat is motion', although he did not believe in a kinetic theory of gases in the modern sense of that phrase. It was often an uncritical veneration for his views that inspired some of the Britons who aspired to make their mark in theoretical physics. Thus a kinetic theory in which the pressure of a gas was ascribed to the bombardment of the walls of the vessel by rapidly moving and widely spaced particles was again put forward. Daniel Bernoulli was overlooked and Newton was the inspiration of John Herapath [7], a teacher turned journalist, and of John James Waterston [8], an engineer. The tragi-comedy of their efforts to publish their kinetic theories is now well known [9]; one of the problems was the attitude reflected in Herschel's address. Nevertheless their ideas slowly reached the wider physical world. The subject was kept alive by James

Joule [10], who was not widely known in the 1840s, and by August Krönig [11], a somewhat isolated figure as a teacher in a technical college in Berlin. These were all men whose vision outran their mathematical skills and much of their work is a confused mixture of real insight and inadequate or even wrong physics. Joule learnt something from Herapath, and Krönig most probably from Waterston. It was Krönig's paper of 1856 that spurred Clausius into action, so the pioneers were not without influence. The subject came to maturity with the work of Clausius and Maxwell, after the development of thermodynamics and a realisation of the central importance of energy. The early work on the kinetic theory of gases is not described here in detail – it is a well-documented story – but the observations that arose from it that are relevant to molecules and their interaction are extracted as they are needed. More will be said later about molecular forces in liquids which is a lesser-known topic and one that does not lend itself so readily to quantitative analysis. The work of Clausius and Maxwell is deferred to Section 4.2, since not only did they put kinetic theory on a sound footing, but they also summarised what could be said (with some confidence by Clausius and with more hesitancy by Maxwell) about molecules and their interactions and about the relation of this synthesis to the experimental behaviour of gases and liquids.

The field that came to be called thermodynamics was also started by those out-side the main stream. Sadi Carnot's brilliant book of 1824 was misleading on one vital point; he held then that heat was a conserved quantity [12]. The book had little influence outside French engineering circles until the 1840s [13]. Then the experiments of Joule on the conversion of work into heat, and the calculations of J.R. Mayer and others [14] convinced physicists that it was energy, not heat, that was conserved. Out of the synthesis of this work and that of Carnot emerged the first and second laws of thermodynamics at the hands of Clausius and William Thomson, with off-beat contributions from W.J.M. Rankine [15]. Helmholtz's pamphlet of 1847, *On the conservation of force* [16], marked an important step in the accep-tance of the doctrine of the conservation of energy (as we now call it). In it he took the mechanical expression of this doctrine to be equivalent to the hypothesis that all forces in nature are attractive or repulsive forces acting along the lines joining the particles of matter, but he did not speculate on the nature of these particles, and he was later to modify this view. He introduced the idea of potential energy [*die Spannkraft*] between the particles, an innovation that recognised the value of this concept outside the fields of gravitation and electrostatics to which it had hitherto been confined, if we except fleeting appearances in Laplace's theory of capillarity and in some of the papers on elasticity.

The acceptance by the pioneer thermodynamicists of the law of the conservation of energy implied a belief that the energy that 'disappears' as heat, and which can emerge again, in part, as work, is an energy of motion, but they were not always explicit about what it was that was moving. Helmholtz and Joule were clear in 1847 that it was a motion of the atoms (initially a rotational motion in Joule's case) that constituted the energy; Clausius shared the same view. Rankine invoked molecular vortices in an aether, but Thomson and Maxwell were more cautious about the implications of the laws of thermodynamics and, from 1860, of the kinetic theory of gases. Thomson came forward eventually in 1867 with his own theory of atoms as vortices in an aetherial fluid, only later to abandon that idea also.

It might be said that what Clausius and Thomson did for thermodynamics around 1850, Clausius and Maxwell did for kinetic theory in 1857–1860; that is, they gave it a proper theoretical foundation and brought out its consequences in a way that was to shed light on the emerging view of the structure of matter. The presence of Clausius's name in both fields is not coincidental for it was thermodynamics that was to rescue kinetic theory from the 'outsiders' and bring it into the mainstream of physics. The phrase 'mechanical theory of heat' was used at first to denote what we now call thermodynamics but it came also to embody the congruence of thermodynamics with the ideas of kinetic theory. This conflation is clear, for example, in Émile Verdet's book *Théorie mécanique de la chaleur* and, in particular, in the valuable bibliography by J. Violle which it includes. Both book and bibliography cover what we now call thermodynamics and kinetic theory [17].

Before following the physicists further let us see briefly what the chemists had contributed to physical theory by 1860. The chemist Lothar Meyer, writing in 1862 [18], from a good grounding in physics [19], acknowledged that Berthollet had had the right idea in wanting to interpret the processes of chemistry by means of interparticle forces, but said that little or no progress had been made in that direction. For most of the 19th century the emphasis was on questions of composition and mass; forces generally received less attention. Berthollet's work was to mark the end of the Newtonian tradition that had started with the *Opticks* and Freind's lectures a hundred years earlier. Once his short-lived influence had waned chemical theories were to evolve on quite different lines. Two of the most striking of these were the electrochemical theories of Davy and Berzelius [20]. In his influential Bakerian Lecture of 1806 Davy brought forward the idea that the formation of chemical compounds from their elements was a consequence of electrical attraction between them. He said of electrical energy that "its relation to chemical affinity is, however, sufficiently evident. May it not be identical with it, and an essential property of matter?" [21]. Berzelius developed this idea further and with more effect since he believed in Daltonian atoms in a way that Davy never did [22]. His creed is summarised in two sentences:

... [in] the *corpuscular theory*, union consists of the juxta-position of the atoms which depends on a force that produces chemical combination between heterogeneous atoms and mechanical cohesion between homogeneous. We shall return later to our conjectures on the nature of this force. [23]

All is revealed thirty pages later where he writes:

... that in all chemical combination, there is a neutralisation of opposite electricities, and that this neutralisation produces fire [feu], *in the same way as it is produced in the discharges of a Leyden jar* [boutielle électrique], *of the electric battery, and of thunder, without being accompanied by chemical combination in the last cases.* [24]

He ordered the known elements into an electrochemical series that ran from oxygen as the most negative, through hydrogen near the middle, to potassium as the most positive. The entities that combined were ranked in orders; in the first order there were simple compounds such as water or sulfuric acid, formed from a radical plus oxygen, in the second simple salts such as calcium sulfate, formed from a positive CaO and a negative SO_3, and in the third and fourth, double salts and salts with water of crystallisation. This scheme did much to rationalise the combinations exhibited by inorganic substances but soon proved less successful with the organic. Belief in its universality never recovered from Dumas's discovery that he could replace the 'positive' hydrogen by the 'negative' chlorine in the methyl group (to use the modern name) without any substantial change in its properties [25]. Whatever the initial hopes of Davy and Berzelius, their scheme contributed nothing to the understanding of cohesion.

Dalton had come to chemistry from meteorology and the study of gases whose properties he interpreted in the same way as Lavoisier and Laplace, that is, as an array of static particles or atoms each surrounded by a sheath of caloric which was attracted to the atoms but which repelled other caloric. To explain the diffusion of gases, and what we now call his law of partial pressures, he had to assume that gas atoms of different chemical species did not repel each other, and he was led from this conclusion and from the differing solubilities of gases in water to some rather inconclusive speculations on the sizes of atoms. From these physical considerations came the notion that atoms had masses in fixed ratios that could be determined, and so to the justification of this theory from the chemical principle of constant combining proportions [26]. He and Davy both made passing mention in their textbooks of the forces of attraction as the origin of cohesion, but their hearts were not in this subject [27]. This attitude persisted for some years in textbooks of chemistry. Thus in 1820 James Millar devoted a chapter of 15 pages (of his 466) to the subject of 'Affinity', which comprised gravitation, adhesion, cohesion, the formation of crystals, and chemical affinity, but this chapter had no discernible influence on the descriptive material that followed [28]. As late as 1847 the young Edward Frankland, in his first lectures at Queenwood College in Hampshire, opened the course proper with 'specific gravities'. He said something on cohesion and repulsion in his 4th and 5th lectures, but his notes show that his understanding was slight and the titles of the remaining lectures suggest that this was no more than a formal bow to Newtonian tradition, although more than most pupils would have learnt of these subjects at the time [29].

In France J.B. Dumas set out his views in the Spring of 1836 in a course of lectures he gave at the Collège de France [30]. He traces the descent of the idea of affinity from the beginning of the 18th century, and then divides the attractive forces into three classes, which may all be different or which may be only modifications of one particular force. The first is the weakest; it is 'the cohesion of the physicists', it acts between particles of the same kind and is capable of infinite replication – a crystal can continue to grow indefinitely if there is an adequate source of material. The second is 'the force of dissolution', which acts between similar bodies; it is stronger than the first force but is limited in its extent – no more solid can be dissolved in a saturated solution. The third is 'affinity', which is the strongest force; it leads to the formation of chemical compounds, but it is the most discriminating in its action. Gay-Lussac had a more committed approach in which he continued to seek physical, and hence 'attractive' explanations of chemical phenomena. In a review of 1839 he implicitly followed Dumas by giving a sympathetic but ultimately critical account of the work of Geoffroy, Bergman and Berthollet [31]. He went on to discuss many of the phenomena that were to become the bread-and-butter of physical chemistry at the end of the century. Thus he noted that the elevation of the boiling point of water on dissolving a salt in it is related to the lowering of the vapour pressure at a given temperature, and that the vapour pressure of a solid at its melting point is equal to that of the liquid then formed. He ascribed this fact to a difference in molecular repulsions, since he believed that the attractive forces are clearly much stronger in the solid. In an earlier paper he had shown that the solubility of a solid is often total at its melting point; that is, there is complete miscibility of solute and solvent [32].

Thus the detachment of chemistry from physics was more marked in Britain than in France where the Laplacian tradition lingered. An early and engaging instance of this is Jane Marcet's book of elementary instruction, *Conversations on chemistry* [33]. She distinguishes between two quite different powers, the attraction of cohesion which acts between particles of the same kind, and the attraction of composition which leads to chemical reaction between particles of different kinds. When a French translation of her book appeared in Geneva in 1809 (she and her husband, a physician and chemist, were both of Swiss descent) it was reviewed by Biot [34]. He chided her for holding the doctrine of elective affinities and for ignoring Berthollet's recent revisions, and he criticised her particularly for her false distinction between the two kinds of attraction. He held then to the orthodox Laplacian view that the forces were the same but were to be distinguished from gravitation.

Chemists almost disappear from our story for much of the 19th century. They felt that they had to defend the autonomy of their subject, and even when they believed in atoms they did not necessarily try to identify these with the particles that some physicists believed in. This feeling was probably more widespread than is apparent from published books and papers, but it can be found in print. It surfaced, for example, in the critical attitude of chemists to the hypothesis of Avogadro and of Ampère that equal numbers of molecules were to be found in equal volumes of gases [35]. An interesting example is furnished by William Prout who, in 1834, expressed some views on heat and light that were old-fashioned by the physical opinions of the day, but then added in a footnote:

We are aware that this opinion is opposed to that of most mathematicians, who favour the undulatory theory of light, and with good reason, so far as they have occasion to consider it; but we are decidedly of the opinion that the *chemical* action of light can be explained only on chemical principles, whatever these may be. Whether these chemical principles will hereafter explain what is now so happily illustrated by undulae, time must determine. [36]

Such a view was not perverse – the chemical action of light was to be a problem for the wave theory until the advent of quantum mechanics – but Prout's conscious detachment of chemistry from physics explains why chemists had so little to contribute to the subject of cohesion.

In 1860 the Karlsruhe Conference led, in principle, to the resolution of the long-standing problems of the chemists over atomic weights and so over the atomic constitution of the simpler gases and organic molecules. In practice it was another decade before some chemists were convinced, but the Conference marked the beginning of the appreciation of the power of Avogadro's hypothesis. With this resolution came the conviction, in the minds of most scientists, that the chemists' molecules, N_2, O_2, CO_2, etc., were also the molecules of the physicists' kinetic theory. Although chemistry still retained its own separateness, the time was not far off when the new subject of physical chemistry would make the boundary between physics and chemistry more one of academic administrative convenience than of internal logic. The hesitant start of the reconciliation of chemistry and physics in the 1860s is reflected in the chemistry textbooks. Thus W.A. Miller of King's College, London, published a book with the title of *Chemical physics*, but even in the fourth edition of 1867, the last before his death in 1870, there is little real chemical engagement with physical principles [37]. He ignores thermodynamics and was probably unaware of the initial attempts in the 1860s of the physicist Leopold Pfaundler and others to interpret the rates of chemical reactions in terms of the collisions of rapidly moving molecules [38]. A contrast is the evolution of Thomas Graham's *Elements of chemistry* [39]. The first 101 pages of Volume 1 of the second edition, published in 1850, are on 'Heat', a subject then regarded as much the province of the chemist as of the physicist. The treatment is still old-fashioned; the section on the nature of heat being essentially unchanged from the first edition of 1842 [40]. There is no mention of Joule's work, indeed there is a positive statement that liquids cannot be heated by friction, and there is support for the Laplacian view of heat as the agent of repulsion between the particles of matter [41]. By 1858, when the second volume appeared, all is changed. In a 'Supplement' there is a clear account of the mechanical theory of heat (but not of the second law) and of the kinetic theory of matter [41]. The treatment of this last subject derived directly from Krönig and from Clausius's great paper of 1857 (discussed below). This was probably the first exposition of the theory in a textbook. Graham was assisted in this volume by Henry Watts, the Editor of the *Journal of the Chemical Society* and a skilled translator from German. It was probably he who was responsible for the inclusion of Clausius's theory [42]. A contrasting pair of German textbooks appeared in 1869. That by Friedrich Mohr has a long and promising title [43], but is quite out of date. The author contents himself with bald statements that lead nowhere, such as "Capillarity is a form of cohesion. One can produce no motion by cohesion. Hardness and difficulty of melting often go in parallel, but not always." In the same year, and from the same publisher (Vieweg), appeared what is probably the first German chemical text to include an up-to-date account of thermodynamics and kinetic theory: Alexander Naumann's *Grundriss der Thermochemie* [44]. Like Graham (or Watts) he follows Clausius in his discussion of molecular motions and interactions [45], and, later in the book, distinguishes between atomic compounds (e.g. H_2O) and molecular compounds with either fixed ratios of components (e.g. $BaCl_2 \cdot 2H_2O$) or variable ratios, as in solutions (e.g. NaCl in H_2O) [46]. His discussion of the heat changes in chemical reactions includes what we should now describe as changes in potential energy in the condensed phases but which he describes in terms of Clausius's 'disgregation' (see below). Perhaps the last word on the detachment of chemistry from physics should rest with Maxwell who attempted a classification of the physical sciences in 1872 or 1873. He wrote:

I have not included Chemistry in my list because, though ~~Physical~~ Dynamical Science is continually reclaiming large tracts of good ground from the one side of Chemistry, Chemistry is extending with still greater rapidity on the other side, into regions where the dynamics of the present day must put her hand upon her mouth. But Chemistry is a Physical Science . . . . [47]

From this brief summary of some of the relevant background in physics and chemistry let us now move to a more detailed account of the experimental work on gases and liquids that is related to cohesion, and to the theoretical deductions that flowed from it. It was work on the bulk properties that proved to be the most productive. Capillary studies, which had played so important a role up to the time of Laplace, were now less important, at least until the 1860s. A mathematically more rigorous version of Laplace's theory by Gauss in 1830 was little more than a tidying-up operation in which he proved, rather than assumed, that a given pair of liquid and solid has a fixed angle of contact at a fixed temperature [48]. It was his first excursion into physics and may have arisen from his work three years earlier on the mathematics of curved surfaces. He distinguished clearly between cohesive and gravitational forces, noting that the relevant integrals diverged for an inverse-square law, but he retained, nevertheless, the notion that the cohesive forces were proportional to the product of the masses of the interacting particles. This was little advance on what had already been done. Solids, as we have seen, became a detached branch of science that had little contact with the study of gases and liquids.

Boyle's law states that the pressure of a gas is inversely proportional to its volume at a fixed temperature; it had been known since the 17th century. At the end of the 18th and early in the 19th, Charles, Gay-Lussac and Dalton showed that the pressure, p, at a fixed volume, V, is a linear function of the temperature as measured, say, on the scale of a mercury thermometer [49]. An extrapolation of that linear relation placed the zero of pressure at about $-270\,°C$. The two laws, Boyle's and Charles's, could be combined into a single equation that described what came to be called the perfect or ideal gas law [49];

$$pV = cT, \qquad (4.1)$$

where T is a temperature measured on a scale whose zero is at about $-270\,°C$, and c is a constant that is proportional to the amount of gas. Avogadro's hypothesis implies that this constant is proportional to the number of molecules in the sample of gas. This equation, to which the common simple gases nitrogen, oxygen and hydrogen conform closely at temperatures near ambient and pressures near atmospheric, was the guiding principle of early workers on the kinetic theory of gases. (Herapath thought, however, that the temperature was a measure of the scalar momentum of the particles, not of their energy, and so wrote $(T^*)^2$ in place of T, where T^* is the 'true' temperature.)

For many years it had been known that the perfect-gas law was not exact; pressures could be a little higher or a little lower than that calculated from this equation. If the molecules had a non-zero size then the effective volume in which each moves is less than the observed volume of the gas, and so the pressure is higher than the ideal pressure, if the kinetic theory be correct. This deduction was made first by Daniel Bernoulli and was repeated in Herapath's work. He was delighted when he found [50] that experiments by Victor Regnault on hydrogen confirmed his prediction. Hydrogen was, according to Regnault, "un fluide élastique plus que parfait" [51]. Other gases, for example, carbon dioxide and steam, had pressures that fell below that calculated from the perfect-gas equation. The implication that this deficit is evidence for (Laplacian?) attraction between the molecules was drawn by Herapath who ascribes the reduction of pressure to an incipient condensation or clustering of the molecules [52]. Nothing quantitative could at that time be deduced from this inference, in the absence of a kinetic theory of interacting molecules.

The French state had provided funds from the 1820s onwards for the experimental study of gases, particularly steam, at high pressures. Dulong and Arago carried out the early work but from the 1840s it became the life work of Victor Regnault. The results in his first full monograph of 1847 [51] were accepted as the authoritative work in the field. They proved difficult to interpret, or even to fit to empirical equations. Regnault tried to do this and it was one of the last tasks that, in 1853, Avogadro set himself [53]; neither had any real success.

A gas that conforms to Boyle's and Charles's law has an energy, U, that is independent of volume at a fixed temperature. In modern notation,

$$(\partial U/\partial V)_T = -p + T(\partial p/\partial T)_V = T^2(\partial/\partial T)_V(p/T), \tag{4.2}$$

and from eqn 4.1 the ratio p/T is c/V. This result is one of classical thermodynamics, that is, it is not dependent on any molecular hypothesis except that which may, according to taste, be used as a theoretical basis of the empirical eqn 4.1. Joule observed, in 1845, that there was no change of temperature on a free expansion of air at 22 atm pressure into an evacuated and thermally-insulated vessel; that is, he found that $\delta T/(V_2 - V_1) = 0$, where V_1 and V_2 are the initial and final volumes [54]. If the system is thermally insulated and if the gas does no work then, by what came to be called the first law of thermodynamics, the expansion is one at constant energy. Joule's result may therefore be expressed, after using eqn 4.2,

$$\left(\frac{\partial T}{\partial V}\right)_U = -\frac{(\partial U/\partial V)_T}{(\partial U/\partial T)_V} = -\frac{1}{C_V}\left(\frac{\partial U}{\partial V}\right)_T$$

$$= \frac{1}{C_V}\left[p - T\left(\frac{\partial p}{\partial T}\right)_V\right] \approx 0, \tag{4.3}$$

where C_V is the heat capacity of the gas at constant volume. This demonstration that the energy was indeed independent of the volume was, therefore, one of the foundations of the first law. It was realised that the energy, U, is a state function, that is, it depends only on the present volume and temperature of a fluid, and not on its past history or how it came to be in its present state. For a perfect gas, the energy depends on the temperature alone.

A more sophisticated series of experiments was carried out by Joule between 1852 and 1853, with the theoretical guidance of William Thomson who had come to accept by 1851 that it was the energy that was conserved in physical changes and not the heat [55]. Joule and Thomson expanded the gas in a continuous flow down a well-insulated pipe in which there was a constriction in the form of a porous plug of cotton wool or, on one occasion, of Joule's silk handkerchief. The pressure falls from p_1 to p_2 on passing the obstruction and Joule observed that there is generally a small fall of temperature of the gas, that is, $\delta T/(p_1 - p_2) < 0$. The fall was negligible with hydrogen, observable with air, and substantial with carbon dioxide; it decreased with increase of initial temperature. This expansion is not one of constant energy because of the work expended in passing the gas through the obstruction. It modern terms it is an expansion at constant enthalpy, where the enthalpy, H, is defined as $U + pV$ [55]. We have, therefore,

$$
\left(\frac{\partial T}{\partial p}\right)_H = -\frac{(\partial H/\partial p)_T}{(\partial H/\partial T)_p} = -\frac{1}{C_p}\left[V - T\left(\frac{\partial V}{\partial T}\right)_p\right]
$$

$$
= -\frac{T^2}{C_p}\left(\frac{\partial}{\partial T}\right)_p\left(\frac{V}{T}\right), \tag{4.4}
$$

where C_p is the heat capacity at constant pressure. This key equation, the basis for modern discussion of the 'Joule–Thomson effect', appears only in the Appendix to the fourth and final paper that they published in the *Philosophical Transactions*. The discussion in the earlier papers, while essentially sound, is less clear owing to the primitive state of development of thermodynamics in the 1850s. For a perfect gas it follows that the differential coefficient $(\partial T/\partial p)_H$ is zero, as in the parallel case of $(\partial T/\partial V)_U$ of eqn 4.3. Hence Joule's observation of cooling, like the deficit of pressure from that required by Boyle's law, is evidence for the existence of attractive forces between the particles. The 1850s were, however, not the time to draw this conclusion. Joule and Thomson were more concerned to use their results to validate the laws of thermodynamics and to establish the absolute scale of temperature. Maxwell, a close friend of Thomson, took little notice of their results, describing the change as "a slight cooling effect" [56]. Clausius ignored the effect, although he introduced in 1862 the concept of 'disgregation' to describe the thermal effects of changing the separation, or more generally, the arrangements of the particles of a fluid. This term has vanished from modern thermodynamics; it became redundant once the concept of entropy was accepted [57]. In modern terms it is the configurational part of the entropy, as was first shown by Boltzmann [58].

It might be asked why it was that, in 1845, Joule found no change of temperature in a free expansion, which measures $(\partial T/\partial V)_U$, but, nine years later, found a cooling in a flowing expansion, which measures $(\partial T/\partial p)_H$. In a real or imperfect gas the first coefficient is zero in the limit of zero pressure, while the second tends to a non-zero limit. This is, however, not the root of the difference, for the two coefficients $(\partial T/\partial p)_U$ and $(\partial T/\partial p)_H$ are of similar size. If we add an empirical correction term, $B(T)$, to the equation of state of a perfect gas we can write (with the modern choice of R for the gas constant)

$$
pV = RT(1 + B(T)/V). \tag{4.5}
$$

In the limit of zero pressure we find that

$$(\partial T/\partial p)_U = (T/C_V)[T(\mathrm{d}B/\mathrm{d}T)],$$
$$(\partial T/\partial p)_H = (T/C_p)[T(\mathrm{d}B/\mathrm{d}T) - B]. \tag{4.6}$$

In general both expressions are non-zero and of similar size. Rankine had proposed an equation of the form of eqn 4.5 in a letter to Thomson of 9 May 1854, with $B(T)$ having the particular form $-\alpha/T$ [59]. The reason that Joule did not detect a non-zero value of $(\partial T/\partial p)_U$ in 1845 was that the thermal capacity of his iron gas-vessels was too large. He himself pointed this out later [60], and it was perhaps a fortunate circumstance that he detected no change of temperature at the time, for such a change would have made more difficult the establishment of the laws of thermodynamics!

Joule and Thomson found in 1862 that for air and carbon dioxide the cooling effect, $(\partial T/\partial p)_H$, was proportional to the inverse square of the absolute temperature. They were therefore able to integrate eqn 4.4 to obtain an equation of state of the form [61]

$$pV = RT - \alpha p/T^2, \tag{4.7}$$

where R is the constant of integration and α is a measure of the strength of the cooling. At low pressures this equation has the same form as eqn 4.5 with $B(T) = -\alpha/T^2$. This is similar to the form proposed by Rankine eight years earlier, but with a stronger dependence on temperature. Previously [60] they had found results that were equivalent to the more complicated form $B(T) = \alpha - \beta/T + \gamma/T^2$, which is closer to our current ideas on the form of this function, for, as we shall see, the coefficient $B(T)$, now called the second virial coefficient, is an important measure of the form and strength of the intermolecular forces and one that was to play an important role in the 20th century.

The study of liquids made less progress than that of gases in the first half of the 19th century since there was no simple limiting law comparable with the perfect-gas law and no simple theory comparable with the struggling kinetic theory of gases. The basic facts were known; liquids have a fixed vapour pressure at a given temperature which is independent of the fraction of the (pure) liquid that is in the vapour state; this vapour pressure rises rapidly with temperature and the density of the liquid falls but more slowly; and the change from liquid to saturated vapour is accompanied by a large intake of heat – 'the latent heat of evaporation'. Solids are more dense than the liquids formed on melting, and this melting is accompanied by the absorption of a smaller latent heat. The exceptional behaviour of ice and water between 0 and 4 °C was well known but no explanation of this behaviour was agreed; it was generally ignored, although John Tyndall made a tentative suggestion that the energy absorbed between 0 and 4 °C went to increasing the speed of rotation of the water molecules [62]. It was known that the 'heavier' vapours such as chlorine, hydrogen sulfide, carbon dioxide and sulfur dioxide could be liquefied by cooling, or compression, or both, and it was freely conjectured that the so-called permanent gases, nitrogen, oxygen and hydrogen, might be liquefied if their temperature could be sufficiently lowered.

Faraday was one of the first to study systematically the liquefaction of gases. After his first experiments in 1823 [63] he became aware of the sporadic efforts of his predecessors and published a short summary of them the next year [64]. He returned to the subject in 1844 and then wrote a long paper [65] in which he reported the condensation of a wide range of gases by a combination of pressures up to about 100 atm and temperatures down to that of a pumped bath of ether and carbon dioxide. He estimated the temperature of this to be about −166 to −173 °F, or −110 to −113 °C, or 160 to 163 K on the later thermodynamic or 'absolute' scale. He failed to liquefy nitrogen, oxygen, and hydrogen, noting presciently that they could probably be liquefied only at lower temperatures, and that increasing the pressure would not suffice. He obtained his solid carbon dioxide from supplies of 220 cu.in. (3.6 litres) of liquid made for him by Robert Addams [66]. The solid had first been prepared in bulk by Thilorier [67] who had realised the usefulness of a mixture of solid carbon dioxide and ether as a refrigerant. Addams improved Thilorier's apparatus.

When a liquid is heated in contact with its saturated vapour it is observed that the pressure rises rapidly, the density of the vapour rises equally rapidly, and the density of the liquid falls more slowly. It is natural to wonder what would happen if the heating were continued. The first answer was provided by Cagniard de la Tour who, in the 1820s, heated ether, alcohol and water in separate sealed glass tubes [68]. He found that a point was reached when the liquid, after a considerable expansion, was apparently converted into vapour. He was also the first to notice what we now call 'critical opalescence', for when his tubes were cooled from the highest temperatures liquid was suddenly formed again in "un nuage très épais". His estimate of this point of apparent vapourisation of ether, a pressure of 37–38 atm and a temperature of 150 °R = 188 °C, is close to what we now call the critical point of ether, 36.1 atm and 194 °C. Faraday, in his 1845 paper [65], wrote, "I am inclined to think that at about 90° Cagniard de la Tour's state comes on with carbonic acid". This estimate, 32.2 °C, is also close to the modern result of 31.04 °C.

Herschel [69] argued on general grounds that Cagniard de la Tour's work was evidence for the lack of a sharp distinction between the three states of matter:

Indeed, there can be little doubt that the solid, liquid, and aëriform states of bodies are merely stages in a progress of gradual transition from one extreme to the other; and that, however strongly marked the distinctions between them may appear, they will ultimately turn out to be separated by no sudden or violent line of demarcation, but shade into each other by insensible gradations. The late experiments of Baron Cagnard de la Tour may be regarded as a first step towards a full demonstration of this (199.).

The reference to § 199 of his book is to

... that general law which seems to pervade all nature – the law, as it is termed, of continuity, and which is expressed in the well known sentence 'Natura non agit per saltum'.

In November 1844 Faraday wrote to William Whewell at Cambridge [70] asking him to suggest a better name for the 'Cagniard de la Tour state'. His description of it is more accurate than anything that has gone before:

... the difference between it [the liquid] & the vapour becomes less & less & there is *a point* of temperature & pressure at which the liquid ether & the vapourous ether are identical in *all their properties.* ... but how am I to name this point at which the fluid & its vapour become one according to a law of continuity? [71]

Whewell replied:

Would it do to call them [the fluids] *vaporiscent*, and this point, the point of *vapor-iscence*[?] ... Or if you wish rather to say that the liquid state is destroyed, you might say that the fluid is *disliquified*. [71]

Faraday was not satisfied with these suggestions:

... for at that point the liquid is vapour & the vapour liquid, so that I am afraid to say the liquid *vaporisces* or that the fluid is *disliquefied*. [71]

In 1861 Mendeleev [72] introduced another name when he wrote:

We must consider that point to be the absolute boiling temperature at which (1) the cohesion of the liquid becomes zero, and $a^2 = 0$, at which (2) the latent heat of evaporation is also zero, and at which (3) the liquid is transformed into vapour, independently of pressure and volume.[†]

His choice of words shows that Mendeleev had an unsymmetrical view of the phenomenon; liquid was changed into vapour.

These confusions were resolved in the 1860s by Thomas Andrews, the first Professor of Chemistry at Queen's College, Belfast [73]. His first results, on carbon dioxide and nitrous oxide, were sent informally to W.A. Miller for inclusion in the third edition of his textbook [74]. By then Andrews had found that the liquid meniscus lost its curvature as the temperature approached that at which the liquid disappeared, 88 °F = 31.1 °C for carbon dioxide. He did not then draw the conclusion that the surface tension vanishes at this point. The flattening of the meniscus

---

[†] The length a is called the 'capillary constant'. The ratio a^2/r is the height to which a fully-wetting liquid rises in a narrow capillary tube of radius r. The capillary constant of water is 3.9 mm at its freezing point.

had been observed previously by Wolf and by Waterston, who attributed it to a failure of the liquid to wet the glass, and not to a vanishing of the surface tension [75]. Andrews saw also the opalescence of the fluid which he described as "moving or flickering striae throughout its entire mass". By 1869 he had mapped out in detail the relations between volume, temperature and pressure, and the boundaries in V, T, p-space of the liquid and gaseous phases of carbon dioxide. He arrived at the important conclusion that it was not the case that liquid was transformed into vapour but, as Faraday had surmised, that both approached a common fluid state at what he christened 'the critical point'. He published this work in his Bakerian Lecture to the Royal Society of June 1869 , 'On the continuity of the gaseous and liquid states of matter' [76]. In the text he demonstrates this continuity by means of a passage in V, T, p-space that passes from a typical liquid state around the critical point to a typical gas state without there ever being a dividing meniscus. He wrote:

The ordinary gaseous and ordinary liquid states are, in short, only widely separated forms of the same condition of matter, and may be made to pass into one another by a series of gradations so gentle that the passage shall nowhere present any interruption or breach of continuity.

And for a fluid above its critical temperature, he added:

. . . but if any one ask whether it is now in a gaseous or liquid state, the question does not, I believe, admit of a positive reply.

He "avoided all reference to the molecular forces brought into play in these experiments", but said enough to show that he thought that there was "an internal force of an expansive or resisting character" and also "a molecular force of great attractive power". He thought that these were "modified" in the passage from gas to liquid.

Others were not so reticent as Andrews and in the years up to 1870 some fragmentary views were expressed on molecular forces and on the cohesion of fluids. These did not form a coherent doctrine and, as with the development of kinetic theory, the first moves came from those outside the main stream.

The increasing interest in electricity led some neo-Laplacians and others to try to interpret cohesion in terms of electrostatic or magnetic forces, rather the gravitational force or a modification of it. These attempts seem to be quite uninfluenced by the earlier electrochemical ideas of Davy and Berzelius. O.F. Mossotti, a professor first in Buenos Aires and then on Corfu, made such an attempt in a pamphlet published in Turin in 1836 [77]. This aroused Faraday's interest, since any attempt to unify electrical and gravitational forces was a theme close to his heart in the 1830s and 1840s. He therefore arranged for an English translation in a new journal to be devoted to foreign memoirs. Mossotti maintained that forces should act only between independent pairs of particles ('two-body forces', in modern jargon), and believed that they changed with temperature. His paper contains an interparticle potential (he does not use that name) that is formed by damping a $(1/r)$ term with an exponential of the form $\exp(-\alpha r)$, where r is the separation of the particles. This form of potential has a long history; it is now called the Yukawa potential, and Mossotti's use of it may be the earliest instance, although Laplace had previously used an exponentially-damped force [78]. It was the Laplace–Poisson equation of electrostatics that led Mossotti to this form of potential. He believed also that in a dense system the attractive forces should lead to a contribution to the pressure proportional to the square of the density. This supposition had also been made previously by Laplace in 1823 [79]; it is one that several simple lines of approximation lead to, and was to recur later in the century. Mossotti's main thesis – an attempt to explain the structure and stability of a dense electrically-neutral system under Coulombic forces – led to a controversy in which he was supported by Philip Kelland, the Professor of Mathematics at Edinburgh (of whose work the young William Thomson had a poor opinion) and in which he was criticised by Samuel Earnshaw and Robert Ellis [80]. The most positive outcome of these exchanges was 'Earnshaw's theorem' that no static system of inverse-square power forces can be at equilibrium.

Waterston later claimed to follow Mossotti in some of his early ideas on 'molecularity', developed before he had fully articulated his kinetic theory. In a book with the unpromising title of *Thoughts on the mental functions* he drew an intermolecular force curve of the kind that we now use regularly (Fig. 4.1), with a positive repulsive branch and a negative attractive branch, the sum of the two leading to a minimum (i.e. greatest energy of attraction) at some particular separation. He believed then that the relative position of the two branches changed with the state of matter, so that the positive or repulsive branch moved to larger separations in the gas, thus making the minimum disappear [81].

Élie Ritter [82] taught mathematics at a school, the Institut Topffer in Geneva. His interests were mainly astronomical but in 1845 he read a paper to the local Physical and Natural History Society, of which he was the Secretary, on 'elastic fluids' [83]. This is entirely in the Laplacian tradition. His particles are static with a mean separation ε and, following explicitly the lead of Laplace [79] and of Poisson [84], he arrives, like Mossotti, at an 'attractive' contribution to the pressure that varies as ε^{-6}, or as the square of the density. His replacement of sums by integrals leads also to minor terms that vary as ε^{2n} where $n = 1, 0, -1, -2$, etc., but he argues these away as unimportant [85]. It is easy to believe that he knew the result he wanted and was not going to be distracted by minor terms even if they seemed to be divergent. For gases at moderate pressures, we have seen that Rankine, Thomson and Joule soon arrived empirically at an equation of state that carries the same implication of an energy that varies as the square of the density. At the end of the century, when

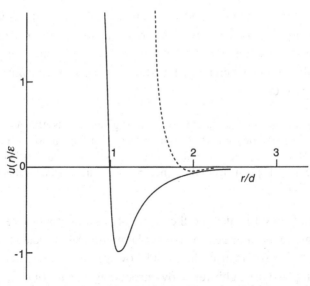

Fig. 4.1 Waterston's view of intermolecular forces, illustrated here by a modern intermolecular potential. The full line shows the potential in the liquid state with an attractive 'bowl' of depth, ε. The force is zero when the two particles are at the separation of the minimum, it is repulsive to the left of it and attractive to the right. In the gas he supposed that the repulsive part of the force or the potential is moved to larger distances. Here this is illustrated by moving the repulsive branch of the potential to the right by half of the original diameter, d. This move eliminates most of the attractive bowl, leaving the force almost wholly repulsive.

van der Waals had established this result as the norm, Émile Sarrau cited the French physicists Poisson and Cauchy as early proponents [86].

By 1860 belief in the reality of atoms and of the physicists' molecules was becoming sufficiently strong for new attempts to be made to estimate the sizes and energies of these particles. The few previous attempts had borne little fruit. Edmond Halley had estimated a maximum size for the atoms of gold from the minimum thickness to which sheets of gold could be hammered out; more could have been made of Franklin's experiments of spreading oil on water; and Young's ingenious estimate of the range of the interparticle forces from the ratio of the surface tension to the cohesive energy was apparently unknown to or ignored by all. None of this work influenced the attempts that grew from the kinetic theory of gases from the middle of the 19th century.

The first 'microscopic' result that was derived from the new kinetic theories was the speed of the molecules in a gas. Herapath showed to his own, and indeed perhaps also to our surprise, that one can get this speed without any 'microscopic' knowledge. The equation that he should have obtained is

$$pV = Mc^2/3,\qquad (4.8)$$

where p is the pressure of a mass M of gas in a volume V, and c is the molecular speed. This equation is correct if we interpret c^2 as the mean of the square of the speeds. He obtained something similar first in 1836 [87] but it apparently only became generally known when he included it in his book *Mathematical physics* in 1847, when he wrote:

At first sight one would imagine that the conditions given are insufficient for the solution of this problem. The size of the particles, the direction of their motions, or something of the kind, seems at an off-handed view to be indispensable; such at first I considered to be necessary. However, it happens from the concurrence of circumstances that nothing of the sort is wanting. [88]

He was, in fact, trying to calculate the speed of sound but since he assumed that sound is transmitted through a gas by molecular motion he expected the value of c to be that of the speed of sound in air; he would, however, have found it to be somewhat larger. His attempt to correct his result by introducing a factor of $(1/\sqrt{2})$, the cosine of 45°, the average angle of collision of a molecule with the wall of the vessel, is quite wrong and was unlikely to have seemed convincing to his contemporaries, but he deserves the credit for the first calculation of what we now call the root-mean-square speed of molecules. He went further and pointed out, not for the first time, that there was a natural zero of temperature at which all motion ceases.

Joule's first thoughts on this subject were not as clear as those of Herapath, since he, like Davy before him, thought at first that the 'heat' in a gas was accounted for by the rotatory motions of the molecules. In a lecture on 28 April 1847, just before the publication of Herapath's book, he made, however, the unsupported statement that the "velocity of the atoms of water, for instance, is at least equal to a mile per second of time." [89] This guess is too high by a factor of three. He returned to the subject after reading Herapath's book, admitted that the attribution of heat to translational molecular motion was a simpler hypothesis than his own, and so arrived at a speed of hydrogen 'atoms' of 1906 m s^{-1} at 15.6 °C (in modern units), a figure close to the now-accepted root-mean-square speed of hydrogen molecules of 1891 m s^{-1} [90]. Waterston had also obtained a correct figure for what he more precisely defined as the 'mean square velocity' in his great manuscript of 1845, but this languished in the stack of rejected papers at the Royal Society until Lord Rayleigh rescued it and published it in 1893 [91].

To go more deeply into the problem and obtain estimates of molecular sizes and energies is more difficult. The first of the new attempts were along lines similar to that followed by Young, although clearly in ignorance of his result [92]. Waterston followed his earlier 'thoughts on molecularity' and his unpublished paper on kinetic theory with some experiments on capillarity. These were carried out in India, where he was teaching naval cadets, but were published only after his return to Scotland in

1857 [75]. His interpretation of the cause of capillary rise is muddled and naive; it resembles most closely the ideas of Jurin and he seems to have had little knowledge of what Young and Laplace had achieved. Nevertheless he stumbles through an argument that parallels that of Young in 1816, using surface tension and latent heat of evaporation for, in effect, Laplace's H and K, to arrive at a figure of 214 778 500 layers of water molecules in a cubic inch of water; that is, a thickness of each layer of 1.2 Å, which therefore becomes his estimate of the diameter of a molecule. Twenty years later N.D.C. Hodges of Harvard followed a similar line of reasoning to arrive again at a diameter of 1.0 Å [93]. Twenty years later again the young Einstein's [94] first paper included another variant of this approach [95]. He then believed that the intermolecular potential function was a universal function of separation, but later retracted this opinion [96]. Meanwhile, as we shall see below, van der Waals had, in 1873, given a more 'modern' and more satisfactory version of Young's argument.

Waterston did not distinguish, as Young and Laplace had been careful to do, between the size of the molecule and the range of the intermolecular force. We now know that the two are of similar magnitude but this was not the view at the beginning of the 19th century and no more evidence had come forward by the middle of the century. The belief that the range of the force greatly exceeded the size was used by the Laplace school as a justification for their 'mean-field' approximation, but there is no evidence that Waterston appreciated this point.

Herapath and Waterston were, perhaps, the last who contributed to the problem of molecules and their interaction without an appreciation of the power and constraints of the new field of thermodynamics. G.-A. Hirn [97] was an engineer from Alsace who, from his early work on steam engines, was one of those who arrived at a value for the mechanical equivalent of heat, and so was led to thermodynamics. In the first edition of his book on heat in 1862 he rejects the new kinetic theory, admitting only that the forces between molecules would cause them to move; he did not clearly say how [98]. This is, of course, very different from the free thermal movement of the kineticists which is independent of the intermolecular forces. In a second edition, three years later, he deals more fully with the intermolecular forces [99]. In Laplacian style, he says that the pressure of a gas is composed of two terms, a 'répulsion calorifique' and an 'ensemble d'actions internes' that he denotes by R. He corrects the volume of a gas by subtraction of Ψ, 'la somme des volumes des atomes d'un corps', and so arrives at an equation of the form,

$$(p + R)(V - \Psi) = \text{constant} \cdot T. \tag{4.9}$$

He speculates on the form of R, saying that it is likely to vary inversely with volume, and that he accepts "as a first approximation that R constitutes a homogeneous sum exclusively a function of V", but then Poisson-like doubts creep in and he covers himself by saying that "In reality, and rigorously speaking, R is almost always heterogeneous and therefore no longer a function only of V".

Those who wrote before van der Waals lacked the insight or courage or, perhaps, the encouragement provided by Andrews's work on the continuity of the states, to apply a common theory to gases and liquids. Hirn was no exception. He had a chapter entitled 'Theory of liquids and solids' [100] but it is a translation of an excerpt from a book by G.A. Zeuner [101] whose approach is entirely macroscopic. Zeuner opens by contrasting what he calls the system of Redtenbacher [102] with that of Clausius, that is, in essence, of the Laplacian versus the kinetic interpretation of the properties of gases. But he does not follow this up; the nearest he comes to a molecular comment is his assertion that the heat of fusion of a solid represents the work done in overcoming cohesion [103]. In a third edition, Hirn notices Andrews's work but draws no inference from it. He now uses Regnault's results to estimate Ψ, the volume of the molecules, and R, which he now calls "la pression interne". He finds this to vary with volume roughly as $V^{-1.3}$ [104].

A route similar to Waterston's was followed by Athanase Dupré, the Professor of Physics at Rennes [105]. In a series of papers in the *Annales de Chimie et de Physique* and in the *Comptes Rendus* of the Academy (of which he was never a member) he explored a number of related problems on the physics of gases and liquids. He received help from his younger colleague, the engineer F.J.D. Massieu [106] who was skilled in thermodynamics. Dupré summarised his work in his book *Théorie mécanique de la chaleur* of 1869 [107]. His work is an advance on Waterston's in that, either because of his wider reading, his innate skill, Massieu's advice, or the mere lapse of time, he was more careful in his handling of thermodynamic functions. He was, however, far from careful in his arithmetic. He discusses gases in the Laplacian manner, that is in terms of forces between static particles, and introduces what we should now call the configurational part of the energy or that part that arises from the intermolecular forces. This he calls φ, "le travail interne", and he shows by a thermodynamic argument that [108]

$$(\partial\varphi/\partial V)_T = T(\partial p/\partial T)_V - p, \tag{4.10}$$

although, as was then customary, he writes the equation with ordinary derivatives not partial ones and does not show the variables to be held constant in the two differentiations. He notes that if φ is a function of volume only then $(\partial^2 p/\partial T^2)_V$ is zero, and that $(\partial p/\partial T)_V$ is also a function of volume only. This leads him, by an argument that is far from rigorous, to what he calls his 'law of co-volumes' [109],

$$p = \alpha T/(V + c), \tag{4.11}$$

where α and c are constants; the latter being what he calls the co-volume. This name has passed into common usage with the understanding that the constant represents the correction of V to allow for the effect of molecular size, a usage that requires c to be negative in the equation as Dupré wrote it. For him it was merely a measure of the departure of a real gas from the perfect-gas law. He does not claim that it is an exact measure for he writes: "In what follows we shall use Mariotte's law as the law of *first approximation*, and that of co-volumes as the law of *second approximation*." [110]

It is significant that he regards his laws as equally applicable to liquids and gases, insisting that he differs from Hirn on this point. When he turns to liquids, however, he uses different methods, and his conviction that both states can be handled by the same law is not followed into practice [111]. He considers first the "attraction au contact", that is the force holding two portions of liquid together, per unit area of their plane surface of contact. To this he gives the symbol A, but it is clearly the same as Laplace's K. He shows that this attraction is proportional to the square of the density – as indeed follows from Laplace's derivation if this is done carefully (see Section 3.2). From the attraction at contact he proceeds to a calculation of the work needed to break up a portion of matter into its separate molecules, 'le travail de désagregation totale', which he shows is the product AV, that is, an energy. The more transparent of the two justifications that he gives for this result is that provided by Massieu, who is responsible also for a derivation of what are, in essence, Laplace's equations of capillarity [112]. Dupré is now in a position to repeat Young's calculation of the range of the intermolecular forces although, since he regards this range and the separation of the molecules as essentially the same, he arrives instead at a minimum value for the number of molecules per unit volume. He quotes numerical values of F, the surface tension of water, as 7.5, and of A of 2.266×10^7 [113]. He gives no units but his usual unit of length is the millimetre and the numbers quoted correspond to modern values of the surface tension and latent heat of 7.35 dyn mm^{-1} and 2.465×10^7 erg mm^{-3} at 15 °C. He takes the latent heat to be the 'work of total disaggregation' but he (or, rather, Massieu) notices correctly that a work term, equal to pV of the gas, should be subtracted from the latent heat, but it is small and he ignores it. He finds that he is led to unacceptable conclusions if he assumes an attractive force proportional to the inverse cube of the molecular separation [113] and turns instead to what is, in effect, Young's method. He shows that the work needed to peel off a layer of liquid one molecule thick leads to a value of N, the number of molecules per unit volume that must exceed $(A/2F)^3$. His figures should therefore give N a minimum value of 3.45×10^{21} molecules per cubic millimetre or, in a more conventional form, 6.21×10^{25} molecules in 18 cm^3 or 1 mole of water. The figure is too large by a factor of 100, and corresponds therefore to an underestimate of the linear separation of the water molecules by a factor of about 5. Unfortunately this is not the result obtained by Dupré. In his paper in *Comptes Rendus* [114] he has 0.125×10^{21} molecules per cubic millimetre, and in his book what seems to be the same calculation leads to 0.225×10^{21}. These are not misprints for each figure is repeated in words, but neither seems to follow from the values of F and A.

It is interesting to compare this result with Young's, which was of course not known to Dupré. Young's 'force at contact' was a pressure of 23 kbar, which is equivalent to a 'work of total disaggregation' of 2.3×10^7 erg mm^{-3}, the same as Dupré's figure, but their arguments are different. Young does not use the energy of the liquid but, insofar as his argument is explicit, relies on his understanding of stress. His figure for the range of the intermolecular force, about 1 Å, is therefore, as should Dupré's have been, too low by a factor of about 5.

Thus Young, Waterston and Dupré followed the same broad route, each using similar figures for water, and each arriving at a distance that we can now see is of the right order of magnitude, although in each case too small. Their arguments are physically sound for rough order-of-magnitude calculations, and are flattered by the taking of the cube root in going from the actual subject of the calculation, a volume, to a length of separation. We now know that a static picture of a liquid is adequate for such rough calculations and so Young and Dupré were not misled by their lack of a kinetic picture of matter. Young took his figure to be the range of the forces, Dupré took it to be the mean separation of the molecules. Both thought that the actual 'size' of the molecules was smaller and could justify the use of a mean-field approximation. Waterston's diagram of the change of intermolecular force with separation shows that he believed that the 'range' and the 'size' differed by only a factor of about two, so his picture would not justify the use of such an approximation, but then neither did he appreciate the need for it.

4.2 Clausius and Maxwell

The return of 'molecular science' to the forefront of physical research was brought about by Clausius and Maxwell. The lines of descent of the kinetic theory of gases are now clear; Herapath influenced Joule, Waterston almost certainly influenced Krönig, Clausius made his own approach to the subject but published nothing until prompted by the appearance of Krönig's paper, while Maxwell knew of Herapath's and Joule's work but did not seriously interest himself in the field until he read Clausius's first two papers. The subject then grew to become, within a few years, an active branch of physics in its own right and one which was to throw much light on molecules and their interactions.

The contributions of Clausius and Maxwell were pivotal not only because they established the kinetic theory of gases on a sound basis and drew quantitative conclusions from it, but also because their wider vision led them to put forward, if only in words, the implications of the molecular–kinetic view of matter for liquids and solids. Here Clausius was the more convinced advocate. Maxwell was always more hesitant and, as we shall see repeatedly, the more conscious of the difficulties and the unresolved problems. Gibbs summarised their styles by saying that Clausius's work was in mechanics and Maxwell's in the theory of probability [57]. Theirs was a synthesis in which, for the first time, we can recognize a description of the microscopic structure of the three phases of matter with which we are wholly comfortable [115]. In this respect it forms a notable contrast with that in the reviews of Joule and Helmholtz of twenty years earlier, written before the development of thermodynamics and kinetic theory [116]. It was a view that was not without its critics, at least until the early years of the 20th century, but it was the dominant view that drove a progressive research programme that has been maintained to this day [117]. Clausius and Maxwell never seriously tackled liquids, however, which remained in the neo-Laplacian limbo of Ritter, Hirn and Dupré until they were rescued by a hitherto unknown Dutch schoolmaster.

Clausius tells that he had been thinking of the relation of heat to molecular motion since the time of his first paper in 1850 on what came to be called thermodynamics [118]. He properly did not wish to compromise his development of thermodynamics, an essentially macroscopic subject, with speculations on its possible molecular foundations. This was a trap that Rankine fell into when he made his thermodynamics depend on a prior assumption of a particular view of matter as molecular vortices; an error of judgement that made his influence on the subject less than it might have been, then and since [119]. Clausius himself criticised Helmholtz's pamphlet of 1847 on the grounds that he had made his conclusions depend on an assumption of a central force acting between the particles of matter [120]. It was only after Clausius had seen Krönig's paper of 1856 [11] that he put forward his own views in the *Annalen der Physik* [115]; he had by then moved to Zürich [121].

His paper falls into into two parts; in the first he explains his ideas on molecular motion, rotation and vibration, and how these movements lead to the existence of matter in gaseous and condensed phases. If the molecules are of minute size and moving rapidly then the pressure caused by their impacts on the walls lead to a gas obeying what we call Boyle's, Charles's and Avogadro's laws. The last law leads him to propose that the common elementary gases have diatomic molecules, a conclusion then novel among the physicists and one that had been discussed, but not always accepted, by the chemists. The known heat capacities of these gases could not be reconciled with the assumption that all their energy of motion was translatory (the *vis viva*); rotation and vibration must also be involved. In solids the molecules continue to move but only about fixed sites. In liquids the motion is similar in the short term to that in solids but the sites about which they move are continually being exchanged so that, although always hemmed in by close neighbours, the neighbours themselves change and the molecules slowly diffuse. This description is followed by a detailed 'kinetic' picture of the evaporation and condensation of a liquid in apparently static equilibrium with its vapour, and of the phenomenon of latent heat.

The second part of his paper puts the kinetic hypothesis into quantitative form for an ideal gas, leading again to the basic equation 4.8. He ends with a calculation of the proportion of molecular energy that is accounted for by the translational motion; it is, in modern notation, $3(C_p - C_V)/2C_V = 3R/2C_V$, where C_p and C_V are the two heat capacities per mole, or "per unit volume", as Clausius puts it. For simple gases such as nitrogen and oxygen this proportion is 0.6315, which implies a ratio of C_p/C_V, denoted by γ, of 1.421.

If the molecules of a gas move at speeds of the order of 500 m s^{-1}, as he had just calculated, why do they not diffuse into one another in milliseconds rather than in minutes? This natural objection to the Krönig–Clausius hypothesis was raised by the Dutch physicist, C.H.D. Buys Ballot of Utrecht, who was best known as a meteorologist [122]. He had earlier worked on capillarity and speculated on a 'unified theory of matter', taking his atoms to be Boscovichian centres of force, but it was just this difficulty over the rate of diffusion that led him to assume that their motion was oscillatory, not translational. In rebutting this criticism Clausius broke new ground in the kinetic theory [123]. He abandoned molecules of infinitesimally small size and assumed instead only that they were small, and so travelled only a finite distance before colliding with another molecule. He could estimate neither their supposed diameter, s, nor the mean free path, l, that they traversed between collisions, but he could show that there were plausible ranges of s and l that were consistent with the gases showing only small departures from Boyle's law and having sufficiently small rates of diffusion. His kinetic theory, in which all molecules were supposed to travel at the same average speed, c, led to an equation that connected s and l;

$$4\pi N l s^2 = 3V, \qquad (4.12)$$

where there are N molecules in a volume V. The assumption that all the molecules had the same speed was clearly a weak point in his derivation of this equation, and one that was soon picked up by Maxwell, who showed, by a less than perfect argument, that there was a wide spread of speeds which followed the well-known 'law of errors' [124]. With this correction, the numerical factor of $(4/3)$ in eqn 4.12 becomes $\sqrt{2}$, but the change is unimportant for the calculations that could be made at the time. Equation 4.12 determines only the product Nls^2; further information is needed if we are to be able to calculate any of the three factors themselves. The first step in this direction was taken by Maxwell in 1860. He used the postulates of kinetic theory (or dynamical theory as it was then usually called) to calculate the rate of transfer of momentum between two layers of gas moving at different speeds, and so obtained an expression for the shear viscosity;

$$\eta = \rho l \langle c \rangle / 3, \tag{4.13}$$

where ρ is the mass density and $\langle c \rangle$ is the mean speed, which he showed is a little less than the root-mean-squared speed, $\langle c^2 \rangle^{1/2}$, which is the speed that properly occurs in eqn 4.8. This equation can then be written,

$$p = \rho \langle c^2 \rangle / 3, \quad \text{where } 3\pi \langle c \rangle^2 = 8 \langle c^2 \rangle. \tag{4.14}$$

A measurement of the viscosity gives, therefore, a direct route to the mean free path, l, if, indeed, the molecules can be treated as hard elastic spheres, as was done in the early versions of the kinetic theory.

Unfortunately the viscosity of a gas is hard to measure. Maxwell asked Stokes for a value for air, and Stokes, relying on some old measurements of the damping of the motion of a pendulum by Francis Baily [125], gave him a figure of $\sqrt{(\eta/\rho)} = 0.116$. This obscure result [126] makes sense only if one knows that the implied units are inch and second. For the viscosity Maxwell uses grains as the unit of mass, where there are 7000 grains in $1\,\text{lb} = 0.454\,\text{kg}$. The density of air was then well known; Maxwell does not say what figure he uses but a modern figure for air at 60 °F or 15.6 °C is $1.220\,\text{kg m}^{-3}$ or $0.3085\,\text{grain in}^{-3}$. The Baily–Stokes result therefore implies a viscosity of $0.004\,15\,\text{grain in}^{-1}\,\text{s}^{-1}$. (Maxwell's figure is 0.004 17.) This is a viscosity of $1.059 \times 10^{-5}\,\text{kg m}^{-1}\,\text{s}^{-1}$ or, in micropoise, 106 μP. A few years later Maxwell, helped by his wife, measured the viscosity of air from the damping of a stack of oscillating discs. He obtained $0.007\,802\,\text{grain in}^{-1}\,\text{s}^{-1}$ or 199 μP [127]. An extensive investigation by O.E. Meyer [128], a physicist at Breslau and the younger brother of the chemist Lothar Meyer, yielded figures of 104, 275 and 384 μP from previous measurements that he quoted, and 305 and 360 μP from his own early measurements. The range of values shows the difficulty of measuring this quantity; the modern value is 179 μP at 16 °C, so Maxwell has proved to be the best experimenter. The value of the mean speed is readily found from eqn 4.14; Maxwell quotes $1505\,\text{ft s}^{-1}\,(=458.7\,\text{m s}^{-1})$ and so, from eqn 4.13 and a viscosity of 106 μP, we get a mean free path, l, of $5.68 \times 10^{-6}\,\text{cm}$, which is Maxwell's figure of $1/447\,000$ in. This he confirmed by a figure of $1/389\,000$ in that he calculated from the rate of diffusion in gases as measured by Thomas Graham [127].

The product $N s^2$ is now calculable but we need another hypothesis before we can calculate each factor separately. This was supplied by Joseph Loschmidt in Vienna in 1865 [129]. He assumed that the liquid formed by condensing a gas is an array of touching spherical molecules. He denoted the ratio of the volume of the liquid to that of the gas by ε, the 'condensation coefficient', and so deduced the relation

$$s = 8\varepsilon l. \tag{4.15}$$

Air had not been liquefied in 1865 and, indeed, cannot be liquefied at ambient temperatures, so he had to estimate its hypothetical volume from the approximate additivity of the atomic volumes of liquids. This additivity had been established some years earlier by Hermann Kopp [130]. He used Kopp's figures, with slight modification, to obtain $\varepsilon = 8.66 \times 10^{-4}$. For l he chose Meyer's value of 1.4×10^{-4} mm, which he preferred to Maxwell's value, and so obtained $s = 9.7 \times 10^{-7}$ mm, or about 10 Å, admitting readily that "this value is only a rough estimate, but it is surely not too large or too small by a factor of ten". He quoted eqn 4.12 in his paper but did not use it explicitly to calculate N, the number of molecules per unit volume which, for a gas at 0 °C and 1 atm pressure, we now call 'Loschmidt's number'. His figures give $N = 1.8 \times 10^{18}$ cm^{-3} at ambient temperature. Had he used Maxwell's measurement of the mean free path his figures would have given $s = 3.9 \times 10^{-7}$ mm, or 4 Å, and $N = 2.7 \times 10^{19}$ cm^{-3}, which is close to the modern figure of 2.54×10^{19} cm^{-3} for an ideal gas at 1 atm and 60 °F.

Loschmidt's work was consolidated by Lothar Meyer [131] who showed that the volume ω of one of the assumed spherical particles [Teilchen] could be expressed, according to the equations found by Clausius and Maxwell,

$$\omega = F(T)m^{3/4}\eta^{-3/2}, \qquad (4.16)$$

where $F(T)$ is a function of temperature that is the same for all gases. He was thus able to show that the ratios of molecular volumes calculated from the viscosity of gases were close to that of the molar volumes of the liquids for a wide range of substances.

The kinetic theory that Maxwell put forward in 1860 was not exact but it was adequate for the calculation of the viscosity of a gas in terms of its molecular characteristics. It was, however, flawed for the calculation of the rate of diffusion and of the thermal conductivity. The root of the problem is the calculation of the distribution of the molecular velocities. At equilibrium these follow the the 'law of errors', as he had found correctly, but by a not wholly convincing argument, in 1860. If, however, the gas or gas mixture is at equilibrium then there is no viscous drag, no diffusion, and no conduction of heat. It is only when the distribution departs from 'Maxwellian' that these processes occur, and he did not know how to calculate this departure. He returned to the problem in 1867 with a much improved treatment [132]. Here he established, for the first time, the modern view of an inhomogeneous gas, and dispensed with the theoretical use of the mean free path. At elastic collisions between hard spherical particles there are three conserved quantities: mass and energy, which are both scalar, and momentum, which is a vector. To each there is a corresponding 'transport property', measured, for a gas of one component, by the coefficients of self-diffusion, D, of thermal conductivity, λ, and the more complex property of viscosity; η is the coefficient of shear viscosity.

Between these properties there are the simple relations,

$$\lambda = k_1 \eta C_V \quad \text{and} \quad D = k_2 \eta / \rho, \tag{4.17}$$

where C_V is the heat capacity at constant volume and k_1 and k_2 are dimensionless constants of the order of unity. In 1867 Maxwell found that $k_1 = 5/3$ and that $k_2 = 6/5$. Boltzmann showed later that k_1 is 5/2 [133].

The experimental predictions of the kinetic theory are surprising. Since l is inversely proportional to the density, ρ, it follows from eqns 4.14 and 4.17 that η and λ are independent of the gas density, and D inversely proportional to it. All vary with the temperature as $T^{1/2}$, if the heat capacity is independent of temperature, as is the case for hard spheres and for air at ambient temperature. It was the first prediction that led Maxwell and his wife to measure the viscosity of air in 1866 and to confirm that this improbable prediction held for pressures between 0.5 and 30 in of mercury (0.02 to 1.0 atm), so providing strong support for the infant theory [127]. The variation with temperature was potentially more interesting. The first experimental results produced a viscosity varying not as $T^{1/2}$ but closer to T^1. One of the more dramatic results of Maxwell's 1867 paper was that the problem of not knowing the departure of the velocity distribution from the equilibrium form could be evaded if the law of force between the molecules was an inverse fifth-power repulsion. For such particles the viscosity varies as the first power of the temperature. Since his experimental results came close to this behaviour he thought for a time that real molecules might have this law of force, although he was always more cautious than Clausius in attributing a real existence to the particles of kinetic theory. He was, however, never committed to the Newtonian view that molecules must have hard cores. Whewell had called this doctrine "an incongrous and untenable appendage to the Newtonian view of the Atomic Theory" [134], and Maxwell shared this opinion; the solidity of matter in bulk did not imply that two atoms could not be in the same place [135]. In his referee's report on Maxwell's 1867 paper Thomson had criticised the use of an inverse fifth-power repulsion between the molecules on the grounds that it was incompatible with the known values of the heat capacities [136]. This criticism could have been made of any system of simple spherical particles. It is interesting that Thomson did not then say that it was also incompatible with the cooling observed in the 'Joule–Thomson' expansion, a cooling that requires the presence of attractive forces between the molecules. This was pointed out by Meyer and by van der Waals in 1873 [137].

Maxwell's theoretical result could be summarised by saying that if we have an intermolecular potential of the form $u(r) = a(r/s)^{-n}$, where r is the separation, then $n = 4$ implies that η varies as T^1, and that the limit $n = \infty$ implies a variation as $T^{1/2}$. These results suggest that we have in the viscosity and other transport properties a new tool for studying intermolecular forces by seeing how their coefficients vary with temperature. This route could not be exploited in the middle of the 19th century since only these two isolated limits could be resolved. A general attack on the problem required a determination of the form of the of the velocity distribution function for a gas not at equilibrium, and that problem was not solved adequately until 1916. Its solution was to lead to the viscosity, in particular, becoming a prime source of information about intermolecular forces in the 20th century. Meanwhile one minor observation whetted the appetite for what might be achieved. In 1900 Rayleigh found that if, as theory and experiment agreed, the viscosity was independent of the gas density, then a dimensional argument shows that a simple repulsive potential with an inverse power of n implies that the viscosity varies with temperature as the power $(n + 4)/2n$; this result includes the two known special cases of $n = 4$ and $n = \infty$ [138]. Meyer had summarised the results for air in 1877 [139] and by 1900 Rayleigh was able to call on his own results for argon, which has a truly spherical molecule. Meyer found that a power of temperature of 0.72 was closer to experiment than Maxwell's power of unity, and Rayleigh found 0.77 for argon. The latter figure is consistent with $n = 7.4$ but, as Rayleigh knew, this assignment is too simplistic since it ignores the effects of the attractive forces.

Maxwell's proposal of a force repelling the molecules as the inverse fifth power of their separation led to further speculations. Stefan, in Vienna, suggested that the continuous repulsions might arise from dense clouds of aether surrounding the hard spherical cores. A continuous repulsive force leads to an effective molecular diameter that decreases with temperature since at high temperatures the molecules collide with a higher average speed of approach. He thought that this effect would increase the apparent power of the temperature with which the transport properties increased [140]. The same thought occurred also to Meyer [141]. Boltzmann, noting the small compressibility of water and the high speed of the molecules, calculated that "two molecules that approach along their line of centres with the speed of the mean kinetic energy approach to a distance that is about $\frac{2}{3}$ of the distance apart of two neighbouring molecules in liquid water." [142] Other contemporary attempts to establish atomic or molecular sizes were made by Stoney [143], Lorenz [144], Thomson [145] and others [146]. Thomson's support of the kinetic theory was influential in Britain, although his short article is typical of his obscurities and reservations on molecular matters. He starts by saying categorically "For I have no faith whatever in attractions and repulsions acting at a distance between centres of force according to various laws", but two pages later seems to be discussing just such forces. No doubt he resolved the apparent contradiction in terms of his favourite picture of atoms as vortices in the aether. He had put this model forward three years earlier and was to support it for another fifteen [147]. It was an idea that attracted both Maxwell and Tait [148]; the former was always uneasy with 'action at a distance' and here was a way of avoiding that problem if one could calculate the force betwen the vortices. Unfortunately that proved to be impossible. Tait's interest was more in the scope that such entities gave for the application of the vector and quaternion calculi and the entry that the subject gave him into the new field of mathematical topology [149]. Maxwell made little or no further use of the inverse fifth-power repulsion; he always had difficulty with any theory of matter that emphasised force at the expense of inertia [150].

By 1870 the experimental basis for the use of gases for the study of intermolecular forces had been truly laid, but could not be exploited because of the primitive state of kinetic theory. If the premises of this theory are accepted then the known departures from Boyle's law and the existence of the Joule–Thomson effect are evidence of interactions, usually of attractions, between the molecules. Indeed both are, in fact, the same evidence since the two effects are linked by macroscopic thermodynamic arguments that are independent of any molecular or kinetic assumptions. If one knows the departures of a gas from Boyle's law over a range of pressure and temperature then one can calculate the isothermal Joule–Thomson coefficient, that is $(\partial H/\partial p)_T$. With rather more difficulty the calculation can also be carried out in the reverse direction. Neither effect is easy to measure but acceptable values were available. The qualitative implications were clear but theory had yet to provide a quantitative link to the intermolecular forces. The three transport properties were also known to be linked to the molecular interactions via the assumptions of kinetic theory but again this theory was not sufficiently developed to exploit the link; indeed the relation was often counter-intuitive, for the viscosity, rate of diffusion, and rate of conduction of heat of a gas of point molecules without interaction are all infinite. Again accuracy was a problem, for none of these properties is easy to measure. Concern over accuracy became a particular interest of Meyer who, as a student of Franz Neumann, had been brought up in a school that was fanatical in its devotion in hunting down errors, probably to the detriment of what might otherwise have been accomplished [151].

There was one worrying problem that hindered the acceptance of the kinetic theory, and this arose not from the interactions of the molecules but apparently from their internal constitutions. If, as was generally assumed, the molecules were modelled by structureless elastic spheres then the heat capacity of a gas at constant volume arises from their translational motion only. Each orthogonal direction of motion contributes $\frac{1}{2}R$ to the molar heat capacity, where R is the universal constant of the perfect-gas law, thus giving a total heat capacity of $(3R/2)$. The heat capacity at constant pressure exceeds that at constant volume by R for all perfect gases. Thus the ratio of the heat capacities, $\gamma = C_p/C_V$, is 5/3 or 1.67. The first experimental confirmation of this figure came in 1875 with the measurement of the speed of sound in mercury vapour [152]. Mercury was known to form a monatomic vapour and its atoms were presumably spherical. This result provided a drop of comfort in the discussion of what was otherwise seen as an insoluble problem, for no common gas conformed to this figure nor, indeed, to any figure for which a generally acceptable explanation could be given. For oxygen and nitrogen, and hence also for air, the ratio γ was found to be 1.40 or 7/5. It was generally accepted by then that these gases had diatomic molecules, O_2 and N_2, which presumably could rotate, but this presumption only led deeper into the mire. Each 'squared term' in the energy, in Hamilton's formulation of mechanics, contributes $\frac{1}{2}R$ to the heat capacity. A diatomic molecule, it was argued, can rotate about each of its three axes of symmetry and so has, in addition to its translational energy, three terms in the square of the angular momentum about each axis. Hence C_V would be $3R$, C_p would be $4R$ and γ would be 4/3 or 1.33, which is smaller than the observed value. It is possible to argue, as Boltzmann did [153], that there is no rotation about the line of centres of a diatomic molecule since the molecule looks 'monatomic' about this axis. This assumption leads to the correct value of 7/5 and is, indeed, the modern interpretation of the anomaly, but in a quantal not classical mechanical framework. Maxwell never accepted this sleight of hand [154] and it was the main ground on which he sometimes doubted the reality of the kinetic theory; in a discussion of 1867 he called it "under probation" [155]. Moreover a diatomic molecule should be able to vibrate since there is no reason to suppose that the bond between the two atoms is wholly rigid. Any departures from perfect rigidity would add more terms to the energy and so reduce the calculated value of γ for air still further. There was evidence that more complicated molecules did have internal motions; for steam, for example the value of the ratio was 1.19. Beyond these problems of rotation and vibration there lay the nightmare of even more complicated internal motions revealed by the rich optical spectra that could be excited in all molecules. These, as Tyndall foresaw [156], were to lead to our deep understanding of atomic and molecular structure, but neither they nor the heat capacity anomalies were to be unravelled until the advent of quantum mechanics. Meanwhile those with less tender consciences than Maxwell wisely decided to put these problems out of their minds and concentrate on what could be achieved with the experimental and theoretical weapons to hand. It is a tactic that most scientists adopt instinctively.

Liquids remained, by comparison with gases, an unknown theoretical territory. By adding thermodynamic arguments to their armoury, but staying within the Laplacian tradition, Ritter and Dupré had deduced that the large internal pressure of a liquid, Laplace's K, depended on the square of the density of the fluid, and they and others had obtained by variants of Young's argument rough estimates of the size of molecules or the range of the attractive forces; the two were not always distinguished. Young's own result re-surfaced in 1890 when it was exhumed by Rayleigh in a paper on capillarity [157]. These estimates were neither as soundly based nor, as we can now see, as accurate as those derived from gas theory. It is significant, however, that there was no correlation of the two types of estimates, in part because those working on liquids were not convinced of the correctness of the kinetic viewpoint. Thomson mentions capillarity in his short paper of 1870 [145] but did not use it constructively as Waterston and Dupré had done. This failure to tackle liquids seriously arose from a general lack of a real conviction that the properties of gases and liquids could be explained in terms of a common molecular model. Even Andrews, who did most to establish experimentally the continuity of the two states, was not convinced of this [76]. Maxwell often wrote as if he were willing to use a common model, notably in his lecture to the British Association in September 1873 [158]. His mature view, however, is in his article 'Atom' of 1875 [135]. He wrote there:

There is considerable doubt, however, as to the relation between the molecules of a liquid and those of its vapour, so that till a larger number of comparisons have been made, we must not place too much reliance on the calculated densities of molecules.

Nevertheless, he was inclined, on balance, to think that the molecules of a gas were the same as those of a liquid. Clausius and Boltzmann had probably the strongest views on the matter before van der Waals, but neither showed much interest in quantitative work on liquids. G.H. Quincke, in Berlin, had made an early and bold claim for the identity of the forces in gas and condensed phases when he opened a paper of 1859 [159] with the italicised premise:

There is therefore a condensation of gaseous substances on to the surfaces of solid bodies that increases proportionally to their area and density, if the law of attraction as a function of separation, is the same for the gas molecule as for the solid.

He clearly believes that this is the case but one sees also here the residuum of the belief, not entirely banished until the 20th century, that intermolecular attractions are linked in some way to gravitational, a view held also at that time, and indeed twenty years later, by Thomson [160].

One publication of 1870 that excited Maxwell's interest three years later, and which may have helped to persuade him that the combination of kinetic theory and attractive intermolecular forces was a key to the understanding of the simple properties of matter, was a remarkable paper of Clausius [161]. It is remarkable because it contains a theorem that nothing then known gave any hint of. Gibbs came also to admire it calling it "a very valuable contribution to molecular science" [162]. Clausius established that the mean kinetic energy of a system of particles is equal to what he called the 'virial'; that is, in modern notation

$$\left\langle \tfrac{1}{2} m_i v_i^2 \right\rangle = \langle r_i \cdot F_i \rangle, \tag{4.18}$$

where m_i, v_i, r_i, and F_i are the mass, speed and position of particle i and the force on it. The theorem applies to systems in which both the positions and the speeds are bounded. If the motions are irregular, as with a molecular system, then the averages are taken over a long enough time for them to become steady. The forces include those exerted by the bounding wall of the vessel which were known to contribute $3pV/2N$ to the term on the right. If the molecules are spherical particles with forces acting between each pair then the contribution of any one pair to the virial of the whole system can be written,

$$r_i \cdot f_i + r_j \cdot f_j = r_i \cdot f_{ij} - r_j \cdot f_{ij} = -r_{ij} f_{ij}, \tag{4.19}$$

where $f_{ij} = f_i = -f_j$ is the mutual force between i and j, which acts in the same direction as $r_{ij} = r_i - r_j$. The virial theorem, as it is now called, can therefore be written

$$\sum \langle m_i v_i^2 \rangle = 3pV + \sum \langle r_{ij} f_{ij} \rangle, \tag{4.20}$$

where the first sum is to be taken over all molecules and the second over all pairs of molecules. Clausius was seeking, as for a time Boltzmann was also, for a purely mechanical basis for the second law of thermodynamics. When he failed to find it in this theorem he apparently took little further interest in it [163]. The equation had, however, other potentialities, for here, at last, was an exact and, indeed, simple equation between the mean kinetic energy of a molecular system, its pressure, and the sum of the forces acting between its molecules. Only one problem remained to be solved before this equation could be exploited to study intermolecular forces – what was the relation between the mean kinetic energy and the temperature? For a perfect gas, for which $f_{ij} = 0$, it was accepted that we have the simple relation

$$\sum \langle m_i v_i^2 \rangle = 3pV = 3RT, \tag{4.21}$$

where T is the absolute temperature, measured on a scale whose zero is at $-273\,°C$, and R is a constant, proportional to the amount of gas, and the same for all gases if V is the volume that contains a mass of gas equal to its 'molecular weight' in grammes. So much was generally accepted in 1870, but it was not obvious then (as it is now) that the same relation between the mean kinetic energy and the absolute temperature holds also for interacting molecules, since the forces between them clearly change the instantaneous value of the molecular speeds. There was, nevertheless, a growing body of opinion that held that the outer part of eqn 4.21 was true for real gases, for liquids, and maybe also for solids. As early as 1851 Rankine, in expounding a 'rotational' theory of the motion of heat, distinguished between the 'real' and the 'observed' specific heats, identifying the former with the motions [164]. More explicitly, Clausius in 1862 distinguished between the 'heat in the body' and the 'disgregation', and wrote in italics that "*The quantity of heat actually present in a body depends only on its temperature, and not on the arrangements of its component particles*" [165]. Sixteen years later, Maxwell, when reviewing Tait's *Thermodynamics*, expressed his amazement at finding this statement of Clausius in a footnote, and described it as "the most important doctrine, if true, in molecular science" [166]. In the concluding paragraphs of his *Theory of heat* of 1871 [56], Maxwell had speculated that the molecules in a liquid might move more slowly than those in its vapour at the same temperature, a speculation that survived in all later editions of the book, down to the tenth, in 1891 which was edited and revised by Lord Rayleigh whose failure to remove it, or at least to comment on it, was perhaps an oversight, although Rayleigh was not wholly willing to commit himself on that point at that time [167].

Maxwell and Rayleigh were not the only agnostics; those arch-enemies Tait and Tyndall had doubts also. Tait upbraided Clausius for muddying the clear waters of thermodynamics by introducing his molecular quantities 'die innere Arbeit' and 'die Disgregation'. He was still arguing the point in a paper of 1891 that he reprinted without comment in 1900 [168]. Tyndall, in a lecture course of 1862, could affirm only that "most well-informed philosophers are as yet uncertain regarding the exact nature of the motion of heat" [169]. Others were more confident about equating the mean kinetic energy and temperature. In 1872, M.B. Pell, the professor of mathematics at Sydney, affirmed without proof, in a Boscovichian description of matter, that in all states "the temperature may be assumed to be proportional to the mean *vis viva*" [170], an assumption that, as we shall see, van der Waals was to make to great effect the next year. Maxwell summarised the doubters' position in a letter to Tait of 13 October 1876:

With respect to our knowledge of the condition of energy inside a body, both Rankine and Clausius pretend to know something about it. We certainly know how much goes in and comes out and we know whether at entrance or exit it is in the form of heat or work, but what disguise it assumes when in the privacy of bodies . . . is known only to R, C, and Co. [171]

From our privileged modern position we can see that the problem of the mean kinetic energy in any state of matter is a trivial one. The translational energy of the molecules at any time is a term in the classical Hamiltonian, or total energy, that is independent of their internal motions of rotation and vibration and of their mutual interactions, and which can be expressed as a sum of squared terms in the instantaneous values of the linear momenta. In the partition function of classical statistical mechanics we can integrate at once over these linear momenta to give a contribution to the total thermodynamic energy that is independent of the state of aggregation. It is therefore equal, in any state, to its value in the dilute gas, or $3RT/2$. This was shown, but not of course in this language, by Boltzmann in 1868–1871 [58], but it was many years before it became a truth universally acknowledged. No doubt Clausius, who was already convinced of the truth, saw no need to comment on these papers of Boltzmann's, while Maxwell probably saw their titles, and since he knew that thermodynamics could not be reduced to mechanics, read no further at that time. But the ways by which this important point was established are still far from clear and could well be a subject for further study.

4.3 Van der Waals's thesis

Johannes Diderik van der Waals was a schoolmaster in The Hague for eleven years from 1866 to 1877 [172]. When he started there he had no university degree but he soon began to attend lectures at Leiden and passed his doctoral examinations in December 1871. Eighteen months later he submitted his thesis *On the continuity of the gaseous and liquid state* [173]. It carries the date 14 June 1873, which was the day of his public defence of it. The 'promotor' was P.L. Rijke, whose speciality was experimental work in electricity and magnetism, so it is clear that the choice of subject was van der Waals's own. Like the early 19th century workers in kinetic theory, he was very much the 'outsider' and brought to the subject a new vision, but unlike them he was well versed in mathematics and physics and so was able to handle his subject in a way that commanded respect even when it attracted criticism.

He tells us at the opening of his thesis, and again in his Nobel lecture of 1910 [174], that his choice of subject was inspired by Clausius's papers on the kinetic theory of gases and a desire to understand the large but mysterious pressure in a liquid that was represented by the integral denoted K by Laplace. He had a clear and simple conviction of the real existence of molecules and wrote that "I never regarded them as a figment of my imagination, nor even as mere centres of force effects" [174]. This conviction led him to a synthesis of the molecular theory of gases and liquids that had escaped his predecessors. There is evidence in the thesis that he had arrived at the form of his famous equation of state by simpler arguments than those that follow from his discussion of the work of Clausius and Laplace, but it was these that he used in his public defence of his derivation.

He has, as he sees it, two problems to solve. First, how to take account of the effect on the pressure of attractive forces of unknown form but, he believes, of essentially short range, that is, of a range comparable with the sizes of the molecules. He and O.E. Meyer [137] were, it seems, the first to emphasise that the cooling of gases on expansion observed by Joule and Thomson was direct evidence for the existence of attractive forces in gases; the statement of this truth is the subject of the first two-page chapter of his thesis (§§ 1–5, see also pp. 70–1). His simplest calculation of the effect of these forces on the pressure comes in Chapter 7 (§ 36); the molecules at the surface of a fluid are pulled inwards and the effect on the pressure, p, is proportional both to the number pulled per unit volume and to the number in the interior doing the pulling. In other words, the corrected pressure to be used in an equation of state is the observed pressure plus a term proportional to the square of the molecular density, $(p + a/V^2)$. A correction term of this form follows also from Laplace's theory when this is carried out carefully and, as we have seen, it was a form that had also been reached by other arguments in the time since Laplace; it would have been surprising if he had arrived at any other form. His second problem is to calculate the amount by which the observed volume must be reduced by the space taken up by the molecules so as to give an effective volume in which they move, and which can be used in the equation of state. He is adamant that there are no repulsive forces; his molecules are hard objects which have size, and he had no sympathy with models such as Maxwell's fifth-power repulsion, although he did not then appreciate fully the contents of Maxwell's papers. Whenever his predecessors had thought of this second problem they had rather casually assumed that the effective volume was the actual volume less the sum of the volumes of the molecules. He showed, by an argument based on Clausius's mean free path in a gas of particles of non-zero size, that the effective volume is $(V - b)$, where b is four times the sum of the volumes of the molecules (Chapter 6). It is to the parameter b that Dupré's name 'co-volume' is now attached, although van der Waals did not use this word.

In his thesis these two justifications of the effects of the attractive forces and of molecular size are preceded by a fuller and more sophisticated discussion of the attractive forces. There are three points to note.

He repeats in full Laplace's derivation of his integrals K and H (Chapters 3 and 4), including correctly the insertion of the factor of the square of the molecular density. This enables him to identify K with his correction term a/V^2 (Chapter 9). The late appearance of this identification and its surprisingly tentative form is not consistent with the opening sentence of the Preface: "The choice of the subject which furnished the material for the present treatise arose out of a desire to understand a magnitude which plays a special part in the theory of capillarity as developed by Laplace". No doubt the emphasis he placed on different parts of the work changed over the years he spent in preparing it, and after he realised what a rich set of results he had produced. He makes no reference to Ritter or Dupré although the work of the latter must have been accessible to him since he cites other papers from the *Annales de Chimie et de Physique*.

The second point to note is that in obtaining Laplace's results he has recourse, as Laplace did also, to integrations over an assumed uniform distribution of molecules in space. In Laplace's day this assumption had been justified by the belief that the forces, although only of microscopic range, were nevertheless long compared with the diameters of the hard cores of the molecules. Van der Waals did not share this belief and, as we shall see, obtained quantitative evidence to rebut it, so this comforting justification of what we call the mean-field approximation was denied to him. He certainly held, however, to the mean-field view itself, writing in words reminiscent of Laplace: "On the particles of a gas no forces act; on the particles within a liquid the forces neutralise each other. In both cases the motion will go on undisturbed so long as no collisions occur." (§ 9) His justification differs from that of Laplace, who had a static picture of matter; for van der Waals it is the molecular motion that produces the averaging over positions needed to justify the approximation. He seems also to ascribe a repulsive effect to this motion, writing: "It is the molecular motion that prevents the further approach of these particles." (§ 23) We now know that both points are incorrect, the first for reasons adduced at the end of the previous Section. The strict separation in classical mechanics of translational motion from configurational interaction means that one cannot simplify expressions for the latter by invoking the former. His inadequate justification of the mean-field approximation was to lead to criticism from Kamerlingh Onnes eight years later and, more forcibly, from Boltzmann some twenty years later. It is one of the few cases where van der Waals's instinct for the correct 'physics' of a problem, even if not always for the correct 'mathematics' with which to handle it, led to a deep flaw in his work. This became apparent many years later in considering the detailed behaviour of fluids near their critical points.

The third point to note in his discussion of his correction to the pressure is his account of Clausius's virial theorem, which he derives and discusses in Chapter 2. He was the first to appreciate the value of this theorem for the study of intermolecular forces, but before he could use it he had to tackle the problem of relating the mean kinetic energy of the molecules in a liquid to the temperature. He makes as little of this difficulty as had Clausius. Indeed, he evades it by saying simply that since the mean energy increases with what is usually called the temperature, it can be replaced by it: "This may be considered to give our definition of temperature." (§ 36) This is an evasion, not a solution, since he does not show that the temperature of a liquid, so defined, is the same as that of the absolute scale of the second law of thermodynamics, or of its equivalent, the perfect-gas scale. Nevertheless his instinct, like that of Clausius, proved to be right when he supposed that "the kinetic energy of the progressive motion is independent of the density; [and] that, for instance, a molecule of water and a molecule of steam at $0\,^\circ$C have the same velocity of progressive motion." (§ 36) He is now in a position to combine the augmented pressure and the effective volume to obtain his well-known equation of state of gases and liquids,

$$(p + a/V^2)(V - b) = RT. \tag{4.22}$$

He knows that the equation is not exact. The co-volume, b, must itself diminish with increasing density since it is equal to four times the sum of the volumes of the molecules only in the dilute gas. Moreover there is chemical and thermal evidence

(Chapter 5) to show that molecules are more complicated entities than the hard spheres that he had assumed. He is more confident about the a/V^2 term.

To test his equation he used first the extensive results that Regnault had published in his monographs of 1847 and 1862 for air, hydrogen, sulfur dioxide and carbon dioxide [175]. His discussion of the last gas is curtailed since he had fortunately become aware of Andrews's results. These were to provide him with a much more convincing demonstration of the power of his equation than he had been able to find from the rather inconclusive comparison with Regnault's results. It is not clear when he first saw Andrews's results. He cites the long abstract in German published in 1871 in a supplement to the *Annalen der Physik* [176]. He had presumably missed the original publication of 1869 [76] and probably the French abstract in the *Annales de Chimie et de Physique* and an English one in *Nature*, both in 1870 [177], although he was later (Chapter 12) to quote from a paper that appeared in the *Annales* in 1872. Once he knew of Andrews's work and the discussion of it by Maxwell in his *Theory of heat* of 1871 [56] he realised its importance, and he borrowed, without acknowledgement, the title of Andrews's Bakerian Lecture for his thesis [178].

Andrews had shown that carbon dioxide has a critical temperature of 31 °C. Above that there is one fluid state with a fixed density for each pressure and temperature. Below the critical temperature there are two densities for each pressure and temperature on the vapour-pressure line, the higher being that of the liquid and the lower being that of the vapour in equilibrium with it. Van der Waals's equation is a cubic in the volume (or density) at a fixed pressure and temperature and so has either one or three real roots. The first case occurs when the temperature is above a value of $(8a/27Rb)$, and the second when it is below this critical value. The lowest and highest real roots correspond to gas and liquid states but the third root at an intermediate density has no real existence for it is a state in which $(\partial p/\partial V)_T$ is positive, and so is mechanically unstable. Such a state, if formed, would spontaneously break up into a mixture of gas and liquid states (Fig. 4.2). It was from Maxwell's book that van der Waals learnt that James Thomson, William's elder brother, had, on seeing Andrews's results, suggested just such a continuous cubic curve to interpolate between gas and liquid [179]. Andrews's results show, of course, not a cubic curve but a straight horizontal line joining the co-existing gas and liquid states at a constant pressure, that is, at the 'vapour pressure' appropriate to the chosen temperature. None of them, Andrews, Thomson, Maxwell or van der Waals, then knew how to use the form of the isothermal curve to decide where this line should be drawn. Maxwell's first attempt at this problem was a failure [180], but he gave the correct answer in a lecture before the Chemical Society in 1875; the line is to be drawn so that it cuts off equal areas above and below the cubic curve [181]. This result rests only on thermodynamic considerations; no molecular arguments are needed.

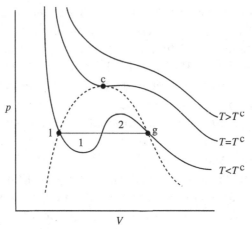

Fig. 4.2 Van der Waals's representation of the relation between pressure and volume of a fluid. Three isotherms are shown: one at a temperature above that of the critical point (marked c), one at the critical temperature, and one below this temperature. The last isotherm shows a maximum and a minimum but what is seen experimentally is the horizontal line joining the liquid state, marked l, and the gas state, marked g. Maxwell showed that this line has to be drawn so that the two areas, 1 and 2, are equal.

Van der Waals chose his parameters a and b for carbon dioxide by fitting his equation to Regnault's results but then used them to calculate the course of the isotherms measured by Andrews. In modern units he chose

$$a = 0.445 \, \text{Pa} \, (\text{m}^3 \, \text{mol}^{-1})^2, \qquad b = 51 \, \text{cm}^3 \, \text{mol}^{-1}.$$

His equation gives for the three critical constants,

$$RT^c = 8a/27b, \qquad V^c = 3b, \qquad p^c = a/27b^2, \tag{4.23}$$

whence

$$T^c = 311 \, \text{K} = 38 \,^\circ\text{C}, \, V^c = 153 \, \text{cm}^3 \, \text{mol}^{-1}, \, p^c = 63.4 \, \text{bar} = 62.5 \, \text{atm}.$$

Van der Waals obtains 306 K, 153 cm^3 mol^{-1}, and "about 61 atm", but the minor discrepancies are a consequence of the two-figure accuracy with which he could estimate a and b. Andrews's experimental results were 30.9 °C, 145 cm^3 mol^{-1}, and "about 70 atm". (Modern figures are 31.0 °C, 94 cm^3 mol^{-1}, and 72.8 atm.) The agreement with Andrews's results is closer than the experimental accuracy of Regnault and Andrews and the approximations inherent in his equation deserve. We can deduce directly from the equation that the critical ratio $(pV/RT)^c$ is 3/8 or 0.375. Andrews's results give 0.40, but the only comment that van der Waals makes (§ 56) is to say that the crude results of Cagniard de la Tour for ethyl ether lead to a ratio of about 0.3, which is closer to the truth; modern values lie in the range 0.22 to 0.29, with carbon dioxide at 0.27.

The importance of van der Waals's achievement lies not so much in the quantitative agreement with Andrews's results as with the fact that, for the first time, the properties of both gases and liquids were derived from a unified theory and related directly to the two essential properties of molecules; they occupy space and they attract each other. The implications of Andrews's observation of the unity of the gas and liquid states and of van der Waals's relating it to the two features of molecular interaction was potentially far-reaching, although neither experiment nor theory was always accepted at first. Maxwell alone had the genius to recognise at once the implications of what was being proposed although, as we shall see, he was not convinced of the rigour of van der Waals's reasoning.

A second important result that flowed from his equation was the information that can be derived from the numerical values of his two parameters a and b (Chapter 10). These are related to the properties of the molecules and their interaction and so complement the information that Maxwell had probably realised was potentially locked up in the transport properties. From the parameter a van der Waals estimated the range of the attractive forces and from b the diameter of the hard core.

He first identifies a/V^2 with Laplace's K and then notes that the surface tension (Laplace's $\frac{1}{2}H$) is the first moment of "the force" which is the integrand of K. The ratio (H/K) is therefore the effective range of the attractive force – a more precise but physically equivalent argument to that of Young. He has no means of measuring the surface tension of liquid carbon dioxide so he turns to the five liquids ethyl ether, ethyl alcohol, carbon bisulfide, water and mercury. We may take the results for ether as typical, and for this the ratio (H/K) yields an effective range of 2.9×10^{-10} m, or 2.9 Å.

From b he can obtain at once the volume of the molecules in a given mass of fluid, but to obtain the volume of one molecule he needs to know Loschmidt's number or its equivalent. He introduces, therefore, Maxwell's estimate of the mean free path in air at 1 atm and 15 °C [124], which he scales appropriately for other gases, and so obtains a diameter of a molecule of ether of 4.0 Å. He comments (§ 68) that:

It is certainly surprising to find s [the diameter] even at all greater than x_1 [the effective range of the attractive force]. In all these calculations, however, we are only dealing with approximate values; and we have been altogether dependent on Maxwell's value of l [the mean free path] for air.

He draws the conclusion that the range of the attractive force is little greater than the size of the core:

By this I do not mean to say that there is no attraction at other distances, but that the attraction at this distance is so much greater, that it is alone necessary to consider it in the calculation.

This was the view that he held for the rest of his life and since his reasoning and his data were essentially correct, his conclusion was also.

The calculation of the molecular diameter gives him also a measure of Loschmidt's number (he does not use that name), and for air at $0\,^\circ$C and 1 atm pressure he deduces a density of 5×10^{19} molecules per cubic centimetre, which is about twice the modern value. He had no way of testing independently the accuracy of his deductions but he was confident that they were more soundly based than earlier estimates of molecular size, as, for example, that of Stoney whose value of Loschmidt's number is 20 times larger [143], or Quincke's estimate of the range of the attractive force from capillary phenomena which is 100 times his [182]; these are the only examples that he cites.

The rest of the thesis is 'thermodynamic' rather than 'molecular'. In Chapter 11 he calculates the cooling of a gas at low densities associated with the Joule and with the Joule–Thomson expansions, that is $(\partial T/\partial p)_U$ and $(\partial T/\partial p)_H$. His equation of state can be arranged to give the second virial coefficient (as we now call it), $B(T)$ of eqn 4.5, as

$$B(T) = b - a/RT, \tag{4.24}$$

whence the expansion coefficients are readily found from eqn 4.6. His calculation of the Joule–Thomson cooling of carbon dioxide is about two-thirds of that found experimentally, a discrepancy larger than he would have expected. He is conscious of the criticisms that have been made of those who drag molecular considerations into thermodynamic arguments but boldly sets out his own view (§ 72):

It is the boast of thermodynamics that its laws do not rest on any assumptions as to the structure of matter, and consequently embody truths which are in so far unassailable. If, however, we are prevented from making more searching investigations into the nature of bodies through fear of leaving the region of invulnerable truths, then it is clear that by so doing we wantonly cut ourselves off from one of the most promising paths to the hidden secrets of nature.

In a resounding peroration he refers to the molecular forces as "nothing but the consequences of a Newtonian law of attraction", but it is clear from what has gone before that he means here only a force that apparently acts at a distance and which varies with the separation, not one that is specifically proportional to the inverse square of the separation. Dutch theses end with a set of *stellingen*, or propositions not directly related to the subject in hand but chosen by the candidate to air his views on cognate matters. Van der Waals had 19 of these [183], one of which was Newton's declaration in his letter to Bentley [184] that action at a distance was "inconceivable". We do not know if the examiners asked him to defend Newton's opinion, nor what he might have replied. He ends his thesis with a quotation from

William Thomson's Presidential Address to the British Association in 1871, in which Thomson, quoting from an anonymous book review by his friend Fleeming Jenkin, the Professor of Engineering at Edinburgh, once again looks forward to that age when the subject of atoms, their motions and their forces, may rival in its precision and richness the field of celestial mechanics.

A Leiden thesis in Dutch by a schoolmaster who was quite unknown outside the Netherlands would have passed unnoticed had it not been circulated to the leaders of the field. Who was the sender, or senders, we do not know; it could have been van der Waals himself or, more likely, his colleague and mentor at the Hague, Johannes Bosscha, or his thesis 'promotor', Professor Rijke. Copies certainly went to Andrews, Maxwell, and the Belgian physicist, J.A.F. Plateau, and probably also to James Thomson and to Clausius, who was now in Bonn [185]. Only Maxwell rose to the challenge with a full review in *Nature* [186] in which he praised the author for his insight and originality but had specific criticisms about the way that he had derived his equation. His first point was that, having introduced Clausius's virial theorem, whose significance Maxwell had not previously appreciated, van der Waals should have used it consistently to treat both the attractive and repulsive forces. Maxwell adopted the modern view that the intermolecular force field is an entity and not something to be split, as van der Waals and most of his predecessors had done, into an attractive field and a space-filling core. Some years later, H.A. Lorentz, the first professor of theoretical physics at Leiden, carried out Maxwell's proposal and treated all forces by means of the virial theorem [187]. Maxwell made his own calculation of the co-volume, b, and found it to be 16 times the volume of the molecules. Whether he obtained this from the virial theorem was not explained here but this seems to be the case from what he wrote in an unpublished manuscript [186]. The result, however, is wrong, and van der Waals, for all the crudity of his calculation from the mean-free path, had arrived at the right answer. Maxwell's second criticism was a re-iteration of his opinion that we are not justified in equating the mean translational energy of the molecules in the liquid state to $3RT/2$. He had not studied Boltzmann in detail but doubts were perhaps beginning to assail him for he was careful to add that "the researches of Boltzmann on this subject are likely to result in some valuable discoveries".

Andrews was asked to give a second Bakerian Lecture in 1876 in which he described further measurements on the equation of state of carbon dioxide [188]. He fitted them only to a simple empirical function of his own devising and ignored van der Waals's equation, perhaps convinced by Maxwell's criticisms that it was flawed. Stokes, the Secretary at the Royal Society, had sent the text to Maxwell to referee before it appeared in print. In his comments Maxwell made it clear that he supported van der Waals's equation as an empirical representation of the results and then he went on to apply the virial theorem to the problem of the equation of state [189]. He followed Boltzmann in writing the probability of finding a molecule at a position in a gas where the energy is Q as proportional to $\exp(-Q/aT)$, where a is "an absolute constant, the same for all gases". If Q arises from the potential energy between a pair of molecules, and if the density is sufficiently low for us to be able to neglect interactions in groups larger than pairs, then he is able to show that the leading correction to Boyle's law is proportional to the integral

$$A_r = 4\pi \int_0^r (e^{-u(r)/aT} - 1)\, r^2 dr, \tag{4.25}$$

where $u(r)$, or Maxwell's Q, is the potential energy of a pair of molecules at a separation r, and where the symbol r also does duty as the upper limit of the integral, where it is the range of the attractive force. An integration by parts leads to an alternative form of the integral in which the force $(-du(r)/dr)$ appears explicitly; $A_r = B_r/3aT$, where

$$B_r = \int_0^r 4\pi r^3 \,[du(r)/dr]e^{-u(r)/aT} dr. \tag{4.26}$$

Clausius's virial theorem now leads to the result that the leading correction to Boyle's law, which we now call the second virial coefficient, $B(T)$ of eqn 4.5, is

$$B(T) = -\tfrac{1}{2}A_r = -B_r/6aT. \tag{4.27}$$

He makes a slip in writing the virial theorem and so obtains a result that is too large by a factor of $(3/2)$, but had he used these results to re-calculate the co-volume, b, he would at least have recognised that his earlier result was seriously wrong. We obtain van der Waals's result by writing $u(r)$ as the potential of a hard core of diameter s;

$$u(r) = \infty \quad (r < s), \qquad u(r) = 0 \quad (r \geq s), \tag{4.28}$$

whence

$$A_r = -4\pi s^3/3 \quad \text{or} \quad B(T) = b = 4[4\pi (s/2)^3/3]. \tag{4.29}$$

But Maxwell never took the calculation this far and never, apparently, retracted his erroneous expression in his review in *Nature*. He had discovered, in eqns 4.25 and 4.26, the most direct connection between an observable physical property, $B(T)$, and the force or potential acting between a pair of molecules. There is a minor problem in fixing the value of the constant a, later to be known as 'Boltzmann's constant', k, and equal to R/N_A, where R is the molar gas constant and N_A is Avogadro's constant, whose value was still uncertain in 1876 but which was then becoming increasingly better known. Boltzmann's constant is, however, needed only to convert the intermolecular energy from a scale of temperature to one in more conventional units. Maxwell's equations were a link that was to be exploited to the full in the 20th century, and Maxwell's failure to use this link or, at least, to publish it so that others could use it, is one of the great missed chances of this field.

If Clausius's paper of 1857 and his popular lecture of the same year [115] mark the birth of the modern molecular–kinetic view of the states of matter then van der Waals's thesis, and Maxwell's formal completion of it in 1875 [181] with his rule for determining the vapour pressure at each temperature, mark its coming of age. For the next thirty years there were critics of the molecular–kinetic interpretation of the properties of matter but henceforth it was the orthodoxy from which physicists departed at peril to their future reputations.

4.4 1873–1900

Maxwell's review in *Nature* ensured that van der Waals's work was soon known in Britain, even if not fully understood or appreciated, but it made its way more slowly in Germany. If Clausius had had a copy in 1873 he cannot have then read it for he calculated the value of the co-volume, b, in November 1874, and made it eight times the volume of the spherical molecules [190]; van der Waals quickly corrected him [191]. Six years later, when he had read the thesis, Clausius published a second paper [192] which contained his own derivation of Maxwell's 'equal-areas rule' for fixing the value of the vapour pressure. At the same time he modified the equation of state, for he supposed that the attractive forces might change with temperature. This supposition arose from a confused discussion of molecules "rushing towards each other" and forming aggregates, a discussion that shows that, notwithstanding his earlier introduction of the concept of 'disgregation', he had not fully appreciated the consequences of Boltzmann's separation of the kinetic and potential energies. He replaced van der Waals's term a/V^2 for the 'internal pressure' with $\alpha/T(V+\beta)^2$. This introduction of a third adjustable parameter allowed him to claim an improved representation of some experimental results, including those of Andrews. The mathematician D.J. Korteweg, a colleague of van der Waals, was later to claim that E.-H. Amagat's results for carbon dioxide, obtained in 1873, were better fitted by van der Waals's original equation than by Clausius's modification of it, that is, the factor of T was not needed and β was best put equal to zero [193]. The apparently greater flexibility of the modified equation and, no doubt, Clausius's greater reputation, meant that it was for some years used more often than the original version. Maxwell had stressed the empirical virtues of van der Waals's equation and had criticised its theoretical basis, so there was little reason not to use a second empirical equation with an even less secure theory behind it.

Boltzmann seems first to have known of the thesis from a long abstract of it that Eilhard Wiedemann published in the first issue of the *Beiblätter* of the *Annalen der*

Physik [194]. One crude measure of the cohesion of a liquid is the height above atmospheric pressure that can be sustained in a barometer tube before the liquid column splits leaving a vacuum [195]. Boltzmann was led from a consideration of this topic to a new determination of the range of the intermolecular attraction [196]. He notes first that the minimum thickness of liquid films appears to give a figure of the order of 500 Å, but then says that van der Waals got a very different result from the ratio of Laplace's two integrals, H and K, and so devised his own method. He estimated the energy needed to separate two molecules by considering the 'unbalanced' force at the surface of a liquid, as measured by the surface tension, and the maximum value of the force between two molecules from the tensile modulus of the solid. Since the energy is an integral of the force with respect to distance, the ratio of these two quantities is a length which he takes to be the effective range. For six metals he thus gets figures that lie between 15 Å for copper and 63 Å for zinc. The corresponding values that he quotes for the internal pressures are equivalent to 4000 atm for copper and 1300 atm for zinc. Neither in method nor, as we can now see, in numerical results, is this an advance on the clearer notions of van der Waals.

In truth, the four great theorists of the developing fields of kinetic theory and statistical mechanics, Clausius, Maxwell, Boltzmann and Gibbs, never gave their full attention to the problem of the attractive forces. Clausius set out the virial theorem but never used it; he turned soon to work on electrical problems where he developed a variant of Weber's theory of central forces between particles that depend on their motions as well as their positions. Maxwell derived the expression for the second virial coefficient, but only when pushed by having to referee Andrews's paper, and he never exploited it. In his last years – he died of cancer in 1879 at the age of 48 – his main interest in this field was the behaviour of highly rarefied gases. Boltzmann's real concern was the newly developing field of statistical mechanics, a generalisation of kinetic theory, and so with the link between mechanics and thermodynamics. This led him into the great problem of irreversibility [197] – how are the time-reversible laws of mechanics compatible with the irreversible operations of thermodynamics? He continued also to worry about the problem of the heat capacities of gases, a problem that was not to be solved in his lifetime. His most influential contribution to the field of cohesion was the perceptive commentary on van der Waals's work in the second volume of his book on gas theory [198]. Gibbs came to the field with his masterly studies of classical thermodynamics in the 1870s and only later turned his attention to statistical mechanics [199]. When he did, his concern was with the foundations of the subject, not with its application to the properties of gases and liquids. Like Maxwell, he was sufficiently worried by the problem of the heat capacity of gases to doubt if his deductions had a wider range of applications than to the formal models that he had set up. There was, moreover, the distraction (as it turned out) of the views of the positivists, the 'energeticists' and the anti-atomists. Their opinions were influential in France and Germany at the end of the 19th century and were not fully overcome until the first decade of the 20th [200]. This movement is relevant to the study of cohesion only in that those who did not accept the need for atoms and molecules could not work on the problem of the forces between them. There is, logically, no reason why they could not have tried to develop a non-particulate theory of the cohesion of liquids, as some of the 'elasticians' had for solids, but none seems to have made the attempt.

The problems that drew the attention of the major theorists were more pressing and more topical than those raised by van der Waals's work; cohesion was not a new topic at the end of the 19th century! Most physicists, then and now, would think also that these other problems were deeper and more important. So for the forty years after the publication of the thesis, the problems it raised became the major concern primarily of the growing Dutch school of physicists. Some of their efforts were defensive, for the Andrews–van der Waals picture of continuity between the gas and liquid states was not everywhere accepted at once. There was resistance particularly in France and in Italy, and new experiments cast doubt on the simple picture. These doubts were reinforced by widespread scepticism about the identity of the 'particles' in the two states, a view that went with the notion that the liquid state persists above the critical point as a solute dissolved in the compressed gas [201]. William Ramsay was one of the first doubters [202] but he later recanted [203] and his collaborator, Sydney Young, made some of the most precise measurements that we have on the relation between pressure, volume and temperature in the critical region [204]. The sources of the errors that seemed to refute Andrews's work were several: impurities, density gradients arising from the great compressibility of fluids near their critical points, and the slowness of these states to reach equilibrium because of the impurities and the high heat capacities of critical fluids. It was well into the 20th century before the situation was clarified, the brunt of the refutation falling on the experimental school established at Leiden by Kamerlingh Onnes [205] who had succeeded Rijke in 1882. A major step in unmasking the effects of impurities was a systematic study of binary mixtures and the development of the theory of their phase behaviour by van der Waals [206]. Even when some measure of agreement about the correctness of the Andrews–van der Waals picture had been restored the identity of the particles in the two phases was not universally accepted. As late as 1904, Émile Mathias, who had done good experimental work in the field, wrote to van der Waals to say that he thought that this idea was flawed: "The great defect, in my view, of your theory of the identity of the liquid and gaseous molecules is that one cannot understand at all the simple phenomenon of the liquefaction of gases." [207]

Van der Waals's equation, when supplemented by Maxwell's equal-area rule, leads in principle to a complete determination of the vapour pressure of the liquid as a function of temperature, and of the co-existing or orthobaric densities of liquid and vapour. In practice, the calculation cannot be made explicitly, as van der Waals soon found out after some trials. (A parametric solution for Clausius's modified equation was found by the young Max Planck in 1881, and is easily adapted to the original equation [208].) In the course of his struggles van der Waals discovered that the vapour pressures and orthobaric densities of different liquids resembled each other more closely than they conformed to the predictions of his equation. This resemblance became apparent if he plotted the dimensionless ratio (p_σ/p^c) as a function of (T/T^c), where p_σ is the vapour pressure. His own equation could be expressed in terms of such ratios in a universal or reduced form. If we define $\pi = p/p^c$, $\omega = V/V^c$, and $\tau = T/T^c$, then his equation can be written

$$(\pi + 3\omega^{-2})(3\omega - 1) = 8\tau. \tag{4.30}$$

(Such a reduction can be made for any equation of state that contains only two adjustable parameters and the universal gas constant, R [209].) Of more value, however, than this explicit form was what came to be called the principle or law of corresponding states, namely that π is a function of ω and τ that is, approximately, the same for all substances; or, formally,

$$\pi = f(\omega, \tau), \tag{4.31}$$

where the function $f(\omega, \tau)$ is a universal function, although not necessarily of the form of eqn 4.30. This law was obtained and applied by van der Waals in 1880 as an outcome of his struggles to fit vapour pressures to his original equation. The long papers in Dutch [210] became more widely known through the abstracts in the *Bleiblätter* [211]. These were the work of Friedrich Roth at Leipzig, who published in the next year a complete translation of the thesis itself, with some revisions by the author [212]. It was from this time that van der Waals's work became to be more fully known outside the Netherlands.

The practical value of the law of corresponding states was immense; one had for the first time a reliable, but not exact, method of predicting any of the thermodynamic properties of a hitherto unstudied substance from a very sparse set of observations, most simply from two of the critical constants, for example, p^c and T^c, but other sets, not necessarily critical, could be used. The law proved invaluable in estimating the conditions needed to liquefy hydrogen and later helium, so that James Dewar, a pioneer in gas liquefaction, called it the most powerful physical principle in the field to be discovered since Carnot's theorem [213]. But what were the theoretical principles that lay behind this powerful law? Kamerlingh Onnes, then a young assistant to van der Waals's friend Johannes Bosscha at Delft, had heard of van der Waals's results by word of mouth and soon perceived that behind this principle of similarity of the macroscopic physical properties there must be a similarity in the underlying molecular force fields [214].

He starts by making three assumptions that are to be found in van der Waals's thesis: the necessary assumption that the temperature is a measure of the mean kinetic energy of the molecules in all states of matter, that the effects of the attractive forces can be subsumed into a pressure of the form a/V^2, and that the molecules can be regarded as miniature solids, by which he and van der Waals [215] understood that they were perfectly elastic bodies that retained their size and shape in all physical encounters. These considerations led him to a generalised form of van der Waals's equation,

$$RT = (p + a/V^2)V\Psi(m, V), \qquad (4.32)$$

where m is the volume of a molecule and V the volume of a fixed amount of substance, e.g. one mole in modern language, and Ψ is an unknown function. It is, however, not a function of m and V separately but only of their ratio, so he wrote it

$$\Psi(m, V) = (1 - rm/V)\chi(m/V), \qquad (4.33)$$

and he proposed that the function χ be expressed as an expansion in powers of the density,

$$\chi(m/V) = 1 + B(m/V) + C(m/V)^2 + \cdots \qquad (4.34)$$

Van der Waals's equation is recovered if one puts $r = 4$ and $\chi(m/V) = 1$. We have here, in this equation of 1881 an incomplete form of what he was to develop twenty years later, the modern 'virial equation of state'. The first general expansion of the pressure in powers of the molecular density was, in fact, made in 1885, by M.F. Thiessen, a German working at the International Bureau of Weights and Measures at Paris [216]. He wrote

$$p = RT\rho(1 + T_1\rho + T_2\rho^2 + T_3\rho^3 + \cdots), \qquad (4.35)$$

where T_i are functions of temperature only. He obtained also expansions of the heat capacities in powers of the density and inverted these to get expansions in powers of the pressure. He estimated T_1, our second virial coefficient, from Regnault's results for carbon dioxide, but made no attempt at a molecular interpretation of his equation which seems to have had little influence.

Kamerlingh Onnes does not, at this stage, try to go beyond eqn 4.34. After a long discussion of the kinetic explanation of evaporation and condensation he comes to what he describes as his "second step" beyond van der Waals. He touches first on the justification for the use of the mean-field approximation, namely that there is an internal pressure of the form a/V^2 only if the range of the attractive force is large compared with the molecular size – the condition that was clear to Laplace and Poisson but which van der Waals had obscured. He notes that van der Waals had provided the evidence that the condition is not fulfilled and adds firmly: "But if the decrease in the law of attraction is so rapid for it to be felt only at a collision, then our argument is no longer applicable". He does not elaborate, perhaps out of respect for the views of van der Waals who was submitting his paper to the Royal Academy of Sciences. He goes on instead to discuss the distinction between physical and chemical association of molecules into groups:

By physical associations I mean those for which we can ignore the mutual interactions of parts of molecules, so that we can consider, to a sufficient approximation, the motion of one molecule with another as the sole result of actions emanating from the similarly situated points [217] in the molecules that we take to be the centres of molecular attraction. Under these circumstances the chemical constitution of the molecule has no effect. On the contrary, in chemical associations – which can be classed with the phenomenon of crystallisation – the points from which the forces emanate that cause the association are no longer those similarly situated points ...

(We note here a persistence of the notion that we first met in the work of Newton and his followers, that crystallisation involves a lack of spherical symmetry; that is, properties of 'sidedness' or 'polarity' are required. There is a confusion here, which the French 'elasticians' would probably not have made at this time, between the fact that a molecule in a crystal is not in a spherically symmetrical environment while, on the average, a molecule in a liquid is, and the erroneous implication that a non-spherical force field is needed to induce crystallisation.) Kamerlingh Onnes continues:

By the nature of our hypothesis we do not consider chemical associations. Thus the law we shall establish will apply only when the molecules can be considered as similar bodies, acting on each other through forces emanating from similarly situated points. So that the departures that we shall observe from this law should be attributed to the fact that the molecules are no longer similar elastic solids of almost constant dimensions, and that their mutual actions are not inversely proportional to a certain power of the separation of the similarly placed points, but the influence of a difference of constitution in different parts of the molecule, and the resultant chemical interactions, make themselves felt in the laws of molecular motions. ... Thus we arrive at the following law: by choosing appropriate units of length, time, and mass, it is possible, according to our new hypothesis concerning the molecular forces, to deduce from the state of motion of one substance an allowed state of motion of the same number of molecules of another substance. The speeds and external pressure should therefore be replaced by corresponding values. If the isotherms have the property of correspondence then the ratios of reduction are equal to the ratios of the pressure, volume, and absolute temperature of the critical state. ... It seems to me, therefore, that in what is said above we have given the simplest explanation of the law discovered by Prof. van der Waals, by means of the principle that similarity of the isotherms and of the [liquid–vapour] boundary curves is the immediate expression of the similarity of the molecular motions. [218]

He then suggests that the principle might be applied to comparisons of capillary constants, viscosity and thermal conductivity of fluids. Thirty years later, when he and his eventual successor at Leiden, W.H. Keesom [219], were writing an article on the equation of state for the *Encyklopädie der mathematischen Wissenschaften*, they expressed his conclusions more concisely and, indeed, more clearly:

First, that the molecules of different substances are completely hard elastic bodies of a common shape; second, that the long range forces that they exert emanate from similarly situated points and are proportional to the same function of the corresponding separation of these; and thirdly, that the absolute temperature is proportional to the mean kinetic energy of the translational motion of the molecules. [220]

Van der Waals at once perceived the value of these ideas and communicated the paper to the Academy on 24 December 1880. He had apparently not known of Kamerlingh Onnes until then, but the contact between them grew into a close personal and professional friendship. When Kamerlingh Onnes went to Leiden in 1882 he established there the leading physics laboratory for the study of fluids and fluid mixtures at high pressures and down to low temperatures. This effort was balanced by the theoretical developments of van der Waals in Amsterdam on the equation of state of pure and mixed fluids and on capillarity.

Ideas similar to those of Kamerlingh Onnes, but more obscurely expressed, were put forward by William Sutherland [221], a free-lance theoretical physicist who worked in Melbourne. As early as 1886, when he was 26, he was writing to his brother: "My head is churning now with theories of molecular force for liquids and solids – hyperbolic and parabolic for gaseous molecules and elliptical for liquids; but in solids the law changes and the question is how?" [222]. He apparently then thought that there were different forces in different states of matter, the view that van der Waals was fighting against. His notions on hyperbolic and parabolic trajectories were to see the light of day twenty years later in a paper that comes closer to Kamerlingh Onnes's position [223]. It is not as clear as even Kamerlingh Onnes's first attempts but it is evidence that the idea that intermolecular forces had a 'family' resemblence to each other was in the air; other similar enquiries into the origin of the law of corresponding states are cited by Kamerlingh Onnes and Keesom [220]. In the intervening years Sutherland had published a long series of papers in the *Philosophical Magazine* in which he had put forward a range of ideas of varying merit. He tried at first to convince his readers that the attractive force varied always as r^{-4}, where r is the separation of the two molecules. He knew that this form of force generated a term in the energy of the fluid that was logarithmic in the volume, and that the laws of thermodynamics did not allow for such a term, but he tried to argue the problem away. Only one of his ideas struck a chord at the time, and indeed is remembered to this day [224]. If molecules have hard spherical cores and are surrounded by attractive fields then, he argued, two molecules in free flight in a gas might be drawn into a collision that would not have occurred in the absence of the attraction. This likelihood is greater the slower the speeds of the molecules and so we expect their apparent collision diameters to increase as the temperature falls. If the attractive forces are weak the viscosity of such a gas can be expressed

$$\eta = (1 + S/T)^{-1}\eta_0, \tag{4.36}$$

where η_0 is the viscosity of a gas of plain hard spheres which was known to vary as $T^{\frac{1}{2}}$, and where S is proportional to the potential energy of a pair of molecules in contact. This result can be written in a different form,

$$(\mathrm{d}\ln\eta/\mathrm{d}\ln T) = \tfrac{1}{2} + S/(S + T), \tag{4.37}$$

to show that the apparent power of T with which η varies, changes from $\frac{1}{2}$ at infinite temperature to 1 at $T = S$. Such a variation comes closer to matching the experimental results than any other expression of its day and, for all its simplicity, is perhaps the most important advance in relating viscosity to temperature that was made from Maxwell's time to the 1920s. A similar proposal, but with the factor for the increased number of collisions in the form $\exp(S/T)$, was made by Max Reinganum, a young German physicist trained at Leiden and Amsterdam who was killed in the First World War [225].

The theory of the equilibrium properties of the imperfect gas advanced as slowly as the kinetic theory of the transport properties, but with less reason since there were no formidable mathematical difficulties in the way. The Dutch school explored the extension of van der Waals's equation to mixtures, a rich field that revealed many fascinating kinds of liquid–liquid–gas phase equilibria and critical lines. The Dutch rarely went beyond the closed form of the van der Waals equation and so were unable to extract any more information about the intermolecular forces than he had done in his thesis. Here his increasing reputation probably inhibited progress. The systematic study of the deviations from the perfect-gas laws at low densities, where the molecules interact only in pairs, would have unlocked new information on the intermolecular forces, as Maxwell had shown in his referee's report on Andrews's paper of 1876, but this route was followed only slowly, with hesitation, and initially by those outside the Netherlands. The very success of van der Waals's equation was again a handicap for it led to most effort being put into improving it and devising other closed-form equations. This was a natural way forward at a time when it was supposed that a sufficiently diligent search would reveal the one true equation of state of gases and liquids. New forms were tried and improvements were made, although many of these were trivial, but it was a long time before it was accepted that there was no universal equation to be found, and that a study of the leading terms of a simple expansion of the pressure in powers of the density would reveal more about the range and intensity of the intermolecular forces.

Maxwell's expression for the second virial coefficient was re-discovered after twenty years, published by Boltzmann in 1896 [226], and exploited by Reinganum [227], who wrote the leading correction to the perfect-gas laws in the form

$$p(V + B) = RT, \tag{4.38}$$

so that his B is the negative of our second virial coefficient. Let us move to the modern convention and write, from Maxwell's first integral;

$$B(T) = -2\pi N \int_0^\infty (\mathrm{e}^{-u(r)/kT} - 1)r^2 \mathrm{d}r, \tag{4.39}$$

where B is the second virial coefficient for N molecules, and $u(r)$ is the intermolecular potential energy of two molecules at a separation r. Reinganum's model was that of hard spherical molecules of diameter σ, surrounded by an attractive force field that varies as r^{-m}. He argued first that m was equal to 4, as Sutherland had done, but then chose m to be equal to $4 + \delta$, where δ is a small positive constant, in order to avoid the logarithmic divergence in the total energy and in B. Let us choose the index more generally and work in terms of the intermolecular potential $u(r)$ rather than its derivative, the force, and so write

$$u(r) = +\infty \; (r < \sigma) \quad \text{and} \quad u(r) = -\alpha r^{-n} \; (r \geq \sigma, n > 3). \tag{4.40}$$

We can now expand the exponential and integrate term by term to get

$$B(T) = \tfrac{2}{3}\pi N \sigma^3 - 2\pi N \sum_{i=1} \frac{(\alpha/kT)^i \sigma^{3-in}}{i! \, (in - 3)}, \tag{4.41}$$

where the first term, van der Waals's b, is four times the volumes of the molecules. Reinganum proceeded slightly differently. For separations greater than σ he separated the integral into two terms, the exponential and the term -1; he then integrated the first by expanding the exponential, and integrating by parts from σ to an upper limit, l. He combined the second term with the integral from 0 to σ. After the upper limit becomes infinite, he obtained

$$B(T) = \tfrac{2}{3}\pi N \sigma^3 \, \mathrm{e}^{-u(\sigma)/kT} - \tfrac{2}{3}\pi N \sum_{i=1} \frac{n(\alpha/kT)^i \sigma^{3-in}}{(i - 1)! \, (in - 3)}. \tag{4.42}$$

Since $u(\sigma)$ is negative he wrote the first term $\exp(c/T)$. He chose this route to emphasise, as Sutherland had done, that molecules are brought into collision by the attractive force and so the positive term in B, the co-volume, is larger at low temperatures. The two expressions for B can be shown to be equivalent by expanding the exponential in eqn 4.42 and re-arranging the terms. The second form is not now used.

Sydney Young had made some precise measurements of the pressure of iso-pentane gas as a function of density [228], from which Reinganum calculated the

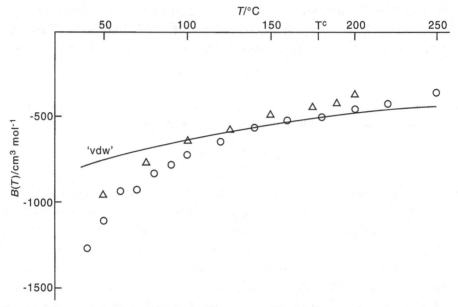

Fig. 4.3 The second virial coefficient of isopentane. The circles are the values calculated by M. Reinganum [227] from the measurements of the pressure by S. Young [228], and the triangles are the more recent measurements of K.A. Kobe and his colleagues [229]. The line is the value calculated from a van der Waals equation that has been fitted to Reinganum's value of $B(T)$ at the critical temperature.

deviations from the perfect-gas laws and compared them with his new theoretical expression, eqn 4.42, and with the corresponding expression that follows from van der Waals's equation, namely

$$B(T) = b - a/RT. \tag{4.43}$$

He observed that the experimental results for B changed more rapidly with temperature than this equation permits. Figure 4.3 shows the values of B calculated from Young's results, and some more modern ones [229]. Reinganum's point can be illustrated by choosing b to be $V^c/3$, as van der Waals's equation requires, and then choosing a to fit the observed value of B at, say, the critical temperature. It is seen that eqn 4.43 does not give a sufficiently rapid variation with temperature. It is only when $u(\sigma)$ is much smaller than kT that eqns 4.41 and 4.42 reduce to the form of eqn 4.43, namely

$$B(T) = (2\pi N\sigma^3/3)\,[1 + 3u(\sigma)/(n - 3)kT], \tag{4.44}$$

and this condition is not satisfied at temperatures as low as the critical. The condition that

$$|u(r)| \ll kT, \qquad (r \geq \sigma) \tag{4.45}$$

is one that ensures the correctness of the mean-field treatment of van der Waals, but as Kamerlingh Onnes and (as we shall see) Boltzmann had already pointed out, it is not one that real molecular systems satisfy at and below the critical temperature.

Reinganum did not try to take the matter further in 1901; in particular, he did not try to fit his theoretical expression, eqn 4.42, to Young's experimental results. Possibly he was deterred by the difficulty of fixing uniquely the three unknown parameters, σ, α and the index n. It was a difficulty that was always going to plague this field. Possibly he also went no further because of a common feature of normal science, as generally carried out by rank-and-file scientists, namely that whenever one makes an advance one is too easily satisfied with that step, and does not consider what further might be done. (The enormous number of papers published today emphasises the small incremental advance made by each of them.) In a later paper he supposed that forces between electric dipoles in the molecules might be the origin of attractive force and tried, without much success, to interpret the viscosity of a gas in terms of a 'Sutherland' factor of $\exp(c/T)$, where c now arises from the dipole–dipole potential [230]. Such electric interactions were to be much discussed in the first twenty years of the new century.

The range of the attractive forces soon again became a matter of discussion and even of controversy. We have seen that van der Waals, arguing from the ratio of Laplace's integrals H and K, had deduced that it was little longer than the size of the molecular core. Kamerlingh Onnes had tentatively pointed out that a mean-field approximation required that $u(r)$ be everywhere less than kT and so that the range had to be large if the integral of $u(r)$, essentially the parameter a of van der Waals's equation, is not to be negligible. Boltzmann made the point more forcibly in 1898, after having disagreed with van der Waals during a visit to the Netherlands [231]. The matter came up again in the context of the thickness of the surface layer of a liquid, a discussion that marked the return to the scene of the phenomenon of capillarity. We have seen that Poisson criticised Laplace's assumption that he could treat the surface of a liquid as a sharp boundary at which the density falls abruptly from that of the liquid to that of the vapour, and had argued, correctly, that the change of density must take place over a distance comparable with the range of the attractive force. Neither he nor Maxwell, who was of the same opinion, contributed anything useful to the problem of determining this thickness [232], which is not easily measurable. The mirror-like surface of a still liquid shows that it is much less than the wavelength of visible light which is around 0.6 μm or 6000 Å for the yellow part of the spectrum. A lower limit was proposed by Quincke [182], who prepared glass slides each coated with two tapering layers of silver of continuously varying thickness, this thickness being almost zero along the line at which the two silver wedges met. By studying the rise of water in the lens-shaped capillary tube formed from two of these slides placed with their silvered faces together he was able to find how thick the intervening layer of silver needed to be before the strong molecular forces between glass and water became negligible. This distance was about 500 Å, and he found similar distances for the minimum range of the forces of other triplets of materials. Van der Waals had quoted, and implicity discarded, these estimates in his thesis, noting that Quincke himself had little confidence in his rather indirect experiments [233]. Maxwell, however, accepted them at face value and dismissed van der Waals's figures as wrong – "so we cannot regard these figures as accurate" [232]. He ignored Boltzmann's estimates which were closer to those of van der Waals [196]. Some years later, A.W. Rücker, a chemist who studied surface films, wrote an extensive review of the field, and backed Quincke and Maxwell [234]. The first clear evidence from thin films that molecular size and, by implication, the range of the forces, were as advocated by van der Waals came in the 1890s when Agnes Pockels [235] and, more explicitly, Lord Rayleigh [236] showed that films of olive oil, etc., on water could be compressed to a point where the the surface tension changed abruptly to an "anomalous" value. The area of the film at this point was recognised by Rayleigh as that at which the surface was covered by a close-packed monomolecular film. He and Pockels both arrived at a thickness of about 10 Å and Rayleigh deduced that this was the size of a molecule of olive oil. Thus we were left with a clash of experimental evidence over the range of the forces, and with the theoretical paradox that van der Waals had experimental evidence for a short range but a valuable equation of state that Boltzmann and also Rayleigh [167] insisted required a long range. Kamerlingh Onnes had tentatively allied himself with Boltzmann but a clearer acceptance of the validity of Boltzmann's criticism came from a member of the Dutch school when P.A. Kohnstamm, then van der Waals's assistant in Amsterdam, wrote in a review in 1905:

If the radius of the sphere of action is large with respect to the molecular diameter, the primitive form of Prof. van der Waals completely retains its validity for high densities; for large volumes, the constant a of the equation of state becomes a function of temperature, tending, as the temperature rises, to a limiting value; the dependence on volume remains however as Prof. van der Waals has established; it is only at intermediate densities that there is a transition region where a depends on volume and temperature. [237]

The first part of this sentence is an acceptance of Boltzmann's point; the second part, the "large volume" limit, shows an awareness of Reinganum's findings but overlooks the fact that these describe the real world, not one in which the attractive forces are of long range, when a is truly a constant. He adds that if the range is not large then the conclusions are no longer exact. And there, for the moment, the matter had to be left without any satisfactory resolution. One by-product was a point of nomenclature; it was Boltzmann who, in 1898, first wrote that "we call this attractive force the van der Waals cohesive force", and the 'van der Waals force' it has remained to this day. Kamerlingh Onnes and Keesom found it necessary, however, to distinguish between the real short-ranged "van der Waals forces" and the hypothetical long-ranged "Boltzmann–van der Waals forces" [238]. Today we recognize the distinction but do not use the second term. It was only after the precise formulation of statistical mechanics by Gibbs in 1902 [239] and its application to fluids by L.S. Ornstein in his Leiden thesis of 1908 [240] that a proper foundation could be laid for calculating the "discontinuous distribution of the attractive centres" [167] needed to tackle the problem of a fluid with short-ranged attractive forces.

Van der Waals himself made only one attempt at guessing the form of the attractive potential. He accepted the correctness of Poisson's argument that the surface of a liquid has a thickness of the range of this potential, and therefore that Laplace's theory, with its sharp interface, was incomplete. He set about developing a theory of the surface tension of a liquid with a diffuse interface. He was not the only one to tackle this problem; Karl Fuchs, the professor of physics at Pressburg (now Bratislava) [241] and Rayleigh [242] had had very similar ideas, but van der Waals's version was the most complete and it was he who worked out the consequences [243]. His work marked a great advance in the theory of capillarity and, as we can now see, in the development of the statistical mechanical theory of non-uniform systems in general. We are concerned here, however, with the particular law of force between molecules to which this work led him. He asked what is the intermolecular potential, $u(r)$, that leads to the field outside a uniform semi-infinite slab of material (e.g. a liquid surface) falling off exponentially with distance z from the face of the slab [244]. It is perhaps surprising that this problem is related to the solution of the Laplace–Poisson equation that Mossotti had studied (see Section 4.1); in both cases the solution is what we now call the Yukawa potential [78]. An intermolecular potential of the form

$$u(r) = -A(\lambda r)^{-1} e^{-\lambda r} \qquad (4.46)$$

between the molecules of the slab generates a field $\phi(z)$ acting on a molecule at height z above the surface of the slab, where

$$\phi(z) = -2\pi A(\rho/\lambda^3) e^{-\lambda z}, \qquad (4.47)$$

and where ρ is the uniform number density of the molecules in the slab. The length λ^{-1} is a measure of the range of both $u(r)$ and $\phi(z)$. If one now calculates Laplace's integrals for a slab with a sharp interface then one finds $K = \lambda H$, thus again showing that λ^{-1} is the range of the potential. The fact that the same form, eqn 4.46, satisfies van der Waals's problem and the Laplace–Poisson equation,

$$(\nabla^2 - \lambda^2) u(r) = 0, \qquad (4.48)$$

is more than a trivial coincidence is shown by the fact that the same identity holds in spaces of all dimensions. The generalised Yukawa potential in a space of dimensionality d is

$$u(r) = -A(2/\pi)^{1/2}(\lambda r)^{-\nu}K_\nu(\lambda r), \qquad (\nu = \tfrac{1}{2}d - 1) \qquad (4.49)$$

where K_ν is the modified Bessel function of order ν [78]. If $\nu = \tfrac{1}{2}$ then eqn 4.49 reduces to eqn 4.46. The Yukawa potential shares with Newton's gravitational potential the property that the total potential between two spheres with this intermolecular potential acts as if all the material were at the centre of the spheres. For a short time van der Waals believed that eqn 4.46 was the true intermolecular potential, but he did not use it again and it is possible that he came to realise that it could not be the answer. His follower Gerrit Bakker used it, however, in a long series of papers on capillarity from 1900 onwards [245] and J.R. Katz used it in studying the adsorption of gases on the surface of a solid [246]. It then fell out of favour but became fashionable again in the second half of the 20th century as an admittedly unrealistic model potential whose attractive mathematical properties make it useful for exploring theoretical ideas.

4.5 The electrical molecule

An intimate relation between electrical forces and chemical bonding had been a commonplace of theoretical discussion throughout the 19th century. It stemmed initially from the experiments and speculations of Davy and Berzelius, but it soon became clear that such forces were only part of the chemical story, applicable to many inorganic compounds but of little use in interpreting the composition and structure of organic compounds. The relationship between atoms and electricity was put on a quantitative footing with Faraday's laws of electrolysis of 1832–1833 [247]. These laws carry the important implication that if matter is composed of discrete atoms then electricity must also be 'atomic'. This was brought out most clearly, at least for British scientists, by Stoney in a paper to the meeting of the British Association in Belfast in 1874 that was published in 1881 [248], and by Helmholtz in his Faraday Lecture to the Chemical Society, also in 1881 [249]. Arrhenius's work extended the understanding of conducting and non-conducting aqueous solutions and it was this fruitful field that led to the coming of age of the new discipline of physical chemistry, which is conventionally marked by the appearance in February 1887 of the first issue of the new international journal, the *Zeitschrift für physikalische Chemie* of Ostwald and van 't Hoff. Such work created a theoretical background in which electrical forces between atoms came to be accepted but it had, at first, little impact on the discussion of cohesive molecular forces. Here the trigger was J.J. Thomson's identification of the electron as a sub-atomic particle in 1897–1899 [250].

In the early years of the 20th century the younger generation of physicists working in statistical mechanics was appreciating the limitations of van der Waals's equation and realising that it was unlikely that any simple closed form of equation would describe fluids exactly. The density expansion of the pressure of a gas, advocated by Kamerlingh Onnes and now written,

$$pV/RT = 1 + B(T)/V + C(T)/V^2 + D(T)/V^3 + \cdots, \tag{4.50}$$

led to a measurable coeffcient, $B(T)$, that was by then known to be rigorously related to an integral of the potential between a pair of molecules. His definitive paper on this 'virial expansion' came at just the right moment to reinforce the point [251]. As we have seen, Reinganum had exploited the link between the second virial coefficient and the intermolecular potential, and in his commentary he supposed that there was an electrical origin to the potential that he wished to measure, but his grounds for doing this were little more than the assertion that if atoms contained electrons they must also contain positively charged entities if they were to be overall neutral. The young Dutch physicists with whom he worked made this connection also and the first twenty years of the new century were marked by a stream of papers in which electrical models of molecules were devised and their validity, or otherwise, tested by comparison of their integral with the observed values of the second virial coefficient. There was one obvious difficulty; the observed coefficient, as a function of temperature, is an integral of the potential over separation, whereas what is really needed is an expression for the potential as an integral (or other function) of the virial coefficient over temperature. There is no simple way of making this inversion and the problem was, apparently, not given any serious attention until the second half of the 20th century. Progress could therefore be made only by guessing the form of the electrical forces involved, calculating the virial coefficient by integration, and seeing if the calculated function had the same magnitude and dependence on temperature as the known experimental results, which were necessarily of less than perfect accuracy and were confined to a limited range of temperature. The danger of this trial-and-error procedure is that there is no guarantee that an incorrectly chosen potential may not yield a coefficient that is sufficiently close to the observations for the potential to be deemed 'in agreement with experiment'.

Another problem which called for some complicated geometrical analysis, was the calculation of the corrections to van der Waals's parameter b at high gas densities, a problem that we now phrase as the calculation of the higher virial coefficients, C, D, etc., of eqn 4.50, for a system of hard spheres. This was tackled by van der Waals himself but his efforts led only to partial results and to errors [252]. The work was completed by Gustav Jäger [253], Boltzmann [226] and J.J. van Laar [254]. At the end of the day the third and fourth coefficients had been calculated correctly for a system of hard spheres; no higher coefficient is known exactly even now. Such results were important for calculating the free or available volume in a fluid of high density but did not touch on the more pressing problem of the origin and form of the attractive forces. Now that the battle over the correctness of the molecular-kinetic theory was being won it became obvious that further advances of the theory required some definite notion of the origin of the intermolecular forces. Van Laar was led from a consideration of the hard-sphere problem to that of the attractive forces. He estimated the second virial coefficient of a gas with an intermolecular force that was attractive at large separations, became repulsive at shorter, and contained within it a hard repulsive shell [255]. His model was more realistic or at any rate more flexible than that of Reinganum but his integrations were carried out more crudely and, like Reinganum, he made no useful comparison of theory and experiment.

Van der Waals's son, also J.D. van der Waals, had succeeded his father as the Professor of Physics at Amsterdam in 1908, having previously held a chair at Groningen [256]. In the same year he took up the electrostatic interaction of molecules. If, as was generally agreed, molecules contained charged sub-atomic particles but were overall electrically neutral, then the simplest picture of the charges was as an electrical doublet or dipole. This comprises a pair of equal and opposite charges separated by a short distance. Its 'strength', μ, is measured by the product of the magnitude of either charge and their separation. The energy of two dipoles at a separation r depends on their mutual orientation, where the direction of a dipole is conventionally represented by the line running from the negative to the positive charge. The mutual energy of two equal dipoles at a centre-to-centre distance r, large compared with the charge separation within each molecule, can be expressed in the modern system of units,

$$u(r) = (\mu^2/4\pi\varepsilon_0 r^3)\,\mathrm{f}(\omega), \qquad (4.51)$$

where ω denotes symbolically the orientations of the two dipoles with respect to the line joining their centres. If the direction of one of the dipoles were to be reversed then u would be changed in sign but unaltered in magnitude. It follows, therefore, that the average of $\mathrm{f}(\omega)$ over all orientations is zero. At first sight it would seem that the dipole–dipole energy would make no net contribution to the second virial coefficient since the integration to give B, eqn 4.39, has to be taken over all orientations as well as all distances for non-spherical potentials. This is not so, however, since $u(r)$ occurs in the exponential (or Boltzmann) factor and so there is a net negative or attractive contribution to B. This much was known to Sutherland and Reinganum and was put forward again by van der Waals Jr as a possible source of the attractive intermolecular potential. The leading term in this potential can be found by expanding the exponential in eqn 4.39 and averaging over all orientations:

$$(e^{-u(r)/kT} - 1) = -\frac{\mu^2 \langle f(\omega) \rangle}{4\pi\varepsilon_0 r^3 kT} + \frac{1}{2}\frac{\mu^4 \langle f(\omega)^2 \rangle}{(4\pi\varepsilon_0)^2 r^6 (kT)^2} - \cdots, \qquad (4.52)$$

where the angle brackets denote the average over all orientations. The first term vanishes since $\langle f(\omega) \rangle = 0$, but the second contributes a negative or attractive term to the potential and so to B. This effective potential falls off rapidly with distance, namely as r^{-6}, but there is a worrying complication in that it also falls off rapidly with increasing temperature, namely as T^{-2}, whereas the van der Waals expression, eqn 4.43, varies only as T^{-1}. Reinganum had shown that higher inverse powers of temperature were needed to fit the experimental results, but it was not expected that there would be no term in the first power. Van der Waals Jr argued that the discrepancy, if there be one, might be misleading since we do not know how, if at all, the dipoles change with temperature [257]. His father, who had submitted his paper to the Academy, was lukewarm in his support for the increasingly popular electrical dipoles. In an unusually metaphysical vein he had characterised the ability of molecules to occupy space as one that was a property necessary and inherent to matter but that the attractive forces, although apparently universal, were not necessary. He said that these forces were not proportional to mass and so there was no reason for the forces between unlike molecules to be the geometric mean of those between like molecules. Experiment bore him out, for the values found for a_{12}^2 were not generally equal to the product $a_{11}a_{22}$, where 1 and 2 denote different molecular species [258]. Such a relation would, however, be required if the forces were due to the interaction of electric dipoles since, from eqn 4.51, we would have u_{12} proportional to the product $\mu_1\mu_2$, while for the like interactions u_{ii} is proportional to μ_i^2.

Keesom tackled the problem of the dipole–dipole energy more systematically. He, like Reinganum, followed Boltzmann's treatment of statistical mechanics, although Ornstein, in his Leiden thesis of 1908 [240], had shown the Dutch school how to use the (to us) more transparent methods of Gibbs. Keesom checked, however, that his results agreed with those of Ornstein that were common to their two methods of working. He found again that a spherically symmetrical attractive potential generates a second virial coefficient in which the leading term is proportional to T^{-1}, and where all higher powers are present, while a dipole–dipole potential leads to an expansion that contains only the inverse even powers of the temperature. He introduced the device, soon to become a standard procedure, of checking the usefulness of theoretical calculations by superimposing experimental and theoretical log–log plots of B as a function of T. It is, however, a device that can be misleading since the strong singularity in such a plot at the Boyle point (the temperature at which $B = 0$) can distort the way in which the eye sees the agreement at other temperatures. The method fell into disuse after the 1930s. In this way he obtained reasonable agreement between the dipole model and the observed virial coefficients of hydrogen and oxygen, but not of nitrogen [259].

It was about this time that serious doubts became irresistible. Molecules are formed of atoms, and atoms contain electrons, but there was at first no agreement on how these electrons were arranged within the atom or where the balancing positive charges were placed. The whole picture became clearer when Rutherford's nuclear model with planetary electrons received impressive support from Bohr's quantal treatment of the optical spectrum of hydrogen atoms. The model was classically unstable, for the rotating planetary electrons, being subject to a continuous centripetal acceleration, would radiate energy, lose speed, and collapse into the nucleus. This problem was dismissed by quantal fiat, to the horror of many. Paul Ehrenfest in Leiden wrote to Lorentz, in May 1913, "If this is the way to reach the goal, I must give up doing physics." [260] Nevertheless the representation was here to stay. Its theoretical implications of spherically symmetrical atoms and cylindrically symmetrical diatomic molecules, such as hydrogen and oxygen, confirmed new incontrovertible experimental evidence from the polarisation of gases in electric fields that such molecules did not possess the supposed electrical dipoles.

The behaviour of matter in electric fields is a difficult problem that had exercised the minds of physicists since the days of Faraday and Mossotti. The efforts of Clausius [261], Lorentz, then in Leiden, and Lorenz, in Copenhagen [262] had led to an equation relating the polarisation of a molecule to the dielectric constant of the material. This constant, ε_r, is the ratio of the electric permittivity of the material, ε, to that of a vacuum, ε_0, and so is readily measured as the ratio of the capacity of a condenser containing the material to that in a vacuum. The Clausius–Mossotti equation, in modern notation, is

$$(\varepsilon_r - 1)/(\varepsilon_r + 2) = N\alpha/3\varepsilon_0 V, \qquad (4.53)$$

where α is the polarisability of a molecule, that is, the ratio of the strength of the dipole moment induced in it to that of the local electric field, and where there are N molecules in a volume V. According to Maxwell's electrodynamics ε_r is equal to the square of the refractive index, n, so that eqn 4.53 can be written

$$(n^2 - 1)/(n^2 + 2) = 4\pi N\alpha_V/3V, \qquad (4.54)$$

where $\alpha_V = \alpha/4\pi\varepsilon_0$ is the polarisability in units of volume. It was this second form of the equation that Lorentz and Lorenz obtained, and they and others confirmed that the function of the refractive index on the left of the equation is proportional to the density of a gas or liquid, and independent of the temperature. The volumes α_V were found to be similar (generally within a factor of two) to the volumes of molecules estimated from kinetic theory or from van der Waals's equation [263].

The first form of the equation holds for some but not all gases and liquids; water and its vapour being notable exceptions. For such fluids the left-hand side of the equation is large and increases further as the temperature falls. These are the materials whose molecules have permanent electric dipoles. Peter Debye, a Dutchman then working in Zürich [264], adapted a treatment that Paul Langevin had used previously for magnetic dipoles to show that if the electric dipoles were free to react independently to the electric field then eqn 4.53 becomes

$$(\varepsilon_r - 1)/(\varepsilon_r + 2) = (N/3\varepsilon_0 V)[\alpha + (\mu^2/3kT)], \qquad (4.55)$$

where μ is the strength of the permanent dipole. The last term in this equation reflects the small average orientation of the permanent dipoles in the applied field, this orientation being opposed by the random thermal motions whose energy is proportional to kT [264]. The term in μ is not present in eqn 4.54, even for molecules with permanent dipoles, since at optical frequencies the dipoles do not have time to re-orient themselves in the electromagnetic field and so do not contribute to the overall polarisation. In liquids the molecules are too close together for their dipoles to react independently to the applied field, but in gases Debye's equation is confirmed and allows one to measure the permanent dipole moments. In this way it was shown that the simpler diatomic molecules, such as hydrogen, oxygen and nitrogen, have no permanent dipole. Heteronuclear diatomic molecules do possess such moments, a large one in hydrogen chloride and a small one in carbon monoxide, for example. It was originally thought that carbon dioxide had a weak permanent moment but we now know that its molecule is linear and centro-symmetric, so it has no moment [265].

These results for the homonuclear diatomic molecules knocked away the foundations of the work of Reinganum, van der Waals Jr and Keesom. The last was not discouraged, however, and returned with an alternative hypothesis – perhaps such molecules have a permanent quadrupole, that is, an array of four equal charges, two positive and two negative, arranged so that the dipole moment of the array is zero. Such an array was compatible with the presumed cylindrical symmetry of the homonuclear diatomic molecules. He showed that two quadrupoles at a separation r have a mutual potential energy of a form similar to that of eqn 4.51, but proportional to r^{-5} and with an orientational function, $f(\omega)$, of different form but one which again averages to zero when integrated over all orientations. This leads by an expansion similar to that of eqn 4.52, to an effective potential proportional to r^{-10}, and again to a leading term in the second virial coefficient proportional to T^{-2} [266]. He noted, moreover, that an empirical expression devised by Daniel Berthelot [267] was of the form $B = \beta - \alpha T^{-2}$. It was found later that this expression when written in terms of the critical constants is remarkably successful in fitting the second virial coefficients of not-too-polar organic vapours [268].

Debye observed that a permanent quadrupole in one molecule would induce a dipole in a nearby polarisable molecule and that the energy of these two charge distributions is always negative. It contributes therefore directly to the second virial coefficient, with a leading term in T^{-1}, without the need to average over a Boltzmann distribution in order to get a non-zero term [269]. In practice, however, this term was found to be smaller than the direct quadrupole–quadrupole term of Keesom [270]. Of more importance is the interaction of a permanent dipole, if present, and an induced dipole; the leading term in B is again proportional to T^{-1}. Such terms were first studied by Debye's student Hans Falkenhagen [271].

There was, however, one great theoretical obstacle in the way of all this work that, for many years, received no recognition from the leading practitioners. Debye had undermined the dipole–dipole interaction as the origin of all intermolecular forces by showing that many simple molecules had no dipoles. He and others then turned to quadrupoles, which could not then be measured directly, but which were plausible and compatible with the known or presumed shapes of the homonuclear diatomic molecules and that of symmetrical linear molecules such as carbon dioxide. These quadrupole moments became measurable in the 1950s. The most direct method was that of David Buckingham and R.L. Disch who measured the optical birefringence induced in carbon dioxide by an electrical field gradient – the quadrupolar analogue of the Kerr effect [272].

It was clear, however, from the time of their discovery that the inert gases, argon, neon and, later, helium [273], could be condensed to liquids and even to solids quite as readily as hydrogen, nitrogen and oxygen. There are therefore attractive forces between their molecules. The second virial coefficients of the inert gases were measured at Leiden from 1907 onwards, and later also elsewhere [274]. Such monatomic molecules have, it was correctly presumed, true spherical symmetry and so no dipole, quadrupole, or any higher multipole, if these electric moments are expressed by traceless tensors of the form needed to describe their electrostatic interactions. None of the electrostatic calculations that had been made could describe the behaviour of these substances. Another flaw in the calculations was that they could not account for the strong cohesion of the liquids. Such success as the gas calculations had had rested on the favourable alignments of each colliding pair. Such alignments are not possible between all pairs in a dense liquid or solid where each molecule can have up to 12 nearest neighbours. The fact was that 'the emperor had no clothes', but this was accepted only slowly and with reluctance. Thus Debye recognised that molecules with what we call 'traceless' quadrupoles could have no electrical interaction potential, and that one would have to go to the next term, that is, an octopole, although he did not then name it [264]. His associate Fritz Zwicky thought that this might be the first non-vanishing moment for argon, ignoring the fact that a spherically symmetrical distribution of charge has *no* non-vanishing moments [275]. Such a regress to ever-higher moments was not a happy route to follow and Debye turned instead to polarisability terms, but without being able to specify the nature of the charge distribution that was doing the polarising. He rightly observed, however, that a spherical distribution of charge would have no repulsive force either and, in 1920, tried at last to remedy the situation with a dynamic model of a hydrogen atom with an electron moving around a nucleus, so that it was only on a time average over its orbit that the atom had spherical symmetry and so no dipole moment [276]. This was a shrewd guess at what turned out to be the ultimate quantal resolution of the problem. It was a later suggestion by Debye that led to this resolution but even his dynamic model could not solve the difficulty in a classical electrical context.

Keesom seems never to have considered seriously the electrostatic impasse posed by the inert gases. In his early work of 1912 he had explored an empirically chosen attractive potential proportional to r^{-n}, where r is the separation, and had found that n was apparently about 4 or 5 for argon [277]. In a footnote ten years later he used this result to argue against the high inverse powers of r required by Debye's multipoles [278], but he never faced the real problem of the inert gases.

Such unwillingness to 'face the facts' is a common and often justifiable tactic of research. Science would advance more slowly if its practitioners worried at each stage about every real or apparent obstacle or inconsistency. We have seen earlier instances of this strategy. In the 18th century and later some worried about 'action at a distance' in both gravitational and cohesive forces; others accepted that it seemed to occur and went on to explore the consequences of this supposition. In the 19th century the inconsistency between the classical law of the equipartition of energy and the observed heat capacities of gases was held by some to be a strong argument against the kinetic theory of gases; others shrugged their shoulders and continued to use the theory. In the early years of the 20th century the 'planetary' structure of the atom was clearly unacceptable in classical electrodynamics, but it seemed to fit the facts and was soon rescued by the early quantum theory, obscure though the basis of that was. Such 'clouds' over classical theory, as Kelvin termed them [279], were eventually to lift, but those studying intermolecular forces with classical electrostatic models were not so lucky; they were facing a real difficulty.

Chemists too had their problems, for the origin of the forces of chemical bonding was as obscure as that of cohesion. Within Nernst's group in Berlin there again arose the Newtonian suggestion that the two might be the same or closely related [280]. Friedrich Dolezalek, a Hungarian-born student of Nernst's, tried to interpret the excess thermodynamic properties of liquid mixtures in terms of chemical bonding between the components [281]. A few of his examples involved what we now call 'hydrogen bonding' [282], but most of his cases were better explained by a lack of balance between the intermolecular or van der Waals forces between the like and unlike molecules. This point was made strenuously by van Laar, perhaps the most combative member of the Dutch school [283]. All these physical and chemical problems were to be resolved by the new quantum mechanics from 1925 onwards.

Disillusion with the electrostatic models led to a partial retreat to a position that Laplace would have appreciated. The evidence was that the cohesive forces were strong compared with the gravitational, weak compared with the Coulombic force between two electrons, and of shorter range than either. The simplest attractive potential that met these criteria is one proportional to $-r^{-m}$, in which the index m is chosen to be large with respect to unity. The non-zero compressibility of liquids and, even more convincing, of solids at low temperatures is evidence not for a hard core but for a Boscovichian repulsive potential proportional to $+r^{-n}$, with $n > m$. The whole potential could therefore by represented by a trial function of the form

$$u(r) = \alpha r^{-n} - \beta r^{-m},\qquad(4.56)$$

where α, β, m and n are four adjustable parameters. For convenience this form is abbreviated to an (n, m) potential.

The first use of this function is commonly ascribed to Gustav Mie in a paper of 1903 [284], although matters are not quite so clear-cut. He proposed a model of liquids and solids in which the monatomic molecules sit on or near the sites of a fixed lattice. The energy of the system is expressed in terms of a Taylor expansion in the displacements from the lattice sites. This leads him, via a repulsive potential of the form of the first term of eqn 4.56, to a contribution to what we should call the negative of the configurational energy of the system [*die innere Verdampfungsenergie*] of the form $-AV^{-n/3}$. He notes that a van der Waals treatment of the contribution of the attractive energies gives a term of the form BV^{-1}, and so writes, in effect, the sum as $-(AV^{-n/3} - BV^{-1})$. He finds that for the heavier metals n seems to be about 5. He does not, however, suggest explicitly that the form of the second term implies that $m = 3$, since he presumably knew that this choice leads to unacceptable consequences – the energy of a solid would depend on its shape and the second virial coefficient of the gas would be infinite. Grüneisen used the same form of the energy in his papers on the relations between the compressibility, heat capacity and coefficient of thermal expansion of metals [285], and he notes explicitly the divergence implied by $m = 3$. Only Simon Ratnowsky, a student of Debye's at Zürich, was rash enough to assume that an energy of the van der Waals form led to an attractive potential of the inverse third power [286]. All were hoping, as Einstein had been earlier [287], that the form of the intermolecular potential would be universal, that is, if it were of the form of eqn 4.56 then the indices m and n

would be the same for all substances. By 1912 Grüneisen was convinced that this was not so, at least for the metals, but it was an idea that was to be resurrected thirty years later in applications of the law of corresponding states to the inert gases and the simpler molecular substances.

Fritz Zwicky made the first attempt at calculating the second virial coefficient for an (n, m) potential in 1920 [288]. He favoured larger values than those working on solid metals, possibly influenced by Debye's multipole models, choosing $m = 8$ and thinking that n was probably about 9 or 10. He made, however, only crude numerical integrations for $m = 8$ and $n = 9$. This model potential came of age when J.E. Lennard-Jones [289] used it more systematically in a series of papers that started in 1924. His first calculation was of the viscosity of a gas with an $(n, 2)$ potential. It had only recently become possible to get at the information on intermolecular forces that was known to be locked up in the transport properties of gases and gas mixtures: viscosity, mass and thermal diffusion, and thermal conductivity. The problem, as we have seen, was that of calculating the departure of the velocity distribution from that of the equilibrium state. For many years no general solution could be found, in spite of some serious effort; even the great mathematician David Hilbert made little progress [290]. During the first World War two independent solutions were found for the general case of an arbitrary intermolecular potential: one by Sydney Chapman, then at Greenwich [291], and one by David Enskog in Uppsala [292]. Chapman's solution derived from Maxwell's work [293] and Enskog's from Boltzmann's [294]; fortunately they agreed, apart from some easily corrected minor errors of Chapman's. The implementation of these solutions required the calculation of the angles of deflection of colliding molecules and the insertion of these angles into some formidable multiple integrals. This was a job that was undertaken only slowly and unsystematically. C.G.F. James, in Cambridge, took a potential of the form (∞, m), now called Sutherland's potential since it is a hard core surrounded by an mth power attractive potential. He calculated the integrals for m from 3 to 8 but only at high temperature, that is, in the limit where $|u(\sigma)/kT| \ll 1$, where $u(\sigma)$ is the energy at contact [295]. Chapman himself calculated the integrals for a purely repulsive potential, with $n = 4, 6, 8, 10, 15$ and ∞ [296]. Lennard-Lones was a junior colleague of Chapman's at Manchester in the early 1920s and it was he who adapted Chapman's solutions to obtain the viscosity of an $(n, 2)$ gas [297]. He chose $m = 2$ because it simplified the calculation, although he knew that such a low index was physically inadmissible for the equilibrium properties of the gas, such as its energy and second virial coefficient. He found that the viscosity of an $(n, 2)$ gas with weak attractive forces varies with temperature in a way that we can express as

$$\eta \approx T^{1/2}\{S/T + [T_0(n)/T]^{2/n}\}^{-1}, \tag{4.57}$$

where $T_0(n)$ is a temperature that changes with n but which remains finite when n becomes infinite, when eqn 4.57 reduces to Sutherland's expression. Much more useful were the results that he obtained in the second part of his paper where he calculated exactly, by a series expansion in reciprocal temperature, the second virial coefficient for an arbitrary (n, m) potential. The powers of temperature in the series are $-(1/n)[3 + j(n - m)]$, where $j = 0, 1, 2$, etc., and so the expansion is less simple than those found for the electrostatic or multipole interactions.

By 1924 the inert gases had become the first choice for testing new gas theories since their molecules are truly spherical and their collisions perfectly elastic. Of these, argon was the most plentiful and the gas for which there was the widest range of experimental results. The second virial coefficient had been measured at low temperatures (below 20 °C) by Kamerlingh Onnes and his student C.A. Crommelin in 1910 [298], and over a wide range by Holborn and Otto in Berlin in 1924 [299]. These last results became available just in time for Lennard-Jones to use them. The viscosity of the gas had been measured in Halle [300] and in Leiden [301], and finally the crystal structure had recently been determined in Berlin; it was a face-centred cubic structure, one of the two close-packed arrays, with a lattice spacing of 5.42 ± 0.02 Å at 40 K [302], a figure that implies a nearest-neighbour distance less by a factor of $\sqrt{2}$, that is a distance of 3.83 ± 0.02 Å.

Lennard-Jones's first conclusion was that the gas properties alone did not determine uniquely the four parameters of an (n, m) potential. He chose $m = 4$, apparently since this was the value favoured by Keesom, and found that n was probably between 10 and 13, two of the values for which he had computed the virial coefficient, with a preference for the higher figure. His viscosity calculations for an $(n, 2)$ potential were of no value to him here. When, however, he introduced Simon's measurement of the nearest-neighbour distance in the crystal, which he took to be 3.84 Å, then he had a firm figure for the minimum of the potential. This is not exactly at this distance but is somewhat greater because of the mutual attractions of the atoms that are not nearest neighbours, but the correction is calculable. The calculation was, however, based on the assumption that the atoms are at rest on their lattice sites at zero temperature, a false assumption that quantum mechanics was soon to destroy. He now found that no potential fitted both Kamerlingh Onnes's values of the virial coefficient and Simon's lattice spacing, but that Holborn and Otto's values and the lattice spacing were consistent with a (13, 4) potential, which, he concluded, was probably close to the true form.

Simon did not try to marry his crystal work with the gas work but tried to extract information about the intermolecular potential of argon by exploiting the methods of Mie, Grüneisen and Max Born [303] for solids composed of atoms with an (n, m) potential that perform (classical) vibrations about their equilibrium lattice sites. Such vibrations are controlled by the curvature of the potential near its minimum but the coefficients of thermal expansion and isothermal compressibility depend in greater detail on the shape of the potential well in the crystal. The two approximate relations that Simon and von Simson drew from Born's analysis are

$$nm = 9V/\kappa U, \tag{4.58}$$

and

$$(n + m + 3)/6 = \gamma \equiv \alpha V/\kappa C_V, \tag{4.59}$$

where U and C_V are the lattice energy and heat capacity of a crystal of volume V, and α and κ are the coefficients of thermal expansion and isothermal compressibility. The dimensionless parameter, γ, defined by the second half of eqn 4.59, is called Grüneisen's constant and is found to change little with temperature for many metals. Simon and von Simson deduced from some measurements by Arnold Eucken of the speed of sound in the crystal that this constant is about 4 or 5 for argon, which is about twice that for a typical metal. The lattice energy could be estimated from the change with temperature of the vapour pressure of the crystal, and κ, rather crudely, from the Einstein frequency of the lattice vibrations determined from the departure of the heat capacity from the classical value of $3R$ of Dulong and Petit. These rough calculations gave them a value of 135 for the product nm which, with Grüneisen's constant, led to the figures $n = 15$ and $m = 9$. These are considerably higher than Lennard-Jones's preferred figures of 13 and 4. Then, and for the rest of the century, this field was often to suffer from calculations that took only a limited range of information and drew conclusions from it that were incompatible with the information from other properties that were known, or should have been known to the authors of the calculations.

The 'reduction' championed by Clausius and van der Waals required that the same molecular entities, with the same forces between them, occurred in all three states of matter. This view became implicitly accepted in the early part of the 20th century with the rout of the anti-atomists. It was reinforced in the 1920s by the careful work of Lennard-Jones. He could do nothing quantitative with liquids, the theory of which had not advanced beyond that of van der Waals, but he was careful to consider all the evidence from the equilibrium and transport properties of the gas and the equilibrium properties of the crystal. His early work on crystals, like the approximations of Born, Grüneisen and others, was based on the two assumptions of the validity of classical mechanics and the ascription of the intermolecular energy to a sum of the pair potentials acting between the molecules. No other course was open to him but neither proved to be adequate after the advent of quantum mechanics, and the consequences of these restrictions are discussed below in Sections 5.4 and 5.2 respectively.

Lennard-Jones's use of the transport properties of gases was hampered by the absence of calculations of the integrals needed for realistic (n, m) potentials. Some progress was made when his colleague H.R. Hassé [304], accepting the preferred value of $m = 4$, calculated the viscosity for $(\infty, 4)$ and $(8, 4)$ potentials [305]. The first is the Sutherland potential, freed from the restriction to high temperatures or weak attractive forces. The second was not chosen for any particular realism in Hassé's or Lennard-Jones's eyes but because 8 is twice 4, a circumstance that simplified the calculations. Both potentials fitted quite well the viscosities of seven gases, but there was an unresolved problem. The parameters of the second, more realistic, potential for argon which fitted the viscosity were not those that fitted the second virial coefficient. The discrepancy was large – about 66% in the strength of the attractive potential. Hassé and Cook noted at the very end of their paper that their method of calculation could be used also for a $(12, 6)$ potential but there was, at that time, no reason to prefer 6 to 4 and it was nearly twenty years before this suggestion was followed up.

A parallel problem to that of the inert gases was that of the physical properties of certain cubic crystals in which the molecular entities were known to be simple charged particles, or ions, that are iso-electronic with the atoms of the closest inert gas in the chemists' Periodic Table, for example, Na^+ with neon, K^+ and Cl^- with argon, and similarly for the doubly-charged ions Ca^{2+} and S^{2-} which are both also iso-electronic with argon. Born and Landé tried first to use Bohr's atom model to explain the structure and properties of these ionic crystals but found that it led to too-high values of the compressibility [306]. They turned therefore to a $(9, 1)$ model in which the attractive term is the strong Coulomb potential between ions of opposite charge; this term is, of course, repulsive between ions of the same sign but these are much farther apart [307]. Polarisation forces between an ion and the dipole it induces in a nearby ion could generally be neglected since their effect is nullified by the high symmetry of the crystal. Born and Landé's repulsive index of 9 conflicted with the value of 14 that Lennard-Jones had deduced from the properties of KCl and CaS [308], and had again found satisfactory for the repulsive potentials of helium and neon [309]. In a later paper he proposed $n = 10$ for the neon-like ions and $n = 9$ for the argon-like, but with the proviso that the effective value rose from 9 to 14 at larger distances in order to bring Ar, K^+, and Cl^- into a common form [310]. At this point he considers briefly $m = 6$, only to dismiss it.

The progress made with the inert gases and with simple ionic crystals did not disguise the fact that the main problem remained unsolved. In spite of many ingenious calculations for Sutherland potentials, for electrical dipole and multipole potentials, and for empirically chosen (n, m) potentials, no sound conclusions had been reached about the form of the van der Waals attractive force or the repulsive force, both of which were a universal feature of molecular systems.

The Faraday Society had always prided itself on its ability to choose for its General Discussions topics that were ripe for a detailed exploration but in November

1927 they made an unfortunate choice by deciding to discuss *Cohesion and related problems* [311] at a time when, had they but known it, the subject was about to be transformed. The papers presented were a miscellany. A few speakers lamented the lack of real progress and fell back on ideas that had been around for many years, such as T.W. Richards with a paper on the internal pressure in fluids [312] and A.W. Porter whose paper [313] on the law of molecular forces used ideas from surface tension in a way that would have seemed crude to van der Waals and his school. Many of the contributors were more interested in the practical problems of the strength of metals and other materials and this part of the meeting provoked the most lively discussion. A few raised the hope that the new quantum mechanics might solve their problems but the subject was still too new and unfamiliar for it yet to be relevant. (A curious instance of this unfamiliarity is in a late note submitted to the discussion by Lennard-Jones, presumably in handwriting, in which he refers to the very recent papers of Heitler and London on the quantum mechanics of the chemical bond [314]. The editor ascribes these papers to the unknown German authors Heitten and Loudon.)

A final commentary on the confusion that prevailed in 1928, on the eve of the first quantal treatment of the problem, was provided by G.A. Tomlinson of the National Physical Laboratory at Teddington [315]. He cited different authorities who had maintained, since 1900, that the attractive potential varied with the inverse of the separation to the powers of 1, 2, 3, 4, 5, 7 or 8. The only number missing from this sequence is 6 which was soon to prove to be the right answer. His own attempt to find the correct solution by a direct measurement of the force of adhesion between two quartz fibres was ingenious but not decisive and, as we shall see, had it been successful it would have given a misleading answer.

Cohesion was not the only unsolved problem of the 1920s; of greater interest was the question of the origin of the forces that led to chemical bonding. These were much stronger than the van der Waals forces but equally mysterious. Indeed, the distinction between chemical and physical forces of attraction was to remain a subject of contention until the clarification brought about by the new quantum theory, as is shown by the long discussion of the point by Irving Langmuir in 1916–1917 [316]. Ignorance of the origin of the chemical forces was, however, not a bar to progress, since for most chemical purposes it sufficed to know that a bond could be formed between two particular atoms and that the strength of that bond could be characterised by a single fixed energy. If it were necessary to know how the energy changed with distance near the minimum then the resources of infra-red and Raman spectroscopy were coming to the rescue. A more detailed knowledge of how the energy changed with distance over wider ranges of separation is needed only if one wishes to study the 'chemical dynamics' of bond formation and breaking, and that was a subject that was only starting to become practicable just before the second World War, and one that only became an active field of research in the second half of the 20th century.

The more delicate problems of gas imperfection, of the transport properties of gases, of the condensation of gases to liquids, of the tension at the surface of liquids, and of the structures and properties of crystals are all ones that demand a detailed knowledge of how the attractive and repulsive forces change with distance over a wide range of separations, and this knowledge was not forthcoming. Classical mechanics and its ad hoc modification by the quantal ideas of Planck, Einstein and Bohr was not up to the job. There seem to be no published attempts to use the 'old' quantum theory to tackle the problem of the intermolecular forces; one reported, but apparently abortive effort, was made by Oskar Klein at the urging of Niels Bohr in 1921 [317]. All these difficulties were to be overcome, at any rate in principle, in the glorious years of 1925 to 1930 when quantum mechanics burst on the molecular scene and revolutionised our understanding, or at least our ability to relate these physical phenomena to a new unified and coherent basis of mechanics.

Notes and references

1 H.v. Helmholtz (1821–1894) R.S. Turner, DSB, v. 6, pp. 241–53.
2 G.R. Kirchhoff (1824–1887) L. Rosenfeld, DSB, v. 7, pp. 379–83.
3 J.C. Maxwell (1831–1879) C.W.F. Everitt, DSB, v. 9, pp. 198–230.
4 M. Faraday (1791–1867) L.P. Williams, DSB, v. 4, pp. 527–40.
5 W.E. Weber (1804–1891) A.E. Woodruff, DSB, v. 14, pp. 203–9. Even in the 20th century, Sommerfeld, as editor of an encyclopaedia, inserted a chapter on electric forces acting at a distance, before Lorentz wrote at much greater length on Maxwell's theory; R. Reiff and A. Sommerfeld, 'Standpunkt der Fernwirkung. Die Elementargesetze', *Encyklopädie der mathematischen Wissenschaft*, Leipzig, v. 5, part 2, chap. 12, pp. 3–62, recd Dec. 1902, pub. Jan. 1904.
6 J.F.W. Herschel (1792–1871) D.S. Evans, DSB, v. 6, pp. 323–8; J.F.W. Herschel, 'Presidential Address of 1845', *Rep. Brit. Assoc.* **15** (1845) xxvii–xliv, see xli.
7 J. Herapath (1790–1868) S.G. Brush, DSB, v. 6, pp. 291–3; For the finished form of his theories, see J. Herapath, *Mathematical physics . . .*, 2 vols., London, 1847.
8 J.J. Waterston (1811–1883) S.G. Brush, DSB, v. 14, pp. 184–6; J.S. Haldane, ed., *The collected scientific papers of John James Waterston*, Edinburgh, 1928, 'Memoir', pp. xiii–lxviii. See also E. Mendoza, 'The kinetic theory of matter, 1845–1855', *Arch. Int. Hist. Sci.* **32** (1982) 184–220.
9 S.G. Brush, *A kind of motion we call heat: a history of the kinetic theory of gases in the 19th century*, 2 vols., Amsterdam, 1976. These two volumes are, together, v. 6 of the series *Studies in statistical mechanics*. An interesting contemporary history is by Maxwell: 'History of the kinetic theory of gases: notes for William Thomson', 1871, reprinted in H.T. Bernstein, 'J. Clerk Maxwell on the kinetic theory of gases', *Isis* **54** (1963) 206–15; and in *The scientific letters and papers of James Clerk Maxwell*, ed. P.M. Harman, v. 2, No. 377, pp. 654–60, Cambridge, 1995.

10 J.S. Rowlinson, 'The development of the kinetic theory of gases', *Proc. Lit. Phil. Soc. Manchester* **129** (1989–1990) 29–38. A short account of the early history of the kinetic theory, as seen in the middle of the 19th century, is in a long footnote that Clausius attached to his paper, 'Ueber die Wärmeleitung gasförmiger Körper', *Ann. Physik* **115** (1862) 1–56, footnote on 2–3; English trans. in *Phil. Mag.* **23** (1862) 417–35, 512–34, footnote on 417–18.

11 A.K. Krönig (1822–1879) E.E. Daub, DSB, v. 7, pp. 509–10; G. Ronge, 'Zur Geschichte der kinetischen Wärmetheorie mit biographischen Notizen zu August Karl Krönig', *Gesnerus* **18** (1961) 45–70; E.E. Daub, 'Waterston's influence on Krönig's kinetic theory of gases', *Isis* **62** (1971) 512–15; A. Krönig, 'Grundzüge einer Theorie der Gase', *Ann. Physik* **99** (1856) 315–22.

12 S. Carnot, *Réflexions sur la puissance motrice du feu*, ed. R. Fox, Paris, 1978; English trans., Manchester, 1986. The original edition was published in 1824.

13 D.S.L. Cardwell, *From Watt to Clausius: The rise of thermodynamics in the early industrial age*, London, 1971; C. Truesdell, *The tragicomical history of thermodynamics, 1822–1854*, New York, 1980, this is v. 4 of the series *Stud. Hist. Math. Phys. Sci.*; P. Redondi, *L'accueil des idées de Sadi Carnot: de la légende à l'histoire*, Paris, 1980; C. Smith, *The science of energy. A cultural history of energy physics in Victorian Britain*, London, 1998.

14 J.R. Mayer (1814–1878) R.S. Turner, DSB, v. 9, pp. 235–40; T.S. Kuhn, 'Energy conservation as an example of simultaneous discovery' in *Critical problems in the history of science*, ed. M. Clagett, Madison, WI, 1959, pp. 321–56; K.L. Caneva, *Robert Mayer and the conservation of energy*, Princeton, NJ, 1993.

15 W.J.M. Rankine (1820–1872) E.M. Parkinson, DSB, v. 11, pp. 291–5.

16 H. Helmholtz, *Über die Erhaltung der Kraft, eine physikalische Abhandlung*, Berlin, 1847; English trans. in *Scientific Memoirs . . .*, ed. J. Tyndall and W. Francis, London, 1853, pp. 114–62, in *Selected writings of Hermann von Helmholtz*, ed. R. Kahl, Middletown, CN, 1971, pp. 3–55, and, in part, in S.G. Brush, *Kinetic theory*, 3 vols., Oxford, 1965–1971, v. 1, pp. 89–110. Helmholtz's later views were added in an Appendix when the pamphlet was reprinted in his *Wissenschaftliche Abhandlungen*, Leipzig, 1882, v. 1, pp. 12–68, 68–75. See also F. Bevilacqua, 'Helmholtz's *Ueber die Erhaltung der Kraft*: The emergence of a theoretical physicist', in *Hermann von Helmholtz and the foundations of nineteenth-century science*, ed. D. Cahan, Berkeley, CA, 1993, chap. 7, pp. 291–333.

17 M. É. Verdet (1824–1866) E. Frankel, DSB, v. 13, pp. 614–15; É. Verdet, *Théorie mécanique de la chaleur*, 2 vols., Paris, 1868, 1870; reprinted in 1868 and 1872 as vols. 7 and 8 of *Oeuvres de É. Verdet*, Paris. Violle's bibliography, v. 2, pp. 267–338, covers the years up to 1870.

18 J.L. Meyer (1830–1895) O.T. Benfey, DSB, v. 9, pp. 347–53; L. Meyer, *Die modernen Theorien der Chemie und ihre Bedeutung für die chemische Mechanik*, 5th edn, Breslau, 1884. The English translation of this edition, *Modern theories of chemistry*, London, 1886, contains the 'Introduction' to the first German edition of 1862, pp. xix–xxvii.

19 He trained for two years in Franz Neumann's celebrated seminar in physics; see K. M. Olesko, *Physics as a calling: discipline and practice in the Königsberg Seminar for Physics*, Ithaca, NY, 1961, pp. 236, 266.

20 H. Davy (1778–1829) D.M. Knight, DSB, v. 3, pp. 598–604; J.J. Berzelius (1779–1848) H.M. Leicester, DSB, v. 2, pp. 90–7; C.A. Russell, 'The electrochemical theory of Sir Humphry Davy', *Ann. Sci.* **15** (1959) 1–25; 'The electrochemical theory of Berzelius', *ibid.* **19** (1963) 117–45.

21 H. Davy, 'The Bakerian Lecture, on some chemical agencies of electricity',
    Phil. Trans. Roy. Soc. **97** (1807) 1–56, esp. Section 8, 'On the relations between
    the electrical energies of bodies, and their chemical affinities', 39–44.
22 J.J. Berzelius, *Traité de chimie*, v. 4, Paris, 1831, pp. 523–641, 'De la théorie des
    proportions chimiques'. This section is a second edition of his *Théorie des proportions
    chimiques . . .* , Paris, 1819 [not seen].
23 Berzelius, ref. 22, p. 538.
24 Berzelius, ref. 22, p. 567.
25 J.B. Dumas, 'Acide produit par l'action du chlore sur l'acide acétique', *Compt. Rend.
    Acad. Sci.* **7** (1838) 474; L.-H.-F. Melsens, 'Note sur l'acide chloracétique', *ibid.* **14**
    (1842) 114–17.
26 H.E. Roscoe and A. Harden, *A new view of the origin of Dalton's atomic theory*,
    London, 1896, pp. 1–5. This book is based on Dalton's unpublished notes which were
    destroyed in an air-raid on Manchester in 1940.
27 J. Dalton, *A new system of chemical philosophy*, Manchester, v. 1, part 1, 1808,
    pp. 148–50. Dalton accepted the doctrine that repulsion was caused by heat.
    H. Davy, *Syllabus of a course of lectures on chemistry delivered at the Royal
    Institution of Great Britain*, 1802, printed in his *Collected works*, 9 vols., London,
    1839–1840, v. 2, pp. 329–436; *Elements of chemical philosophy*, London, 1812,
    Part 1, v. 1, pp. 68–9, reprinted as v. 4 of his *Collected works*.
28 J. Millar (1762–1827) DNB; J. Millar, *Elements of chemistry*, Edinburgh, 1820.
29 E. Frankland (1825–1899) W.H. Brock, DSB, v. 5, pp. 124–7; C.A. Russell,
    Edward Frankland: Chemistry, controversy and conspiracy in Victorian England,
    Cambridge, 1996, p. 46. I thank Colin Russell for a copy of Frankland's notes for
    his 4th and 5th lectures.
30 J.-B.-A. Dumas (1800–1884) S.C. Kapoor, DSB, v. 4, pp. 242–8; J.B. Dumas,
    Leçons sur la philosophie chimique, Paris, 1837.
31 J.L. Gay-Lussac, 'Considérations sur les forces chimiques', *Ann. Chim. Phys.* **70**
    (1839) 407–34.
32 J.L. Gay-Lussac, 'Premier mémoire sur la dissolubilité des sels dans l'eau',
    Ann. Chim. Phys. **11** (1819) 296–315.
33 J. Marcet (1769–1858) DNB; S. Bahar, 'Jane Marcet and the limits of public science',
    Brit. Jour. Hist. Sci. **34** (2001) 29–49; [J. Marcet], *Conversations on chemistry . . .* ,
    2 vols., London, 1806, see v. 1, pp. 10–14, and v. 2, pp. 1–13.
34 J.B. Biot, 'Conversations sur la chimie . . . , Genève,1809'. This review in the
    Mercure de France of 1809 is reprinted in his *Mélanges scientifiques et littéraires*,
    3 vols., Paris, 1858; see v. 2, pp. 97–107, and especially the footnote on
    pp. 103–4.
35 A. Avogadro (1776–1856) M.P. Crosland, DSB, v. 1, pp. 343–50; M. Morselli, *Amedeo
    Avogadro, a scientific biography*, Dordrecht, 1984, chaps. 3–5; J.H. Brooke,
    'Avogadro's hypothesis and its fate: a case-study in the failure of case-studies', *Hist.
    Sci.* **19** (1981) 235–73; N. Fisher, 'Avogadro, the chemists and historians of
    chemistry', *ibid.* **20** (1982) 77–102, 212–31; M. Scheidecker-Chevallier, 'L'hypothèse
    d'Avogadro (1811) et d'Ampère (1814): la distinction atome/molécule et la théorie de
    la combinaison chimique', *Rev. d'Hist. Sci.* **50** (1997) 159–94. For the autonomy of
    chemistry, see Meyer, ref. 18, and D.M. Knight, *The transcendental part of chemistry*,
    Folkestone, 1978.
36 W. Prout (1785–1850) W.H. Brock, DSB, v. 11, pp. 172–4; W. Prout, *Chemistry,
    meteorology and the function of digestion, considered with reference to natural
    theology*, London, 1834, p. 49. This is the eighth of the Bridgwater Treatises.

37 W.A. Miller (1817–1870) J.D. North, DSB, v. 9, pp. 391–2. He is not to be confused with J. Millar, ref. 28, nor with W.H. Miller (1801–1880), the mineralogist. W.A. Miller, *Elements of chemistry; theoretical and practical*, 3 parts, London, 3rd edn, 1863–1867, 4th edn, 1867–1869, see Part 1, *Chemical physics.*

38 L. Pfaundler (1839–1920), Professor of Physics at Innsbruck, Pogg., v. 3, p. 1033; v. 4, p. 1151; v. 5, p. 966. See also J. Berger, 'Chemische Mechanik und Kinetik: die Bedeutung der mechanischen Wärmetheorie für die Theorie chemischer Reaktionen', *Ann. Sci.* **54** (1997) 567–84.

39 T. Graham (1805–1869) G.B. Kauffman, DSB, v. 5, pp. 492–5; T. Graham, *Elements of chemistry*, 2nd edn, 2 vols., London, 1850, 1858.

40 Graham, ref. 39, 1st edn, London, 1842, pp. 85–7.

41 Graham, ref. 39, 2nd edn, v. 1, p. 101; v. 2, Supplement 'Heat', pp. 421–57.

42 H. Watts (1815–1884) DNB.

43 C.F. Mohr (1806–1879) F. Szabadváry, DSB, v. 9, pp. 445–6. Mohr was one of those who has a claim to have contributed to the discovery of the conservation of energy, see Kuhn, ref. 14. F. Mohr, *Allgemeine Theorie der Bewegung und Kraft, als Grundlage der Physik und Chemie. Ein Nachtrag zur mechanischen Theorie der chemischen Affinität*, Braunschweig, 1869, p. 22. American readers were no better served by an old-fashioned book from Harvard, J.P. Cooke, *Elements of chemical physics*, Boston, 1860.

44 A. Naumann (1837–1922) F. Szabadváry, DSB, v. 9, pp. 619–20; A. Naumann, *Grundriss der Thermochemie, oder der Lehre von den Beziehungen zwischen Wärme und chemischen Erscheinungen vom Standtpunkt der mechanischen Wärmetheorie dargestellt*, Braunschweig, 1869.

45 Naumann, ref. 44, pp. 23–38.

46 Naumann, ref. 44, pp. 78–81. This section sems to derive from the similar views of his mentor, Hermann Kopp (1817–1892) H.M. Leicester, DSB, v. 7, pp. 463–4.

47 J.C. Maxwell, 'Remarks on the classification of the physical sciences', Ms. printed in *Scientific letters and papers,* ref. 9, v. 2, No. 432, pp. 776–82. This manuscript was used for his posthumous article, 'Physical sciences', in the 9th edn of *Encyclopaedia Britannica* in 1885, where there is the same comment on chemistry.

48 C.F. Gauss , 'Principia generalia theoriae figurae fluidorum in statu aequilibrii', *Comm. Soc. Reg. Sci. Göttingen* **7** (1830) 39–88; translated in Ostwald's *Klassiker*, No.135, Leipzig, 1903, as 'Allgemeine Grundlagen einer Theorie der Gestalt von Flüssigkeiten im Zustand des Gleichgewichts'. See also L. Boltzmann, 'Über die Ableitung der Grundgleichungen der Kapillarität aus dem Prinzipe der virtuellen Geschwindigkeit', *Ann. Physik* **141** (1870) 582–90, reprinted in his *Wissenschaftliche Abhandlungen*, 3 vols., Leipzig, 1909 [hereafter cited as WA], v. 1, pp. 160–7.

49 J.-A.-C. Charles (1746–1823) J.B. Gough, DSB, v. 3, pp. 207–8. Gay-Lussac and Dalton deserve some credit for this law, but it has long been known as Charles's law in the English-speaking world.

50 Herapath, ref. 7, v. 1, p. 276.

51 H.V. Regnault (1810–1878) R. Fox, DSB, v. 11, pp. 352–4; see also Fox's book, *The caloric theory of gases from Lavoisier to Regnault*, Oxford, 1971, chap. 8; V. Regnault, 'Sur la loi de la compressibilité des fluides élastiques', *Compt. Rend. Acad. Sci.* **23** (1846) 787–98, see p. 796; 'Relation des expériences . . . pour déterminer les principales lois et les données numériques qui entrent dans le calcul des machines à vapeur', *Mém. Acad. Sci. Inst. France* **21** (1847) 1–767.

52 Herapath, ref. 7, v. 1, p. 270.

53 Morselli, ref. 35, pp. 339–44.

54 J.P. Joule, 'On the changes of temperature produced by the rarefaction and condensation of air', *Phil. Mag.* **26** (1845) 369–83; reprinted in the *Scientific papers of James Prescott Joule*, London, 1884, pp. 172–89. His equipment still exists; there is a photograph of it in Plate 23 of Cardwell's book, ref. 13. Essentially the same experiment had been carried out by Gay-Lussac many years earlier, J.L. Gay-Lussac, 'Premier essai pour déterminer les variations de température qu'éprouvent les gaz en changeant de densité, et considérations sur leur capacité pour le calorique', *Mém. Phys. Chim. Soc. d'Arcueil* **1** (1807) 180–203.

55 J.P. Joule and W. Thomson, 'On the thermal effects experienced by air rushing through small apertures', *Phil. Mag.* **4** (1852) 481–92. This preliminary paper was read at the meeting of the British Association on 3 September 1852. W. Thomson and J.P. Joule (or Joule and Thomson), 'On the thermal effects of fluids in motion, Parts 1–4', *Phil. Trans. Roy. Soc.* **143** (1853) 357–65; **144** (1854) 321–64; **150** (1860) 325–36; **152** (1862) 579–89. A parallel series in *Proc. Roy. Soc.* is mainly abstracts of these papers. All are reprinted in the *Joint scientific papers of James Prescott Joule*, London, 1887. See also C. Sichau, 'Die Joule-Thomson-Experimente: Anmerkungen zur Materialität eines Experimentes', *Int. Zeit. Ges. Ethik Naturwiss., Tech. u. Med.* **8** (2000) 222–43.

56 J.C. Maxwell, *Theory of heat*, London, 1871, pp. 194–5.

57 For a near-contemporary discussion, see J.W. Gibbs, 'Rudolf Julius Emanuel Clausius', *Proc. Amer. Acad. Arts Sci.* **16** (1889) 458–65, reprinted in Gibbs's *Collected works*, New York, 1928, v. 2, pp. 261–7. For more modern discussions, see Cardwell, ref. 13, pp. 269–73; E. Daub, 'Atomism and thermodynamics', *Isis* **58** (1967) 293–303; and M.J. Klein, 'Gibbs on Clausius', *Hist. Stud. Phys. Sci.* **1** (1969)127–49.

58 L. Boltzmann (1844–1906) S.G. Brush, DSB, v. 2, pp. 260–8; L. Boltzmann, 'Studien über das Gleichgewicht der lebendigen Kraft zwischen bewegten matierellen Punkten', *Sitz. Math. Naturwiss. Classe Kaiser Akad. Wissen. Wien*, Abt.2 **58** (1868) 517–60; 'Über das Wärmegleichgewicht zwischen mehratomigen Gasmolekülen', *ibid.* **63** (1871) 397–418; 'Einige allgemeine Sätze über Wärmegleichgewicht', *ibid.* 679–711; 'Analytischer Beweis des 2. Hauptsatzes der mechanischen Wärmetheorie aus den Sätzen über das Gleichgewicht der lebendigen Kraft', *ibid.* 712–32 see 728, reprinted in WA, ref. 48, v. 1, pp. 49–96, 237–58, 259–87, 288–308 see 303.

59 The letter is printed in Part 2 of Joule and Thomson's papers, ref. 55, and in Joule's *Joint scientific papers*, ref. 55, pp. 269–70.

60 Joule and Thomson, ref. 55, Part 2.

61 Joule and Thomson, ref. 55, Part 4. Unfortunately the final equation is misprinted in a form that requires p^2, not p, in the final term.

62 J. Tyndall (1820–1893) R. MacLeod, DSB, v. 13, pp. 521–4; J. Tyndall, *Heat considered as a mode of motion*, London, 1863, Lecture 3; 2nd edn, 1865, pp. 98–9.

63 M. Faraday, 'On fluid chlorine', *Phil. Trans. Roy. Soc.* **113** (1823) 160–64; 'On the condensation of several gases into liquids', *ibid.* 189–98. After each paper Davy inserted an addendum to describe his own part in these and related experiments, 164–5, 199–205.

64 M. Faraday, 'Historical statement respecting the liquefaction of gases', *Quart. Jour. Sci.* **16** (1824) 229–40. This paper and those in refs. 63 and 65 are reprinted in his *Experimental researches in chemistry and physics*, London, 1859, pp. 85–141. For a fuller history see W.L. Hardin, *The rise and development of the liquefaction of gases*,

New York, 1899, and for the 20th century, R.G. Scurlock, ed., *History and origins of cryogenics*, Oxford, 1992.

65 M. Faraday, 'On the liquefaction and solidification of bodies generally existing as gases', *Phil. Trans. Roy. Soc.* **135** (1845) 155–77.

66 Little is known about Robert Addams, although a man who could prepare nearly a gallon of liquid carbon dioxide in 1844 is surely worthy of some notice. In 1825 he took out a patent for improving carriages (No. 5310) and later called himself 'Lecturer on Chemistry and Natural Philosophy', see *Phil. Mag.* **6** (1835) 415. He was twice mentioned by Faraday as a lecturer whom he knew and had heard, see *The correspondence of Michael Faraday*, ed. F.A.J.L. James, London, v. 1, 1991, Letter 453 of 1830, and v. 3, 1996, Letter 1365 of 1841. For one year, at least, he was a member of the British Association, see the *List of members, 1838*, p. 17, bound into the *Report* **6** (1837). His address was then 20 Pembroke Square, Kensington. He gave a brief paper at the Newcastle meeting of the Association in 1838, 'On the construction of apparatus for solidifying carbonic acid, and on the elastic force of carbonic acid gas in contact with the liquid form of the acid, at different temperatures', *Rep. Brit. Assoc.* **7** (1838) 'Transactions of the Sections', pp. 70–1.

67 Charles Saint-Ange Thilorier (1797– ?) He, like Addams, is overlooked by Poggendorff. He was at the École Polytechnique from 1815 to 1816, and twenty years later was described as the 'ausgezeichneten Mechaniker' in an anonymous article on his apparatus; 'Apparat zur Verdichtung der Kohlensäure', *(Liebig's) Ann. Pharm.* **30** (1839) 122–6, Tables 1 and 2. By operating this apparatus seven times he could produce 4 litres of liquid carbon dioxide. See D.H.D. Roller, 'Thilorier and the first solidification of a 'permanent' gas (1835)', *Isis* **43** (1952) 109–13; J. Pelseneer, 'Thilorier', *ibid.* **44** (1953) 96–7. A. Thilorier, 'Propriétés de l'acide carbonique liquide', *Ann. Chim. Phys.* **60** (1835) 427–31; 'Solidification de l'acide carbonique', *ibid.* 432–4; 'Sur l'acide carbonique solide', *Compt. Rend. Acad. Sci.* **3** (1836) 432–4. John Mitchell, an American doctor, used liquid carbon dioxide therapeutically; his apparatus was a variant of that of Thilorier, see J.K. Mitchell, 'On the liquefaction and solidification of carbonic acid', *(Silliman's) Amer. Jour. Sci. Arts* **35** (1839) 346–56; see also 301–2, 374–5.

68 C. Cagniard de la Tour, 'Exposé de quelques résultats obtenus par l'action combinée de la chaleur et de la compression sur certain liquides, tels que l'eau, l'alcool, l'éther sulphurique et l'essence de la pétrole rectifiée', *Ann. Chim. Phys.* **21** (1822) 127–32, 178–82; 'Note sur les effets qu'on obtient par l'application simultanée de la chaleur et de la compression à certains liquides', *ibid.* **22** (1823) 410–15. There is an annotated translation of the first paper in *Phil. Mag.* **61** (1823) 58–61. The translator was Philip Taylor (1786–1870), the brother of Richard Taylor, the publisher of the journal. Philip was an enthusiast for the use of high pressures in steam engines.

69 J.F.W. Herschel, *Preliminary discourse on the study of natural philosophy*, London, 1830, §§ 199, 252.

70 W. Whewell (1794–1866) R.E. Butts, DSB, v. 14, pp. 292–5.

71 *Correspondence of Michael Faraday*, ref. 66, v. 3, 1996, Letter 1646, 9 November; Letter 1648, 12 November; and Letter 1650, 14 November 1844.

72 D.I. Mendeleev (1834–1907) B.M. Kedrov, DSB, v. 9, pp. 286–95; D. Mendelejeff, 'Ueber die Ausdehnung der Flüssigkeiten beim Erwärmen über ihren Siedepunkt', *Ann. Chem. Pharm.* **119** (1861) 1–11. See also his 'Sur la cohésion moléculaire de quelque liquides organiques', *Compt. Rend. Acad. Sci.* **50** (1860) 52–4; **52** (1860) 97–9. For an account of early work on the critical point, see Y. Goudaroulis,

'Searching for a name: the development of the concept of the critical point (1822–1869)', *Rev. d'Hist. Sci.* **47** (1994) 353–79.

73 T. Andrews (1813–1885) E.L. Scott, DSB, v. 1, pp. 160–1; Memoir by P.G. Tait and A. Crum Brown in T. Andrews, *The scientific papers*, London, 1889, pp. ix–lxii.

74 Miller, ref. 37, 3rd edn, 1863, Part 1, *Chemical physics*, pp. 328–9. Andrews had previously made a short communication to the British Association on the liquefaction of gases, *Rep. Brit. Assoc.* **31** (1861) 'Transactions of the Sections', pp. 76–7.

75 C. Wolf, 'De l'influence de la température sur les phénomènes qui se passent dans les tubes capillaires', *Ann. Chim. Phys.* **49** (1857) 230–81; J.J. Waterston, 'On capillarity and its relation to latent heat', *Phil. Mag.* **15** (1858) 1–19, reprinted in *Scientific papers*, ref. 8, pp. 407–28.

76 T. Andrews, 'On the continuity of the gaseous and liquid states of matter', *Phil. Trans. Roy. Soc.* **159** (1869) 575–90, see 587–8.

77 O.F. Mossotti (1791–1863) J.Z. Buchwald, DSB, v. 9, pp. 547–9; O.F. Mossotti, *Sur les forces qui régissent la constitution intérieure des corps, aperçu pour servir à la détermination de la cause et des lois de l'action moléculaire*, Turin, 1836; trans. in (*Taylor's*) *Scientific Memoirs* **1** (1837) 448–69.

78 J.S. Rowlinson, 'The Yukawa potential', *Physica A* **156** (1989) 15–34.

79 P.-S. Laplace, *Traité de mécanique céleste*, v. 5, Paris, 1823, 'Sur l'attraction des sphères, et sur la répulsion des fluides élastiques', Book 12, chap. 2, pp. 100–18.

80 P. Kelland (1809–1879) Pogg., v. 3, p. 712; [Anon.] *Proc. Roy. Soc.* **29** (1879) vii–x; S. Earnshaw (1805– ?) Pogg., v. 3, pp. 395–6; R.L. Ellis (1817–1859) DNB. P. Kelland, 'On molecular equilibrium, Part 1', *Trans. Camb. Phil. Soc.* **7** (1839–1842) 25–59; S. Earnshaw, 'On the nature of the molecular forces which regulate the constitution of the luminiferous ether', *ibid.* 97–112; R.L. E[llis]., 'Remarks on M. Mossotti's theory of molecular action', *Phil. Mag.* **19** (1841) 384–6.

81 J.J. Waterston, *Thoughts on the mental functions*, Edinburgh, 1843; 'Note on molecularity', reprinted in *Scientific papers*, ref. 8, pp. 167–82.

82 É. Ritter (1810–1862) Pogg., v. 2, cols. 654–5, 1438–9. A. de Candolle wrote a short memoir of Ritter, see 'Rapport sur les travaux de la Société', *Mém. Soc. Phys. d'Hist. Nat. Genève* **16** (1861) 437–57, see 450–2. The Institut Topffer is now remembered for the delightful accounts of the rambles of its pupils in the Alps; Rodolfe Topffer (or Töpffer, 1799–1846), *Voyages en Zigzag*, Paris, 1844, and later volumes.

83 É. Ritter, 'Note sur la constitution physique des fluides élastiques', *Mém. Soc. Phys. d'Hist. Nat. Genève* **11** (1846–1848) 99–114. Ritter devised also an equation of state for solids, based on the caloric theory, but similar to that of Grüneisen in 1926, see E. Mendoza, 'The equation of state for solids 1843–1926', *Eur. Jour. Phys.* **3** (1982) 181–7.

84 S.-D. Poisson, 'Sur les équations générales de l'équilibre et du mouvement des corps solides, élastiques, et des fluides', *Jour. École Polytech.* 20me cahier, **13** (1831) 1–174, see p. 33ff.

85 A modern account of Ritter's derivation is given by Brush, ref. 9, v. 2, pp. 397–401.

86 É. Sarrau (1837–1904), an authority on explosives, wrote the Preface to the French translation of J.D. van der Waals's thesis, *La continuité des états gazeux et liquide*, Paris, 1894, see p. x. His reference to Poisson is presumably the long article in *Jour. École Polytech.* for 1831 that Ritter had used, see ref. 84.

87 J. Herapath, 'Exact calculation of the velocity of sound', *Railway Magazine, New Series* **1** (1836) 22–8. He became editor of this journal in 1835, when he started a new series of volume numbers and added the sub-title *and Annals of Science*. He used it as a vehicle in which to publish papers he could not or did not wish to publish in more regular journals. The same organisation published his book in 1847, ref. 7.

88 Herapath, ref. 7, v. 2, p. 60.

89 J.P. Joule, *On matter, living force, and heat*, a lecture at St Ann's Church, Manchester, 1847, reported in the *Manchester Courier*, and printed in *Scientific papers*, ref. 54, pp. 265–76, see p. 274; and in Brush, ref. 16, v. 1, pp. 78–88, see p. 86.

90 J.P. Joule, 'On the mechanical equivalent of heat, and on the constitution of elastic fluids', *Rep. Brit. Assoc.* **18** (1848), 'Transactions of the Sections', pp. 21–2, reprinted in *Scientific papers*, pp. 288–90. This abstract was followed by the full paper, read on 3 October 1848, 'Some remarks on heat, and the constitution of elastic fluids', *Mem. Lit. Phil. Soc. Manchester* **9** (1851) 107–14, reprinted, after a complaint by Clausius that he had not been able to see a copy of this journal, in *Phil. Mag.* **14** (1857) 211–16, and in *Scientific papers*, ref. 54, pp. 290–7.

91 J.J. Waterston, 'On the physics of media that are composed of free and perfectly elastic molecules in a state of motion', *Phil. Trans. Roy. Soc. A* **183** (1893) 5–79, and Rayleigh's introduction, 1–5. The paper is reprinted in *Scientific papers*, ref. 8, pp. 207–319. An abstract had been published by the Royal Society in its *Proceedings* **5** (1846) 604.

92 This account of the work of Waterston and Dupré draws on the account by S. Richardson, *The development of the mean-field approximation*, an unpublished dissertation for Part 2 of Chemistry Finals examination at Oxford, 1988.

93 N.D.C. Hodges 'On the size of molecules', (*Silliman's*) *Amer. Jour. Sci. Arts* **18** (1879)135–6.

94 A. Einstein (1879–1955) M.J. Klein and N.L. Balazs, DSB, v. 4, pp. 312–33; A. Pais, '*Subtle is the Lord . . .*': *The science and life of Albert Einstein*, New York, 1982, chaps. 4 and 5.

95 A. Einstein, 'Folgerungen aus den Capillaritätserscheinungen', *Ann. Physik* **4** (1901) 513–23; reprinted in *The collected papers of Albert Einstein*, Princeton, NJ, v. 2, 1989, pp. 9–21. See also the Introduction to this volume, 'Einstein on the nature of molecular forces', pp. 3–8. The paper is translated in the *English translation* of *The collected papers*, v. 2, pp. 1–11; J.N. Murrell and N. Grobert, 'The centenary of Einstein's first scientific paper', *Notes Rec. Roy. Soc.*, **56** (2002) 89–94. The main purpose of Einstein's paper was to represent the surface tension as a sum of contributions from each atom in the molecule. He was not the first to try to do this, see R. Schiff, 'Ueber die Capillaritätsconstanten der Flüssigkeiten bei ihrem Siedepunkt', (*Leibig's*) *Ann. Chem.* **223** (1884) 47–106. Schiff's results were discussed by W. Ostwald in his *Lehrbuch der allgemeinen Chemie*, v. 1, *Stöchiometrie*, 2nd edn, Leipzig, 1891, pp. 526–31, and it is from this source that Einstein takes his figures. Such attempts to relate physical properties to the constituent atoms in a molecule reached its climax with Sugden's 'parachor', which was the molar volume of a liquid multiplied by the fourth root of the surface tension. This was used for some years to try to predict molecular structures from physical properties, see S. Sugden, *The parachor and valency*, London, 1930, but the method has no sound basis and was soon abandoned when better spectroscopic and crystallographic results became available.

96 A. Einstein, 'Bemerkung zu dem Gesetz von Eötvös', *Ann. Physik* **34** (1911) 165–9; reprinted in *The collected papers*, ref. 95, v. 3, pp. 401–7 and in the *English translation*, v. 3, pp. 328–31.

97 G.A. Hirn (1815–1890) R.S. Hartenberg, DSB, v. 6, pp. 431–2.

98 G.-A. Hirn, *Exposition analytique et expérimentale de la théorie mécanique de la chaleur*, Paris and Colmar, 1862, pp. 498–9, 531–58, 599–600.

99 G.-A. Hirn, *Théorie mécanique de la chaleur, Première partie*, 2nd edn, Paris, 1865, pp. 191–6, 224–32.

100 Hirn, ref. 99, chap. 5, pp. 233–52; see also ref. 98, part 4, pp. 133–299.
101 G.A. Zeuner (1828–1907) O. Mayr, DSB, v. 14, pp. 617–18; G. Zeuner, *Grundzüge der mechanischen Wärmetheorie . . .*, Freiburg, 1860; 2nd edn, Leipzig, 1866.
102 F.J. Redtenbacher (1809–1863) O. Mayr, DSB, v. 11, pp. 343–4; F. Redtenbacher, *Das Dynamiden-System, Grundzüge einer mechanischen Physik*, Mannheim, 1857.
103 Zeuner, in Hirn, ref. 99, p. 242.
104 Hirn, ref. 99, 3rd edn, 2 vols., Paris, 1875, 1876; v. 2, pp. 212–23, 282.
105 A.L.V. Dupré (1808–1869) R. Fox, DSB, v. 4., p. 258.
106 F.J.D. Massieu (1832–1896) Pogg., v. 3, p. 881.
107 A. Dupré, *Théorie mécanique de la chaleur*, Paris, 1869.
108 Dupré, ref. 107, eqn 64, p. 51.
109 Dupré, ref. 107, p. 61.
110 Dupré, ref. 107, p. 80.
111 Dupré, ref. 107, p. 144ff.
112 Massieu in Dupré, ref. 107, pp. 152–7, 213–26.
113 Dupré, ref. 107, p. 261, 403–4. The printed figure for *A*, with a long row of zeros, requires 10^8, but the calculation that follows and the known value of the latent heat, require 10^7.
114 A. Dupré, 'Note sur le nombre des molécules contenues dans l'unité de volume', *Compt. Rend. Acad. Sci.* **62** (1866) 39–42.
115 For reviews of this field, see R. Clausius, 'Ueber die Art der Bewegung, welche wir Wärme nennen', *Ann. Physik* **100** (1857) 353–80; English trans. in *Phil. Mag.* **14** (1857) 108–27, reprinted in Brush, ref. 16, v. 1, pp. 111–34; and the popular lecture that Clausius gave the same year in Zürich, *Ueber das Wesen der Wärme, verglichen mit Licht und Schall*, Zürich, 1857. E. Garber, 'Clausius and Maxwell's kinetic theory of gases', *Hist. Stud. Phys. Sci.* **2** (1970) 299–312. For Maxwell, see ref. 56 and *Maxwell on molecules and gases*, ed. E. Garber, S.G. Brush and C.W.F. Everitt, Cambridge, MA, 1986. Clausius's early papers were collected in two volumes entitled *Abhandlungen über die mechanische Wärmetheorie*, Braunschweig, 1864, 1867, which is hereafter cited as *Abhandlungen*. The first volume contains the papers on thermodynamics, Abhandlung I to IX, and the second those on electricity, Abhandlung X to XIII, and on molecular physics, XIV to XVIII. The paper above, of 1857, is Abhandlung XIV. The reprints often contain long notes that are not in the original papers. The English translation, edited by T.A. Hirst, *Mechanical theory of heat*, London, 1867, contains only the first nine Memoirs, that is, those on thermodynamics. A French translation by F. Folie, *Théorie mécanique de la chaleur*, 2 vols., Paris, 1868, 1869, contains all but the last memoir, XVIII, on oxygen, which he omitted because of its overlap with XVII, on ozone.
116 Joule's lecture of 1847, ref. 89, and Helmholtz's pamphlet of the same year, ref. 16.
117 The idea that the end of the 19th century was marked by a stagnation in kinetic theory was put forward by P. Clark, 'Atomism versus thermodynamics', in *Method and appraisal in the physical sciences; the critical background to modern science, 1800–1905*, ed. C. Howson, Cambridge, 1976, pp. 41–105; and was opposed by C. Smith, 'A new chart for British natural philosophy: the development of energy physics in the nineteenth century', *Hist. Sci.* **16** (1978) 231–79.
118 R. Clausius, 'Ueber die bewegende Kraft der Wärme und die Gesetze, welche sich daraus für die Wärmelehre selbst ableiten lassen', *Ann. Physik* **79** (1850) 368–97, 500–24; English trans. in *Phil. Mag.* **2** (1851) 1–21, 102–19; *Abhandlungen*, I, ref. 115.

119 W.J.M. Rankine, *Miscellaneous scientific papers*, London, 1881.
120 R. Clausius, 'Ueber einige Stellen der Schrift von Helmholtz, "über die Erhaltung der Kraft"', *Ann. Physik* **89** (1853) 568–79, and Helmholtz's reply, 'Erwiderung auf die Bemerkungen von Hrn. Clausius', *ibid.* **91** (1854) 241–60 and in his *Wissenschaftliche Abhandlungen*, ref. 16, v. 1, pp. 76–93. The point is discussed by L. Koenigsberger in his biography, *Hermann von Helmholtz*, Oxford, 1906, pp. 115–20. See also Bevilacqua, ref. 16.
121 His time in Zürich, 1856–1867, has been described by G. Ronge, 'Die Züricher Jahre des Physikers Rudolf Clausius', *Gesnerus* **12** (1955) 73–108.
122 C.H.D. Buys Ballot (1817–1890) H.L. Burstyn, DSB, v. 2, p. 628; K. van Berkel, A. van Helden and L. Palm, *A history of science in the Netherlands*, Leiden, 1999, pp. 429–31. [C.H.D.] Buijs-Ballot, 'Ueber die Art von Bewegung, welche wir Wärme und Elektricität nennen', *Ann. Physik* **103** (1858) 240–59.
123 R. Clausius, 'Ueber die mittlere Länge der Wege, welche bei der Molecularbewegung gasförmiger Körper von den einzelnen Molecülen zurückgelegt werden; nebst einigen anderen Bemerkungen über die mechanische Wärmetheorie', *Ann. Physik* **105** (1858) 239–58; English trans. in *Phil. Mag.* **17** (1859) 81–91, and Brush, ref. 16, v. 1, pp. 135–47; *Abhandlungen*, XV, ref. 115.
124 J.C. Maxwell, 'Illustrations of the dynamical theory of gases', *Phil. Mag.* **19** (1860) 19–32; **20** (1860) 21–37, reprinted in Brush, ref. 16, v. 1, pp. 148–71.
125 F. Baily, 'On the correction of a pendulum for the reduction to a vacuum, . . .', *Phil. Trans. Roy. Soc.* **122** (1832) 399–492; G.G. Stokes, 'On the effect of the internal friction of fluids on the motion of pendulums', *Trans. Camb. Phil. Soc.* **9** (1856) 8–106, see 17 and 65, and in brief in *Phil. Mag.* **1** (1851) 337–9.
126 *Memoir and scientific correspondence of the late Sir George Gabriel Stokes, Bart.*, ed. J. Larmor, Cambridge, 1907, v. 2, pp. 8–11; Maxwell's *Scientific letters and papers*, ref. 9, v. 1, No.157, pp. 606–11.
127 J.C. Maxwell, 'On the viscosity or internal friction of air and other gases', *Phil. Trans. Roy. Soc.* **156** (1866) 249–68. The common-sense view that the viscosity would be less at low pressures goes back at least to Newton. In Query 28 of the fourth edition of his *Opticks* (1730) he wrote that "in thinner air the resistance is still less", saying that he had seen performed the experiment of a feather dropping as fast as a metal ball in a vacuum. Stokes had earlier told Maxwell that Graham's experiments on the flow of air through fine tubes were consistent with the viscosity being independent of the density, but not with it being proportional to the density; see Maxwell's letter to H.R. Droop of 28 January 1862, printed in his *Scientific letters and papers*, ref. 9, v. 1, No. 193, p. 706. For Maxwell's calculation of the mean free path from Graham's measurements of the rate of diffusion in gases, see ref. 124, v. 20, p. 31.
128 O.E. Meyer (1834–1909) Pogg., v. 3, pp. 907–8; v. 4, pp. 996–7; O.E. Meyer, 'Ueber die innere Reibung der Gase', *Ann. Physik* **125** (1865) 177–209, 401–20, 564–99; **127** (1866) 253–81, 353–82; **143** (1871) 14–26; **148** (1873) 1–44, 203–36; and with F. Springmühl, **148** (1873) 526–55.
129 J.J. Loschmidt (1821–1895) W. Böhn, DSB, v. 8, pp. 507–11; J. Loschmidt, 'Zur Grösse der Luftmolecüle', *Sitz. Math. Naturwiss. Classe Kaiser Akad. Wissen. Wien, Abt.2* **52** (1865) 395–413. See also R.M. Hawthorne, 'Avogadro's number: early values by Loschmidt and others', *Jour. Chem. Educ.* **47** (1970) 751–5. Maxwell later extended Loschmidt's calculations by using Loschmidt's measurements of diffusion to estimate molecular diameters, on the assumption that, as for spheres, the cross-diameter for unlike molecules is the arithmetic mean of the like diameters,

see J.C. Maxwell, 'On Loschmidt's experiments on diffusion in relation to the kinetic theory of gases', *Nature* **8** (1873) 298–300.

130 H. Kopp, 'Beiträge zur Stöchiometrie der physikalischen Eigenschaften chemischer Verbindungen', *Ann. Chem. Pharm.* **96** (1855) 1–36, 153–85, 303–35. His life's work on molar volumes is summarised in 'Ueber die Molecularvolume von Flüssigkeiten', *(Liebig's) Ann. Chem.* **250** (1889) 1–117.

131 L. Meyer, 'Ueber die Molecularvolumina chemischer Verbindungen', *Ann. Chem. Pharm.* Suppl. 5 (1867) 129–47.

132 J.C. Maxwell, 'On the dynamical theory of gases', *Phil. Trans. Roy. Soc.* **157** (1867) 49–88, reprinted in Brush, ref. 16, v. 2, pp. 23–87.

133 L. Boltzmann, 'Weitere Studien über das Wärmegleichgewicht unter Gasmolekülen', *Sitz. Math. Naturwiss. Classe Kaiser Akad. Wissen. Wien, Abt. 2* **66** (1872) 275–370, reprinted in WA, ref. 48, v. 1, pp. 316–402; English trans. in Brush, ref. 16, v. 2, 88–175. Boltzmann later listed other incorrect values that had been proposed for k_1 which ranged from $\pi^2/8$ (O.E. Meyer) to 25/12 (Stefan), see L. Boltzmann, 'Zur Theorie der Gasreibung, I', *ibid.* **81** (1880) 117–58; WA, ref. 48, v. 2, pp. 388–430.

134 W. Whewell, *The philosophy of the inductive sciences, founded upon their history,* 2 vols., London, 1840, v. 1, p. 416.

135 J.C. Maxwell, art. 'Atom', *Encyclopaedia Britannica*, 9th edn, London, 1875.

136 *The correspondence between Sir George Gabriel Stokes and Sir William Thomson, Baron Kelvin of Largs*, ed. D.B. Wilson, 2 vols., Cambridge, 1990, Letter 249, v. 1, pp. 327–31.

137 Meyer, ref. 128, 1873, see 205, and O.E. Meyer, *Die kinetische Theorie der Gase,* Breslau, 1877, p. 6; English trans. of 2nd edn, London, 1899, p. 7. For van der Waals, see Section 4.3. Thomson later made this deduction in his address to the British Association in 1884; see 'Steps toward a kinetic theory of matter', *Rep. Brit. Assoc.* **54** (1884) 613–22; reprinted in his *Popular lectures and addresses*, 2nd edn, London, 1891, v. 1, pp. 225–59.

138 [J.W. Strutt] Lord Rayleigh (1842–1919) R.B. Lindsay, DSB, v. 13, pp. 100–7; Lord Rayleigh, 'On the viscosity of argon as affected by temperature', *Proc. Roy. Soc.* **66** (1900) 68–74.

139 Meyer, ref. 137, 1877, pp. 157–60.

140 J. Stefan, 'Über die dynamische Theorie der Diffusion der Gase', *Sitz. Math. Naturwiss. Classe Kaiser Akad. Wissen. Wien, Abt. 2* **65** (1872) 323–63, see 339–40.

141 Meyer, ref. 128, 1873, pp. 203–36, see § 4.

142 L. Boltzmann, 'Über das Wirkungsgesetz der Molecularkräfte', *Sitz. Math. Naturwiss. Classe Kaiser Akad. Wissen. Wien, Abt. 2* **66** (1872) 213–19, reprinted in WA, ref. 48, v. 1, pp. 309–15.

143 G.J. Stoney (1826–1911) B.B. Kelham, DSB, v. 13, p. 82; G.J. Stoney, 'The internal motions of gases compared with the motions of waves of light', *Phil. Mag.* **36** (1868) 132–41.

144 L.V. Lorenz (1829–1891) M. Pihl, DSB, v. 8, pp. 501–2; L. Lorenz, 'Zur Moleculartheorie und Elektricitätslehre', *Ann. Physik* **140** (1870) 644–7; English trans. in *Phil. Mag.* **40** (1870) 390–2.

145 W. T[homson]., 'The size of atoms', *Nature* **1** (1870) 551–3. Thomson had been interested in such estimates for some time, possibly prompted by his earlier interest in contact electricity and a letter from Maxwell of 17 December 1861, asking what was the maximum breadth of an atom, see J. Larmor, 'The origins of Clerk Maxwell's electric ideas, as described in familiar letters to W. Thomson', *Proc.*

Camb. Phil. Soc. **32** (1936) 695–750, esp. 731–3, or Maxwell's *Scientific letters and papers*, ref. 9, v. 1, No. 190, pp. 699–702. Thomson wrote to Joule at about this time saying that he hoped to be able to fix an upper limit "for the sizes of atoms, or rather, as I do not believe in atoms, for the dimensions of molecular structures". An extract from this letter was read in Manchester on 21 January 1862, see *Proc. Lit. Phil. Soc. Manchester* **2** (1860–1862) 176–8. A second letter to the Society was printed in *Nature* **2** (1870) 56–7. Thomson's later lecture on 'The size of atoms', a Friday evening Discourse at the Royal Institution on 3 February 1883, adds little to the paper of 1870, see *Proc. Roy. Inst.* **11** (1884) 185–213 or *Popular lectures and addresses*, ref. 137, pp. 154–224.

146 For references to other attempts to estimate molecular sizes, see J.R. Partington, *An advanced treatise on physical chemistry*, v. 1, *Fundamental principles. The properties of gases*, London, 1949, pp. 243–5, and Brush, ref. 9. A review of the contribution of gases to the understanding of molecular properties at the end of the 19th century is in Part III, 'On the direct properties of molecules', pp. 297–352 of the English trans. of Meyer's book, ref. 137.

147 For Thomson's vortex atoms, see C. Smith and W.N. Wise, *Energy and Empire, a biographical study of Lord Kelvin*, Cambridge, 1989, chap. 12.

148 P.G. Tait (1831–1901) J.D. North, DSB, v. 13, pp. 236–7.

149 M. Epple, 'Topology, matter, and space, I , Topological notions in 19th-century natural philosophy', *Arch. Hist. Exact Sci.* **52** (1998) 297–392.

150 J.C. Maxwell, Letter to Mark Pattison, 13 April 1868, printed in *Maxwell on heat and statistical mechanics*, ed. E. Garber, S.G. Brush and C.W.F. Everitt, Bethlehem, PA, 1995, pp. 189–94 and in *Scientific letters and papers*, ref. 9, v. 2, No. 287, pp. 362–8. P.M. Harman has discussed this matter further in *The natural philosophy of James Clerk Maxwell*, Cambridge, 1998, pp. 182–7 and 195–6.

151 Olesko, ref. 19, pp. 280–5. Quincke's work on capillarity was similarly affected, see pp. 371–4.

152 A. Kundt and E. Warburg, 'Ueber die specifische Wärme des Quecksilbergases', *Ann. Physik* **157** (1876) 353–69.

153 L. Boltzmann, 'Über die Natur der Gasmoleküle', *Sitz. Math. Naturwiss. Classe Kaiser Akad. Wissen. Wien, Abt. 2* **74** (1876) 553–60, reprinted in WA, ref. 48, v. 2, pp. 103–10.

154 J.C. Maxwell, 'The kinetic theory of gases' [A review of H.W. Watson's book of that title], *Nature* **16** (1877) 242–6.

155 J.C. Maxwell, Contribution to a discussion on atomic theory at the Chemical Society, 6 June 1867, printed in *Scientific letters and papers*, ref. 9, v. 2, No. 270, pp. 304–5.

156 M. Yamalidou, 'John Tyndall, the rhetorician of molecularity', *Notes Rec. Roy. Soc.* **53** (1999) 231–42, 319–31.

157 Lord Rayleigh, 'On the theory of surface forces', *Phil. Mag.* **30** (1890) 285–98, 456–75.

158 J.C. Maxwell, 'A discourse on molecules', *Phil. Mag.* **46** (1873) 453–69, and his notes for this lecture in *Scientific letters and papers*, ref. 9, v. 2, No. 478, pp. 922–33.

159 G.H. Quincke (1834–1924) F. Fraunberger, DSB, v. 11, pp. 241–2; "... for theories he had little affection", see A. Schuster, 'Prof. G.H. Quincke, For. Mem. R.S.', *Nature* **113** (1924) 280–1; *Proc. Roy. Soc. A* **105** (1924) xiii–v; G. Quincke, 'Ueber die Verdichtung von Gasen und Dämpfen auf der Oberfläche fester Körper', *Ann. Physik* **108** (1859) 326–53.

160 W. Thomson, 'Note on gravity and cohesion', *Proc. Roy. Soc. Edin.* **4** (1857–1862) 604–6, reprinted in *Popular lectures and addresses*, ref. 137, pp. 59–63, as App. B to 'Capillary attraction', a Friday evening Discourse at the Royal Institution, 29 January 1886, *Proc. Roy. Inst.* **14** (1887) 483–507, reprinted in *Popular lectures and addresses*, ref. 137, pp. 1–55. The idea dies hard; my first research student told me in 1951 that he had been taught as an undergraduate in the Physics Department at Manchester that intermolecular forces were gravitational in origin.

161 R. Clausius, 'Ueber einen auf die Wärme anwendbaren mechanischen Satz', *Ann. Physik* **141** (1870) 124–30; English trans. in *Phil. Mag.* **40** (1870) 122–7, reprinted in Brush, ref. 16, v. 1, pp. 172–8.

162 J.W. Gibbs (1839–1903) M.J. Klein, DSB, v. 5, pp. 386–93; J.W. Gibbs, ref. 57, see p. 462.

163 M.J. Klein, 'Historical origins of the van der Waals equation', *Physica* **73** (1974) 28–47.

164 W.J.M. Rankine, 'On the centrifugal theory of elasticity, as applied to gases and vapours', *Phil. Mag.* **2** (1851) 509–42, see Section III.

165 R. Clausius, 'Ueber die Anwendung des Satzes von der Aequivalenz der Verwandlungen auf die innere Arbeit', *Ann. Physik* **116** (1862) 73–112, see 95; English trans. in *Phil. Mag.* **24** (1862) 81–97, 201–13, see 201; *Abhandlungen*, VI, ref. 115.

166 J.C. Maxwell, 'Tait's "Thermodynamics" ', *Nature* **17** (1878) 257–9, 278–80, see 259. Maxwell accepted Boltzmann's derivation of this result five months later; 'On Boltzmann's theorem on the average distribution of energy in a system of material points', *Trans. Camb. Phil. Soc.* **12** (1878) 547–70.

167 Lord Rayleigh, 'On the virial of a system of hard colliding bodies', *Nature* **45** (1891) 80–2. He was convinced by 1900, see 'The law of partition of kinetic energy', *Phil. Mag.* **49** (1900) 98–118. J.J. Thomson accepted Boltzmann's and, later, van der Waals's view but the derivation in his *Applications of dynamics to physics and chemistry*, London, 1888, pp. 89–93 is unsatisfactory. P.W. Bridgman was still in doubt about the relation between kinetic energy and temperature in 1913; 'Thermodynamic properties of twelve liquids . . .', *Proc. Amer. Acad. Arts Sci.* **49** (1913–1914) 1–114, see 109–10.

168 P.G. Tait, 'Reply to Professor Clausius', *Phil. Mag.* **43** (1872) 338; 'Foundations of the kinetic theory of gases, Part IV', printed in his *Scientific papers*, 2 vols., Cambridge, 1898, 1900, v. 2, pp. 192–208.

169 Tyndall, ref. 62, Lecture 3, p. 62. Helmholtz was expressing similar doubts in a lecture at Karlsruhe the same year, 'On the conservation of force', in his *Popular lectures on scientific subjects*, London, 1873, pp. 317–62, see p. 350.

170 M.B. Pell (1827–1879) I.S. Turner, *Australian dictionary of biography*, Melbourne, 1974, v. 5, pp. 428–9. Pell was appointed the first professor of mathematics and natural philsophy at Sydney in 1852. M.B. Pell, 'On the constitution of matter', *Phil. Mag.* **43** (1872) 161–85.

171 J.C. Maxwell, Letter to Tait of 13 October 1876, printed by Garber *et al.*, ref. 150, pp. 267–9, and in *Scientific letters and papers*, ref. 9, v. 3, No. 623, in press.

172 J.D. van der Waals (1837–1923) J.A. Prins, DSB, v. 14, pp. 109–11; A. Ya. Kipnis, B.E. Yavelov and J.S. Rowlinson, *Van der Waals and molecular science*, Oxford, 1996.

173 J.D. van der Waals, *Over de continuiteit van den gas- en vloeistoftoestand,* Thesis, Leiden, 1873. This is now most easily accessible in an English translation, *On the continuity of the gas and liquid states*, ed. J.S. Rowlinson, Amsterdam, 1988. The

book is v. 14 of the series *Studies in statistical mechanics*. All references to chapters or paragraphs of the thesis are to this translation.

174 J.D. van der Waals, 'The equation of state', in *Nobel lectures in physics*, Amsterdam, 1967, pp. 254–65.

175 Regnault, ref. 51 (1847); H.V. Regnault, 'Recherches sur les chaleurs spécifiques des fluides élastiques', *Mém. Acad. Sci. Inst. France* **26** (1862) 3–924.

176 T. Andrews, 'Ueber die Continuität der gasigen und flüssigen Zustände der Materie', *Ann. Physik, Ergänzband* **5** (1871) 64–87.

177 T. Andrews, 'Sur la continuité de l'état gazeux et liquide de la matière', *Ann. Chim. Phys.* **21** (1870) 208–35; J. Thomson, 'On the continuity of the gaseous and liquid states of matter', *Nature* **2** (1870) 278–80.

178 There is a minor curiosity here. Andrews and van der Waals's German translator, Eilhard Wiedemann, wrote naturally of the continuity of the gaseous and liquid *states*, in the plural. Van der Waals himself, however, used the singular, *state*. The plural is used here, as it was in the English translation, ref. 173.

179 J. Thomson (1822–1892) DNB; J.T.B[ottomley]., *Proc. Roy. Soc.* **53** (1893) i–x; 'Biographical sketch' in J. Thomson, *Collected papers in physics and engineering*, Cambridge, 1912, pp. xiii–xci. J. Thomson, 'Considerations on the abrupt change at boiling or condensation in reference to the continuity of the fluid state of matter', *Proc. Roy. Soc.* **20** (1871) 1–8.

180 J.C. Maxwell, Letter to James Thomson, 24 July 1871, in *Scientific letters and papers*, ref. 9, v. 2, No. 382, pp. 670–4, and in Garber *et al.*, ref. 150, pp. 212–15. See also Maxwell, ref. 56, pp. 124–6.

181 J.C. Maxwell, 'On the dynamical evidence of the molecular constitution of bodies', *Jour. Chem. Soc.* **13** (1875) 493–508; *Nature* **11** (1875) 357–9, 374–7.

182 G. Quincke, 'Ueber die Entfernung, in welcher die Molekularkräfte der Capillarität noch wirksam sind', *Ann. Physik* **137** (1869) 402–14.

183 Nine of these are quoted by Kipnis *et al.*, ref. 172, p. 50.

184 See Section 2.1, and ref. 48 of Chapter 2.

185 For the evidence, see Kipnis *et al.*, ref. 172, pp. 51–2, 55 and 58.

186 J.C. Maxwell, 'Van der Waals on the continuity of the gaseous and liquid states', *Nature* **10** (1874) 477–80. A partial derivation of his faulty expression in this review for the second virial coefficient for a system of hard spheres is in a manuscript printed by Garber *et al.*, ref. 150, pp. 309–13, and in *Scientific letters and papers*, ref. 9, v. 3, No. 522, in press.

187 H.A. Lorentz (1853–1928) R. McCormmach, DSB, v. 8, pp. 487–500; Van Berkel, *et al.*, ref. 122, pp. 514–18. H.A. Lorentz, 'Ueber die Anwendung des Satzes vom Virial in der kinetischen Theorie der Gase', *Ann. Physik* **12** (1881) 127–36, 660–1; 'Bemerkungen zum Virialtheorem', in *Festschrift Ludwig Boltzmann gewidmet zum sechzigsten Geburtstage*, Leipzig, 1904, pp. 721–9.

188 T. Andrews, 'On the gaseous state of matter', *Phil. Trans. Roy. Soc.* **166** (1876) 421–49.

189 The original report by Maxwell is in v. 7 of the Royal Society's Referees' Reports, and the copy sent to Andrews is in the archives of Queen's University, Belfast, with the papers of Thomas Andrews, MS2/16-1. The expressions for the second virial coefficient are in J.S. Rowlinson, 'Van der Waals and the physics of liquids', the Introduction to the 1988 edition of van der Waals's thesis, ref. 173. The whole report has been published by Garber *et al.*, ref. 150, pp. 298–305 and in *Scientific letters and papers*, ref. 9, v. 3, No. 604, in press. Boltzmann's constant, k, was first so expressed by Planck in his famous lecture to the German Physical Society of 14

December 1900 on the theory of black-body radiation. He thereby obtained the best value to date for Avogadro's constant, $R/k = 6.175 \times 10^{23}$ mol^{-1}. The lecture introduced also the expression for the entropy, S, in terms of $\mathfrak{N}_0$, the number of arrangements of his resonators for a given energy; $S = k \ln \mathfrak{N}_0$, see M. Planck, 'Zur Theorie des Gesetzes der Energieverteilung im Normalspectrum', *Verhand. Deutsch. Phys. Gesell.* **2** (1900) 237–45; English trans. in D. ter Haar, *The old quantum theory*, Oxford, 1967, pp. 82–90. A fuller account of the lecture a year later introduced the more familiar equation between entropy and the number of complexions: $S = k \ln W$, see 'Ueber das Gesetz der Energieverteilung im Normalspectrum', *Ann. Physik* **4** (1901) 553–63.

190 R. Clausius, 'Ueber den Satz vom mittleren Ergal und seine Anwendung auf die Molecularbewegungen der Gase', *Ann. Physik, Ergänzband* **7** (1876) 215–80, see 248ff. This had been published in Bonn in 1874 and was translated into English in *Phil. Mag.* **50** (1875) 26–46, 101–17, 191–200, see 104ff.

191 J.D. van der Waals, 'Sur le nombre relatif des chocs que subit une molécule suivant qu'elle se meut au milieu de molécules en mouvement ou au milieu de molécules supposées en repos, et sur l'influence que les dimensions des molécules, dans la direction du mouvement relatif, exercent sur le nombre de ces chocs', *Arch. Néerl.* **12** (1877) 201–16. This paper had previously appeared in Dutch in *Versl. Med. Konink. Akad. Weten. Afd. Natuur.* **10** (1876) 321–36.

192 R. Clausius, 'Ueber das Verhalten der Kohlensäure in Bezug auf Druck, Volumen und Temperatur', *Ann. Physik* **9** (1880) 337–57.

193 D.J. Korteweg (1848–1941) D.J. Struik, DSB, v. 7, pp. 465–6; D.J. Korteweg, 'Ueber den Einfluss der räumlichen Ausdehnung der Molecüle auf den Druck eines Gases', *Ann. Physik* **12** (1881) 136–46.

194 J.D. van der Waals, 'Ueber den Uebergangszustand zwischen Gas und Flüssigkeit', *Beiblätter Ann. Physik* **1** (1877) 10–21.

195 J. Moser, 'Ueber die Torricelli'sche Leere', *Ann. Physik* **160** (1877) 138–43.

196 L. Boltzmann, 'Über eine neue Bestimmung einer auf die Messung der Moleküle Bezug habenden Grösse aus der Theorie der Capillarität', *Sitz. Math. Naturwiss. Classe Kaiser Akad. Wissen. Wien, Abt.* 2 **75** (1877) 801–13, reprinted in WA, ref. 48, v. 2, pp. 151–63.

197 C. Cercignani, *Ludwig Boltzmann: the man who trusted atoms*, Oxford, 1998, 'A short biography', pp. 5–49.

198 L. Boltzmann, *Vorlesungen über Gastheorie*, 2 vols., Leipzig, 1896, 1898. English trans. by S.G. Brush, in one volume, *Lectures on gas theory,* Berkeley, CA, 1964.

199 J.W. Gibbs, *Elementary principles in statistical mechanics*, New Haven, CT, 1902.

200 M. v. Smoluchowski, 'Gültigkeitsgrenzen des zweiten Hauptsatzes der Wärmetheorie', in M. Planck *et al.*, ed., *Vorträge über die kinetische Theorie der Materie und der Elektrizität*, Leipzig, 1914, pp. 87–121, see p. 87. This was one of a series of lectures given under the auspices of the Wolfskehlstiftung.

201 These doubts and the disputes that they gave rise to have been reviewed in detail by J.M.H. Levelt Sengers, 'Liquidons and gasons; controversies about the continuity of states', *Physica A* **98** (1979) 363–402.

202 W. Ramsay (1852–1916) T.J. Trenn, DSB, v. 11, pp. 277–84; W. Ramsay, 'On the critical state of gases', *Proc. Roy. Soc.* **30** (1880) 323–9; 'On the critical point', *ibid.* **31** (1880) 194–205.

203 W. Ramsay and S. Young, 'On the thermal behaviour of liquids', *Phil. Mag.* **37** (1894) 215–18, 503–4.

204 S. Young (1857–1937) T.J. Trenn, DSB, v. 14, pp. 560–2; S. Young, 'The influence of the relative volumes of liquid and vapour on the vapour-pressure of a liquid at

constant temperature', *Phil. Mag.* **38** (1894) 569–72; 'The thermal properties of isopentane', *Proc. Phys. Soc.* **13** (1894–1895) 602–57; S. Young and G.L. Thomas, 'The specific volumes of isopentane vapour at low pressures', *ibid.* 658–65.

205 H. Kamerlingh Onnes (1853–1926) J. van der Handel, DSB, v. 7, pp. 220–2; K. Gavroglu and Y. Goudaroulis, 'Heike Kamerlingh Onnes' researches at Leiden and their methodological implications', *Stud. Hist. Phil. Sci.* **19** (1988) 243–74; *Through measurement to knowledge: The selected papers of Heike Kamerlingh Onnes, 1853–1926*, ed. K. Gavroglu and Y. Goudaroulis, Dordrecht, 1991; Van Berkel *et al.*, ref. 122, pp. 491–4.

206 Kipnis *et al.*, ref. 172, pp. 106–16, 249–86.

207 É. Mathias to J.D. van der Waals, 7 May 1904. The letter is quoted in translation by Levelt Sengers, ref. 201, pp. 390.

208 M.K.E.L. Planck (1858–1947) H. Kangro, DSB, v. 11, pp. 7–17; M. Planck, 'Die Theorie des Sättigungsgesetzes', *Ann. Physik* **13** (1881) 535–43.

209 G. Meslin, 'Sur l'équation de Van der Waals et la démonstration du théorème des états correspondants', *Compt. Rend. Acad. Sci.* **116** (1893) 135–6.

210 J.D. van der Waals, 'Onderzoekingen omtrent de overeenstemmende eigenschappen der normale verzadigden- damp- en vloeistoflijen voor de verschillende stoffen en omtrent een wijziging in den vorm dier lijnen bij mengsels', *Verhand. Konink. Akad. Weten. Amsterdam* **20** (Aug. and Sept. 1880) No. 5, 32 pp.; 'Over de coëfficiënten van uitzetting en van samendrukking in overeenstemmende toestanden der verschillende vloeistoffen', *ibid.* **20** (Nov. 1880) No. 6, 11 pp.; 'Bijdrage tot de kennis van de wet der overeenstemmende toestanden', *ibid.* **21** (Jan. 1881) No. 5, 10 pp.

211 See *Beiblätter Ann. Physik* **5** (1881) 27–8, 250–9, 567–9.

212 J.D. van der Waals, *Die Continuität des gasförmigen und flüssigen Zustandes*, trans. T.F. Roth, Leipzig, 1881.

213 J. Dewar (1842–1923) A.B. Costa, DSB, v. 4, pp. 78–81; J. Dewar, Presidential address, *Rep. Brit. Assoc.* **72** (1902) 3–50, see 29.

214 H. Kamerlingh Onnes, 'Algemeene theorie der vloeistoffen', *Verhand. Konink. Akad. Weten. Amsterdam* **21** (Dec. 1880 and Jan. 1881) No. 4, in three parts, 24 pp., No. 5, 14 pp., No. 6, 9 pp. There was later a partial translation into French, 'Théorie générale de l'état fluide', *Arch. Néerl.* **30** (1897) 101–36.

215 Van der Waals, ref. 173, § 27.

216 M.F. Thiesen (1849–1936) Pogg., v. 3, p. 1336; v. 4, p. 1490; v. 5, p. 1250; v. 6, p. 2645. M. Thiesen, 'Untersuchungen über die Zustangsgleichung', *Ann. Physik* **24** (1885) 467–92.

217 'similarly situated points' seems to be the best rendering of the Dutch 'gelijkstandige punten'. In later writings in French and German, Kamerlingh Onnes, or his translators, uses the less transparent phrases 'points homologues' and 'homologen Punkte'.

218 Kamerlingh Onnes, ref. 214, pp. 3–5 of the 3rd section in No. 6, or pp. 131–3 of the French translation.

219 W.H. Keesom (1876–1956) J.A. Prins, DSB, v. 7, pp. 271–2; Van Berkel *et al.*, ref. 122, pp. 498–500.

220 H. Kamerlingh Onnes and W.H. Keesom, 'Die Zustandsgleichung', in *Encyklopädie*, ref. 5, v. 5, part 1, chap. 10, pp. 615–945, recd Dec. 1911, pub. Sept. 1912, see p. 694; reprinted as *Comm. Phys. Lab. Leiden*, No. 11, Suppl. 23 (1912), see p. 80. In this monograph they suggested that van der Waals's parameter b should be called the 'core volume' [*Kernvolum*] and that the name 'co-volume' be used for $(V-b)$,

see *Encyklopädie*, p. 671, or *Comm. Leiden*, p. 57. The suggestion is logical but it has not been adopted.

221 W. Sutherland (1859–1911) T.J. Trenn, DSB, v. 13, pp. 155–6; W.A. Osborne, *William Sutherland: a biography*, Melbourne, 1920. This book contains a list of Sutherland's papers. For a sympathetic modern account of his work, see H. Margenau and N.R. Kestner, *Theory of intermolecular forces*, Oxford, 1969, pp. 5–8.

222 Osborne, ref. 221, p. 41.

223 W. Sutherland, 'The principle of dynamical similarity in molecular physics', in Boltzmann's *Festschrift*, ref. 187, pp. 373–85.

224 W. Sutherland, 'The viscosity of gases and molecular force', *Phil. Mag.* **36** (1893) 507–31. For a modern account of Sutherland's model, see S. Chapman and T.G. Cowling, *The mathematical theory of non-uniform gases*, Cambridge, 1939, pp. 182–4, 223–6.

225 M. Reinganum (1876–1914) Pogg., v. 4, p. 1226; v. 5, p. 1035. There is an obituary by E. Marx in *Phys. Zeit.* **16** (1915) 1–3. M. Reinganum, 'Über die Theorie der Zustandsgleichung und der inneren Reibung der Gase', *Phys. Zeit.* **2** (1900–1901) 241–5.

226 L. Boltzmann, 'Über die Berechnung der Abweichungen der Gase vom Boyle–Charles'schen Gesetz und der Dissociation derselben', *Sitz. Math. Naturwiss. Classe Kaiser Akad. Wissen. Wien, Abt. 2a* **105** (1896) 695–706, reprinted in WA, ref. 48, v. 3, pp. 547–57. The result was reproduced in his book, ref. 198, English trans. pp. 356–8.

227 Reinganum's first results were in his Göttingen thesis [not seen], and appeared again in his first two papers: M. Reinganum, 'Über die molekuläre Anziehung in schwach comprimirten Gasen', in *Recueil de travaux offerts par les auteurs à H.A. Lorentz, Professeur de Physique à l'Université de Leiden, à l'occasion du 25me anniversaire de son doctorat*, The Hague, 1900, pp. 574–82. (The Lorentz Festschrift is a supplementary volume of the *Archives Néerlandaises*.) The work on the 'second virial coefficient' followed in M. Reinganum, 'Zur Theorie der Zustandsgleichung schwach comprimirte Gase', *Ann. Physik* **6** (1901) 533–48; 'Beitrag zur Prüfung einer Zustandsgleichung schwach comprimirte Gase', *ibid.* 549–58.

228 Young, ref. 204 (1894–1895).

229 I.H. Silberberg, J.J. McKetta and K.A. Kobe, 'Compressibility of isopentane with the Burnett apparatus', *Jour. Chem. Eng. Data* **4** (1959) 323–9.

230 M. Reinganum, 'Über Molekularkräfte und elektrische Ladungen der Moleküle', *Ann. Physik* **10** (1903) 334–53.

231 Boltzmann, ref. 98, pp. 220 and 375 of the English translation. See also Kipnis *et al.*, ref. 172, p. 224. Rayleigh had made the same point some years earlier, ref. 167.

232 J.C. Maxwell, art. 'Capillary action', *Encyclopaedia Britannica*, 9th edn, London, 1876.

233 Van der Waals, ref. 173, chap. 10.

234 A.W. Rücker (1848–1915) T.E.T[horpe]., *Proc. Roy. Soc. A* **92** (1915–1916) xxi–xlv; A.W. Rücker, 'On the range of molecular forces', *Jour. Chem. Soc.* **53** (1888) 222–62. Rücker describes Quincke's experiment on 233–4.

235 A. Pockels (1862–1935) Pogg., v. 6, pp. 2034–5; C.H. Giles and S.D. Forrester, 'The origin of the surface film balance', *Chem. Indust.* (1971) 43–53. A. Pockels, 'Surface tension', *Nature* **43** (1891) 437–9; 'On the relative contamination of the water-surface by equal quantities of different substances', *ibid.* **46** (1892) 418–19.

236 Lord Rayleigh, 'Investigations in capillarity, . . . ', *Phil. Mag.* **48** (1899) 321–37.

237 P.A. Kohnstamm (1875–1951) Pogg., v. 5, pp. 663–4; v. 6, pp. 1364–5; Kipnis *et al.*, ref. 172, pp. 122–4; P. Kohnstamm, 'Les travaux récents sur l'équation d'état', *Jour. Chim. Phys.* **3** (1905) 665–722, see 703. The first part of this review is a stout defence of the 'molecular' school against the 'energetics' of Ostwald and Duhem, who were arguing that one should not speculate beyond the bounds of classical thermodynamics.

238 Kamerlingh Onnes and Keesom, ref. 220; *Encyklopädie*, p. 705, *Comm. Leiden*, p. 91.

239 Gibbs, ref. 199. Similar and independent work was published by Einstein in 1902–1904, ref. 95, *Collected papers*, v. 2, pp. 41–108; *English translation*, v. 2, pp. 30–77.

240 L.S. Ornstein (1880–1941) P. Forman, DSB, v. 10, pp. 235–6. His former students published *L.S. Ornstein, A survey of his work from 1908 to 1933*, Utrecht, 1933, which contains a list of his papers to 1933, pp. 87–121; Van Berkel *et al.*, ref. 122, pp. 550–1; L.S. Ornstein, *Toepassing der statistische mechanica van Gibbs op molekulair-theoretische vraagstukken*, Leiden, 1908. There is an augmented French translation of this thesis in *Arch. Néerl.* **4** (1918) 203–303.

241 K. Fuchs, Pogg., v. 5, p. 402 (no dates given). K. Fuchs, 'Ueber Verdampfung', *(Exner's) Reportorium Physik* **24** (1888) 141–60, and later papers, 298–317, 614–47; 'Über die Oberflächenspannung einer Flüssigkeit mit kugelförmiger Oberfläche', *Sitz. Math. Naturwiss. Classe Kaiser Akad. Wissen. Wien, Abt. 2a* **98** (1889) 740–51; 'Directe Ableitung einiger Capillaritätsfunctionen', *ibid.* 1362–91.

242 Lord Rayleigh, 'On the theory of surface forces-II. Compressible fluids', *Phil. Mag.* **33** (1892) 209–20.

243 J.D. van der Waals, 'Thermodynamische theorie der capillariteit in de onderstelling van continue dichtheidsverandering', *Verhand. Konink. Akad. Weten. Amsterdam* **1** (1893) No. 8, 1–56. (He had published a preliminary note as early as May 1888, see Kipnis *et al.*, ref. 172, pp. 116–19.) The paper was soon translated into German, *Zeit. phys. Chem.* **13** (1894) 657–725, and into French, *Arch. Néerl.* **28** (1895) 121–209, and later into English, *Jour. Stat. Phys.* **20** (1979) 197–244. The German and French versions have five appendices that are not in the Dutch original; the English version has the first of these.

244 It appears in Appendix 5 of the German and French versions, ref. 243.

245 This work is summarised in his book; G. Bakker, *Kapillarität und Oberflächenspannung*, which is v. 6 of the *Handbuch der Experimentalphysik*, ed. W. Wien, F. Harms and H. Lenz, Leipzig, 1928. Carl Neumann also made great use of the Yukawa potential in his *Allgemeine Untersuchungen über das Newton'sche Princip der Fernwirkungen*, ... , Leipzig, 1896.

246 J.R. Katz, 'The laws of surface-adsorption and the potential of molecular attraction', *Proc. Sect. Sci. Konink. Akad. Weten. Amsterdam* **15** (1912) 445–54. For a survey of this field, see S.D. Forrester and C.H. Giles, 'The gas–solid adsorption isotherm: a historical survey up to 1918', *Chem. Industry* (1972) 831–9.

247 M. Faraday, *Experimental researches in electricity*, London, 1839, v. 1, Sect. 5–8.

248 G.J. Stoney, 'On the physical units of nature', *Phil. Mag.* **11** (1881) 381–90.

249 H. Helmholtz, 'On the modern development of Faraday's conception of electricity', *Jour. Chem. Soc.* **39** (1881) 277–304.

250 J.J. Thomson (1856–1940) J.L. Heilbron, DSB, v. 13, pp. 362–72; J.J. Thomson, *Conduction of electricity through gases*, Cambridge, 1903, esp. pp. 131–2.

251 H. Kamerlingh Onnes, 'Expression of the equation of state of gases by means of series', *Proc. Sect. Sci. Konink. Akad. Weten. Amsterdam* **4** (1901–1902) 125–47.

252 J.D. van der Waals, 'Eine bijdrage tot de kennis der toestandsvergelijking', *Versl. Konink. Akad. Weten. Amsterdam* **5** (1896–1897) 150–3; there is an extended French translation in *Arch. Néerl.* **4** (1901) 299–313. 'Simple deduction of the characteristic equation for substances with extended and composite molecules', *Proc. Sect. Sci. Konink. Akad. Weten. Amsterdam* **1** (1898) 138–43.

253 G. Jäger, 'Die Gasdruckformel mit Berücksichtigung des Molecularvolumens', *Sitz. Math. Naturwiss. Classe Kaiser Akad. Wissen. Wien, Abt. 2a* **105** (1896) 15–21.

254 J.J. van Laar (1860–1938) Pogg., v. 4, p. 1552, v. 5, pp. 1295–7, v. 6, pp. 1439–40; E.P. van Emmerik, *J.J. van Laar (1860–1938). A mathematical chemist*, Thesis, Delft, 1991; J.J. van Laar, 'Calculation of the second correction to the quantity b of the equation of condition of van der Waals', *Proc. Sect. Sci. Konink. Akad. Weten. Amsterdam* **1** (1898–1899) 273–87, and, in more detail, in *Arch. Musée Teyler* **6** (1900) 237–84. For a modern account of the work on the fourth virial coefficient, see J.H. Nairn and J.E. Kilpatrick, 'Van der Waals, Boltzmann, and the fourth virial coefficient of hard spheres', *Amer. Jour. Phys.* **40** (1972) 503–15.

255 J.J. van Laar, 'Sur l'influence des corrections à la grandeur b dans l'équation d'état de M. van der Waals, sur les dates critiques d'un corps simple', *Arch. Musée Teyler* **7** (1901–1902) 185–218, see 212–17.

256 J.D. van der Waals, Jr (1873–1971) Pogg., v. 5, p. 1292; v. 6, p. 2785; v. 7b, p. 5843; S.R. de Groot in *Biografisch woordenboek van Nederland*, v. 1, 's Gravenhage, 1979, pp. 637–8.

257 J.D. van der Waals, Jr, 'On the law of molecular attraction for electrical double points', *Proc. Sect. Sci. Konink. Akad. Weten. Amsterdam* **11** (1908–1909) 132–8, and a correction, prompted by a communication from Reinganum, *ibid.* **14** (1911–1912) 1111–12.

258 J.D. van der Waals, 'Contribution to the theory of binary mixtures. VII', *Proc. Sec. Sci. Konink. Akad. Weten. Amsterdam* **11** (1908–1909) 146–57. Forces between different molecules are clearly needed in any discussion of the properties of mixtures but these are not treated here; for the early history of this topic, see J.M.H. Levelt Sengers, *How fluids unmix: Discoveries by the school of Van der Waals and Kamerlingh Onnes*, Amsterdam, in press. It is natural to take the parameters of the van der Waals equation in a mixture to be a quadratic function of the mole fractions since the forces arise from collisions in pairs, see Lorentz, ref. 187 (1881). The proposal that the cross-parameter in a binary mixture, a_{12}, could be put equal to the geometric mean of the like parameters, a_{11} and a_{22} was made by D. Berthelot, 'Sur le mélange des gaz', *Compt. Rend. Acad. Sci.* **126** (1898) 1703–6, 1857–8, and was promptly challenged by van der Waals in a letter with the same title: *ibid.* 1856–7. Lorentz had proposed the less controversial assumption that the cube root of b_{12} be the arithmetic mean of the cube roots of b_{11} and b_{22}, an assumption that follows naturally if the three co-volumes arise from the excluded volumes of spherical hard cores, as Maxwell had observed in 1873, ref. 129. The name 'Lorentz–Berthelot relations' for these two assumptions is modern and due to W. B[yers]. Brown, 'The statistical thermodynamics of mixtures of Lennard-Jones molecules', *Phil. Trans. Roy. Soc. A* **250** (1957) 175–220, 221–46, see 207. For a repulsive potential of the form br^{-n}, R.A. Buckingham suggested that $b_{12}^{1/n}$ be taken as the arithmetic mean of the corresponding like terms, see R.H. Fowler, *Statistical mechanics*, Cambridge, 2nd edn, 1936, p. 307. Attempts to determine the cross-energy, ε_{12}, in terms of the like energies, ε_{11} and ε_{22}, were a popular pastime in the 1950s and 1960s and led to a vast amount of work on the non-trivial task of measuring the thermodynamic properties of mixing of volatile liquids. The consensus was that the cross-energy is

usually a little less than the geometric mean of the like energies. Much of the effort put into this problem was, however, inspired more by the fun of overcoming the experimental difficulties than any real importance of the answers. This now unfashionable field is almost abandoned in the leading scientific countries but still has a small following elsewhere. For a summary, or obituary, see J.S. Rowlinson, *Liquids and liquid mixtures*, London, 1959, 3rd edn, with F.L. Swinton, 1982.

259 W.H. Keesom, 'On the deduction of the equation of state from Boltzmann's entropy principle', *Proc. Sect. Sci. Konink. Akad. Weten. Amsterdam* **15** (1912–1913) 240–56; 'On the deduction from Boltzmann's entropy principle of the second virial-coefficient for material particles (in the limit rigid spheres of central symmetry) which exert central forces upon each other and for rigid spheres of central symmetry containing an electric doublet at their centre', *ibid.* 256–73; 'On the second virial coefficient for di-atomic gases', *ibid.* 417–31.

260 M.J. Klein, 'Not by discoveries alone: the centennial of Paul Ehrenfest', *Physica A* **106** (1981) 3–14.

261 R. Clausius, *Abhandlungen*, ref. 115, Zusatz zu Abhandlung X, 1866, pp. 135–63; *Die mechanische Behandlung der Electricität*, Braunschweig, 1879, Abschnitt III, 'Behandlung dielectrischer Medien', pp. 62–97. This is v. 2 of a second revised edition of the *Abhandlungen* of 1864 and 1867, issued in three volumes in 1876, 1879 and 1891.

262 H.A. Lorentz, 'Ueber die Beziehung zwischen der Fortpflanzungsgeschwindigkeit des Lichtes und der Körperdicht', *Ann. Physik* **9** (1880) 641–65; L. Lorenz, 'Ueber die Refractionsconstante', *ibid.* **11** (1880) 70–103. Both Lorentz and Lorenz wrote other papers on the subject but these are the the usual sources, cited, for example, by R. Gans in his review, 'Elekrostatik und Magnetostatik', *Encyklopädie*, ref. 5, v. 5, part 2, chap. 15, pp. 289–349, see p. 330, recd Oct. 1906, pub. March 1907. For a short account of the confusing history of these equations, with references, see B.K.P. Scaife, *Principles of dielectrics*, Oxford, 1989, pp. 177–81.

263 See, for example, the table in chap. 5 of successive editions of J.H. Jeans, *The mathematical theory of electricity and magnetism*, Cambridge, 1907 to 1925.

264 P.J.W. Debye (1884–1966) C.P. Smyth, DSB, v. 3, pp. 617–21; M. Davies, *Biog. Mem. Roy. Soc.* **16** (1970) 175–232. P. Debye, 'Einige Resultate einer kinetischen Theorie der Isolatoren', *Phys. Zeit.* **13** (1912) 97–100, 295; English translation in Debye's *Collected papers*, New York, 1954, pp. 173–9. See also J.J. Thomson, 'The forces between atoms and chemical affinity', *Phil. Mag.* **27** (1914) 757–89.

265 H. Weight, 'Die elektrischen Momente des CO- und CO_2- Moleküls', *Phys. Zeit.* **22** (1921) 643.

266 W.H. Keesom, 'The second virial coefficient for rigid spherical molecules, whose mutual attraction is equivalent to that of a quadruplet placed at their centre', *Proc. Sect. Sci. Konink. Akad. Weten. Amsterdam* **18** (1915–1916) 636–46; W.H. Keesom and C. van Leeuwen, 'On the second virial coefficient for rigid spherical molecules carrying quadruplets', *ibid.* 1568–71.

267 D. Berthelot, 'Sur les thermomètres à gaz et sur la reduction de leurs indications à l'échelle absolue des températures', *Trav. Mém. Bureau Int. Poids et Més.* **13** (1907) B, 1–113.

268 J.D. Lambert, G.A.H. Roberts, J.S. Rowlinson and V.J. Wilkinson, 'The second virial coefficients of organic vapours', *Proc. Roy. Soc. A* **196** (1949) 113–25.

269 P. Debye, 'Die van der Waalsschen Kohäsionskräfte', *Phys. Zeit.* **21** (1920) 178–87; English trans. in *Collected papers*, ref. 264, pp. 139–57.

270 W.H. Keesom, 'Die van der Waals Kohäsionskräfte', *Phys. Zeit.* **22** (1921) 129–41, 643–4; 'The cohesion forces in the theory of van der Waals', *Proc. Sect. Sci. Konink. Akad. Weten. Amsterdam* **23** (1922) 943–8 [The paper is dated 27 November 1920]; 'On the calculation of the molecular quadrupole-moments from the equation of state', *ibid.* **24** (1922) 162–7; 'Die Berechnung der molekularen Quadrupolmomente aus der Zustandsgleichung', *Phys. Zeit.* **23** (1922) 225–8.

271 H. Falkenhagen, 'Kohäsion und Zustandsgleichung bei Dipolgasen', *Phys. Zeit.* **23** (1922) 87–95.

272 A.D. Buckingham, 'Direct method of measuring molecular quadrupole moments', *Jour. Chem. Phys.* **30** (1959) 1580–5; A.D. Buckingham and R.L. Disch, 'The quadrupole moment of the carbon dioxide molecule', *Proc. Roy. Soc. A* **273** (1963) 275–89. Buckingham was then in Oxford and Disch was an American working at the National Physical Laboratory, Teddington, where the experiment was made. An earlier but less direct method was devised by N.F. Ramsey at Harvard, and applied to hydrogen, see N.F. Ramsey, 'Electron distribution in molecular hydrogen', *Science* **117** (1953) 470; and *Molecular beams*, Oxford, 1956, pp. 228–30.

273 Helium was liquified in 1908 but solidified only in 1926, by applying a pressure of more than 25 atm to the liquid at low temperatures, see W.H. Keesom, 'Solid helium', *Proc. Sect. Sci. Konink. Akad. Weten. Amsterdam* **29** (1926) 1136–45. The standard reference for all early work on helium is W.H. Keesom, *Helium*, Amsterdam, 1942.

274 H. Kamerlingh Onnes, 'Isotherms of monatomic gases and their binary mixtures. I. Isotherms of helium between $+100\,°C$ and $-217\,°C$', *Proc. Sect. Sci. Konink. Akad. Weten. Amsterdam* **10** (1907–1908) 445–50; ' . . . II. Isotherms of helium at $-253\,°C$ and $-259\,°C$', *ibid.* 741–2.

275 F. Zwicky (1898–1974) K. Hufbauer, DSB, v. 18, pp. 1011–13; F. Zwicky, 'Der zweite Virialkoeffizient von Edelgasen', *Phys. Zeit.* **22** (1921) 449–57.

276 P. Debye, 'Molekularkräfte und ihre elektrischer Deutung', *Phys. Zeit.* **22** (1921) 302–8; English trans. in *Collected papers*, ref. 264, pp. 180–92.

277 W.H. Keesom, 'On the second virial coeffcient for monatomic gases, and for hydrogen below the Boyle-point', *Proc. Sect. Sci. Konink. Akad. Weten. Amsterdam* **15** (1912) 643–8.

278 Keesom, ref. 270, 'On the calculation of the molecular quadrupole-moments . . .', footnote on p. 162.

279 Lord Kelvin, 'Nineteenth century clouds over the dynamical theory of heat and light', *Phil. Mag.* **2** (1901) 1–40, a Friday evening Discourse at the Royal Institution, 27 April 1900. For a discussion of the unease felt by some physicists at the end of the 19th century, see H. Kragh, *Quantum generations: A history of physics in the twentieth century*, Princeton, NJ, 1999, chap. 1, and sources cited there.

280 H.W. Nernst (1864–1941) E.N. Hiebert, DSB, v. 15, pp. 432–53; W. Nernst, 'Kinetische Theorie fester Körper', in Planck *et al.*, ref. 200, pp. 61–86, see p. 64.

281 F. Dolezalek (1873–1920) Pogg., v. 5, p. 301; v. 6, pp. 586–7; Obituary by H.G. Möller, *Phys. Zeit.* **22** (1921) 161–3; F. Dolezalek, 'Zur Theorie der binären Gemische und konzentrieten Lösungen', *Zeit. phys. Chem.* **64** (1908) 727–47; **71** (1910) 191–213.

282 This term is discussed in Section 5.3.

283 J.J. van Laar, 'Über Dampfspannung von binären Gemische', *Zeit. phys. Chem.* **72** (1910) 723–51. The argument continued for some years, see J.H. Hildebrand, *Solubility*, New York, 1924, pp. 72–84.

284 G. Mie (1868–1957) J. Mehra, DSB, v. 9, pp. 376–7; G. Mie, 'Zur kinetischen Theorie der einatomigen Körper', *Ann. Physik* **11** (1903) 657–97. His 'monatomic

bodies' were metals, not the inert gases. A few years later P.W. Bridgman also supposed that an intermolecular potential proportional to separation to the inverse 4th power led to an internal energy proportional to $V^{-4/3}$, ref. 167, 95–9.

285 E.A. Grüneisen (1877–1949) Pogg., v. 4, p. 540; v. 5, p. 456; v. 6, pp. 965–6; v. 7a, •p. 295. E. Grüneisen, 'Zur Theorie einatomiger fester Körper', *Verhand. Deutsch. Phys. Gesell.* **13** (1912) 836–47; 'Theorie des festen Zustandes einatomiger Elemente', *Ann. Physik* **39** (1912) 257–306, and many other papers from 1908 onwards. For a review of this and earlier work on solids, see Mendoza, ref. 83. Lorentz repeated the point that a term in the energy proportional to V^{-1} does not imply an intermolecular potential proportional to r^{-3} in the discussion of Grüneisen's paper at the 1913 Solvay Conference, *La structure de la matière*, Paris, 1921, p. 289.

286 S. Ratnowsky (1884–1945) Pogg., v. 5, p. 1023; v. 6, p. 2176; v. 7a, p. 682. S. Ratnowsky, 'Die Zustandsgleichung einatomiger fester Körper und die Quantentheorie', *Ann. Physik* **38** (1912) 637–48.

287 Einstein, see ref. 95 for his early belief in universality, and ref. 96 for his disillusion with it.

288 Zwicky, ref. 275.

289 J.E. Lennard-Jones (1894–1954) S.G. Brush, DSB, v. 8, pp. 185–7; N.F. Mott, *Biog. Mem. Roy. Soc.* **1** (1955) 175–84. J.E. Jones added the name Lennard in 1925, after his marriage to Kathleen Lennard.

290 D. Hilbert, 'Begrundung der kinetische Gastheorie', *Math. Ann.* **72** (1912) 562–77. There is an English translation in Brush, ref. 16, v. 3, pp. 89–102. Max Born claimed at a meeting in Florence in 1949 that Hilbert's results anticipated those of Chapman and Enskog, but this claim is hard to justify in terms of useful results; M. Born, [no title], *Nuovo Cimento* **6** , Suppl. 2 (1949) 296.

291 S. Chapman (1888–1970) T.G. Cowling, DSB, v. 17, pp. 153–5; *Biog. Mem. Roy. Soc.* **17** (1971) 53–89.

292 D. Enskog (1884–1947) S.G. Brush, DSB, v. 4, pp. 375–6; M. Frudland, 'International acclaim and Swedish obscurity: The fall and rise of David Enskog' in *Center on the periphery. Historical aspects of 20th-century Swedish physics*, ed. S. Lindqvist, Canton, MA, 1993, pp. 238–68.

293 S. Chapman, 'On the law of distribution of velocities, and on the theory of viscosity and thermal conduction, in a non-uniform simple monatomic gas', *Phil. Trans. Roy. Soc. A* **216** (1916) 279–348; 'On the kinetic theory of a gas. Part II – A composite monatomic gas: diffusion, viscosity, and thermal conduction', *ibid.* **217** (1917) 115–97.

294 Enskog's results were set out in his dissertation at Uppsala in 1917, *Kinetische Theorie der Vorgänge in mässig verdünnten Gasen* [not seen], of which there is an English translation in Brush, ref. 16, v. 3, pp. 125–225. The first part of Brush's volume contains an account of the development of the Chapman–Enskog theory and its use for the determination of intermolecular forces. Chapman's own account is set out in Chapman and Cowling, ref. 224, see especially the 'Historical summary', pp. 380–90, and in a lecture of 1966, reprinted by Brush, ref. 16, v. 3, pp. 260–71.

295 C.G.F. James, 'The theoretical value of Sutherland's constant in the kinetic theory of gases', *Proc. Camb. Phil. Soc.* **20** (1921) 447–54. See also Fowler's unsuccessful attempt to reconcile Sutherland's constant, S, and van der Waals's constant, a, in R.H. Fowler, 'Notes on the kinetic theory of gases. Sutherland's constant S and van der Waals' a and their relations to the intermolecular field', *Phil. Mag.* **43** (1922) 785–800. For Fowler (1889–1944), see S.G. Brush, DSB, v. 5, pp. 102–3, and E.A. Milne, *Obit. Notices Roy. Soc.* **5** (1945–1948) 61–78.

296 S. Chapman, 'On certain integrals occurring in the kinetic theory of gases', *Mem. Lit. Phil. Soc. Manchester* **66** (1922) No. 1, 1–8.

297 J.E. [Lennard-]Jones, 'On the determination of molecular fields – I. From the variation of the viscosity of a gas with temperature; II. From the equation of state of a gas; III. From crystal measurements and kinetic theory data', *Proc. Roy. Soc. A* **106** (1924) 441–62, 463–77, 709–18.

298 H. Kamerlingh Onnes and C.A. Crommelin, 'Isotherms of monatomic gases and of their binary mixtures. VII. Isotherms of argon between +20 °C and −150 °C', *Proc. Sec. Sci. Konink. Akad. Weten. Amsterdam* **13** (1910–1911) 614–25.

299 L. Holborn and J. Otto, 'Über die Isothermen einiger Gase zwischen +400° und −183°' [−100 °C for argon] *Zeit. f. Physik* **33** (1924) 1–11.

300 K. Schmitt, 'Über die innere Reibung einiger Gase und Gasgemische bei verschiedenen Temperaturen', *Ann. Physik* **30** (1909) 393–410.

301 H. Kamerlingh Onnes and S. Weber, 'Investigation of the viscosity of gases at low temperatures. III. Comparison of the results obtained with the law of corresponding states', *Proc. Sec. Sci. Konink. Akad. Weten. Amsterdam* **15** (1912–1913) 1399–1403.

302 F.E. Simon (1893–1956) K. Mendelssohn, DSB, v. 12, pp. 437–9; F. Simon and C. von Simson, 'Die Krystallstruktur des Argon', *Zeit. f. Physik* **25** (1924) 160–4.

303 Max Born wrote a monograph for the *Encyklopädie*, ref. 5, v. 5, part 3, chap. 25, pp. 527–781, which was reprinted the same year, without change of title or pagination, as *Atomtheorie des festen Zustands*, Leipzig, 1923. In this, § 28, 'Entwicklung der Lehre von Zustandsgleichung', is a summary of the work of Mie and Grüneisen in which he sets out clearly all the assumptions made; see also Mendoza, ref. 83.

304 H.R. Hassé (1884–1955) Pogg., v. 6, pp. 1043–4. Hassé was Professor of Mathematics at Bristol where, in 1927, Lennard-Jones was Reader in Physics. W.R. Cook was a research student who worked with both men.

305 H.R. Hassé and W.R. Cook, 'The viscosity of a gas composed of Sutherland molecules of a particular type', *Phil. Mag.* **3** (1927) 977–90; 'The determination of molecular forces from the viscosity of a gas', *Proc. Roy. Soc. A* **125** (1929) 196–221.

306 M. Born and A. Landé, 'Kristallglitter und Bohrsches Atommodel', *Verhand. Deutsch. Phys. Gesell.* **20** (1918) 202–9.

307 M. Born and A. Landé, 'Über die Berechnung der Kompressibilität regulärer Kristalle aus der Gittertheorie', *Verhand. Deutsch. Phys. Gesell.* **20** (1918) 210–16.

308 [Lennard-]Jones, ref. 297, Part III.

309 J.E. [Lennard-]Jones, 'On the atomic fields of helium and neon', *Proc. Roy. Soc. A* **107** (1925) 157–70.

310 J.E. Lennard-Jones and P.A. Taylor, 'Some theoretical calculations of the physical properties of certain crystals', *Proc. Roy. Soc. A* **109** (1925) 476–508. Lennard-Jones summarised the state of this field in 1929 in a chapter he contributed to the first edition of Fowler's *Statistical mechanics*, 1929, ref. 258, see chap. 10, 'Interatomic forces'.

311 The General Discussion was published in *Trans. Faraday Soc.* **24** (1928) 53–180, and as a separate booklet.

312 T.W. Richards, 'A brief review of a study of cohesion and chemical attraction', *Trans. Faraday Soc.* **24** (1928) 111–20. For Richards (1868–1928), see S.J. Kopperl, DSB, v. 11, pp. 416–18. His earlier work in this field from 1898 is summarised in 'A brief history of the investigation of internal pressures', *Chem. Rev.* **2** (1925–1926) 315–48.

313 A.W. Porter, 'The law of molecular forces', *Trans. Faraday Soc.* **24** (1928) 108–11.

314 J.E. Lennard-Jones [no title], ref. 311, p. 171.

315 G.A. Tomlinson, 'Molecular cohesion', *Phil. Mag.* **6** (1928) 695–712.

316 I. Langmuir, 'The constitution and fundamental properties of solids and liquids. I. Solids', *Jour. Amer. Chem. Soc.* **38** (1916) 2221–95; '. . . II. Liquids', *ibid.* **39** (1917) 1848–1906.
317 O. Klein, in an interview in September 1962, as reported by A. Pais, 'Oskar Klein', in *The genius of science: A portrait gallery*, Oxford, 2000, pp. 122–47, see p. 128. There is no report of any other attempt at a calculation of the attractive force in ter Haar, ref. 189, nor in A. d'Abro, *The decline of mechanism*, New York, 1939, 2 vols., nor in any of the short articles on the early quantum theory in *Science* **113** (1951) 75–101, nor in v. 1 of J. Mehra and H. Rechenberg, *The historical development of quantum mechanics*, New York, 1982.

5

Resolution

5.1 Dispersion forces

The understanding of cohesion has two main strands; first, what are the forces between the constituent particles of matter and, second, how does the operation of these forces give rise to the transformation of gases into liquids, liquids into solids, and to all other manifestations of cohesion, of which the elasticity of solids and the surface tension of liquids have, throughout the years, been the two that have attracted most attention. We have seen that in the 18th century there were some interesting speculations about the form of the forces, in particular that they fell off with r, the separation of the particles, as r^{-n}, where n is greater than 2, its value for the law of gravitation. The second strand received some attention at this time but little progress was made. The situation was reversed by Laplace who found that he had to dismiss speculation about the nature or form of the forces with the dictum that all we could know of them was that they were 'insensible at sensible distances'. He made, however, a substantial contribution to the second strand of the problem with his theory of capillarity and, in the hands of his followers, his ideas proved fruitful, if controversial, in the interpretation of the elastic properties of solids. No further progress could be made until the kinetic theory and the laws of thermodynamics had been established. The time was then ripe for van der Waals to resume the Laplacian programme; first, to advance our understanding of the condensation of gases to liquids and, second, to make the first real advance in the theory of surface tension since the time of Laplace. The success of van der Waals's programme re-awakened interest first among his Dutch followers, and then more widely, into the origin of the forces themselves, to which Boltzmann soon attached van der Waals's name. Classical mechanics and electromagnetism proved unable to explain why the simplest substances, the monatomic inert gases, should cohere, and provided only unconvincing suggestions to explain the coherence of substances such as hydrogen, nitrogen and oxygen. This failure was only one aspect of a much wider problem; why do some pairs of atoms exhibit only the weak cohesive 'van der Waals' attraction while other pairs are violently attracted and form strong chemical bonds? Theoretical physics and chemistry could make little progress until such questions could be answered. In 1895 Boltzmann wrote:

For a long time the celebrated theory of Boscovich was the ideal of physicists. According to his theory, bodies as well as the ether, are aggregates of material points, acting together with forces, which are simple functions of their distances. If this theory were to hold good for all phenomena, we should still be a long way off what Faust's *famulus* hoped to attain, viz. to know everything. But the difficulty of enumerating all the material points of the universe, and of determining the law of mutual force for each pair, would only be a quantitative one; nature would be a difficult problem, but not a mystery for the human mind. [1]

Boltzmann's mystery was resolved in the early years of the 20th century, although not in a way that he or Boscovich would have suspected. The realisation that classical mechanics was inappropriate for atomic systems grew steadily after first Planck and later Einstein, Bohr and others, found that the quantisation of energy removed many of the 'clouds' (to use again Kelvin's term) that were obscuring the understanding of the optical, electrical, mechanical and thermal properties of matter. The rules for quantisation were at first ad hoc, each was invented to rationalise a particular phenomenon, but a coherent basis for a new mechanics was developed in 1925 and 1926. The most fruitful form – Erwin Schrödinger's wave mechanics – was applied with astonishing speed and success to a wide range of physical and chemical phenomena in the next five years. As early as 1929 Paul Dirac made a claim that echoed Boltzmann's expectations. He wrote:

The underlying physical laws necessary for a mathematical theory of a large part of physics and the whole of chemistry are thus completely known, and the difficulty is only that the exact application of these laws leads to equations much too complicated to be soluble. [2]

Since 1929 the history of quantum mechanics, as applied to most of physics and all of chemistry, has been the search for ever better solutions of Schrödinger's wave equation. Implicit in this programme is the formal abandonment of the particle models that had come down to us from Newton and Boscovich. Heisenberg's 'uncertainty principle' and the 'Copenhagen' interpretation of quantum mechanics require that we think about electrons and, at least formally, also about atomic nuclei in new ways, as both waves and particles. Fortunately for many problems, including the calculation of the cohesive forces, we can use the fact that the large masses of the nuclei, compared with that of the electrons, means that we can conceptually place the nuclei in fixed positions and confine the quantal calculations to the solution of the wave equation for the electrons as they move around the fixed nuclei. This simplification is called the Born–Oppenheimer approximation [3]. Once this has been done and we know the forces as a function of intermolecular separation and orientation then we can usually use this information in a purely classical way to calculate the properties of matter. Only for the lightest molecules, hydrogen and helium, must we use quantal methods also for the calculation of these properties, and then only at low temperatures or when we need high accuracy. All this, of course, is in an ideal world in which the quantal calculation of the forces and the classical calculation of the properties can actually be made. We consider both problems in this chapter.

The first advance that is directly relevant to the problem of intermolecular forces arose from a suggestion made by Debye on a visit to New York in 1927. John Slater [4] wrote later that he had been told by H.A. Kramers that Wolfgang Pauli had earlier made a similar suggestion in his lectures, but it was Debye's that bore fruit. We have seen that Debye had thought that electrons oscillating about a positive nucleus might be the mechanism by which atoms attracted each other, but a classical electrostatic calculation shows that the net effect of the interaction of two such systems is zero. At Columbia University he met a research student, S.C. Wang [5], whom he persuaded to repeat the calculation with the new wave mechanics. Wang proposed a crude model of a pair of hydrogen atoms as two electron oscillators confined to a common plane [6]. With this he obtained the important result that there is indeed an attractive force at (atomically) large distances, which is proportional to r^{-7}, where r is the atomic separation. The potential energy of this force can be written

$$u(r) = -C_6 r^{-6}, \tag{5.1}$$

where his estimate of C_6 was 8.2×10^{-79} J m^6 or, in the so-called 'atomic units', $C_6 = 8.6$ a.u. These units are convenient to use in this field since not only do they remove the inconveniently high positive and negative powers of ten needed with conventional units, but the actual calculations are made in them. The atomic unit for C_6 is $(e^2 a_0^5 / 4\pi\varepsilon_0) = 0.9574 \times 10^{-79}$ J m^6. Here e is the charge on the electron, a_0 is the Bohr radius of the hydrogen atom, $a_0 = \varepsilon_0 h^2 / \pi m_e e^2 = 0.529\,18$ Å, ε_0 is the permittivity of free space, $4\pi\varepsilon_0 = 1.112\,65 \times 10^{-10}$ C^2 J^{-1} m^{-1}, h is Planck's constant, 6.6261×10^{-34} J s, and m_e is the mass of the electron, 9.1094×10^{-31} kg. Wang saw that his value of C_6 was of the right order of magnitude since the energy at a separation of 2 Å is about three times the translational energy of a molecule at 0 °C, but it is, as we now know, not quite the correct result for two hydrogen atoms [7]. He offered no more in the way of interpretation but his result was important since it showed, for the first time, that two atomic systems with no permanent electric multipoles should, according to the rules of the new quantum mechanics, attract each other with a force that was apparently strong enough to explain the phenomenon of cohesion.

At the same time as Wang was tackling the problem of the long-range forces between hydrogen atoms, Fritz London [8] was working with Walter Heitler in Zürich on what turned out to be a different kind of force at much shorter separations, although their original aim had also been to understand the van der Waals attractive force [9]. They made the dramatic discovery that the short-range force is repulsive if the electrons on the two hydrogen atoms have their spins in a parallel orientation, but changes sign and is attractive if they are anti-parallel [10]. At extremely short distances there is an even stronger repulsion in both cases which could be explained as the classical Coulomb repulsion between the two positively charged nuclei when they are so close that they are no longer shielded by the orbiting electrons. The attractive force with the anti-parallel electrons arises from a term in the interaction that represents the possibility of the electrons switching from movement around one nucleus to movement around the other. It has no classical analogue; they called it the 'exchange energy' [*Austauschenergie*] and found that it leads to a deep minimum in the potential energy as a function of separation which is comparable with the energy of the covalent chemical bond between the two atoms in the hydrogen molecule. For helium, where each atom has two electrons with no net spin on the atom, there is no possibility of forming a chemical bond. They had therefore solved at last, in principle, two major theoretical problems. First, they had shown how, and under what circumstances, two atoms could share a pair of electrons and so form a covalent bond. Chemists had known empirically for ten years that sharing a pair of electrons is the essence of covalent bonding but had not been able to explain how this came about [10]. Second, they had shown that where there are no available electrons with anti-parallel spins then the energy is large and positive, a consequence of Pauli's exclusion principle of quantum mechanics that forbids the overlap of electron clouds with no anti-parallel pairing. This positive energy or repulsive force explains why many atoms and most molecules repel each other at short distances, or, in simpler terms, why they have size. This repulsive energy dies away exponentially with distance and so is ultimately less in magnitude than the universal attractive energy in r^{-6} discovered by Wang. The total energy, $u(r)$, as a function of r, has therefore a weak minimum at (atomically) moderately large distances for all chemically unreactive pairs of atoms and molecules, as is required to explain the cohesive properties of all matter.

To produce an attraction between atoms with anti-parallel electron spins Heitler and London had used first-order quantal perturbation theory, in which the mutual Coulombic energies between the electrons and protons on different atoms are treated as a perturbation of the energies of the isolated atoms. The consequences of this perturbation are found by averaging it over the known wave function (i.e. the electron distribution) found by solving Schrödinger's equation for the isolated or unperturbed atoms. The weaker effect discovered by Wang does not appear at this order of approximation. London, by then in Berlin, first mentioned Wang's work in a review he wrote for an issue of *Naturwissenschaft* commemorating the 50th anniversary of Planck's doctorate [11]. He quoted from a later paper of Wang's and said that the calculated depth of the energy minimum in a hydrogen molecule was -3.8 eV at a separation of 0.75 Å and added, but without giving the source of his estimate, that for a pair of atoms with parallel spins a "more exact calculation shows a much weaker attraction of some thousandths of a[n electron] volt at a separation of about 5 Å". (The thermal energy, kT, at 25 °C, is 0.0257 eV; 1 eV $= 1.6021 \times 10^{-19}$ J.)

In Berlin, London met Robert Eisenschitz [12] who was working at the laboratories of the Kaiser-Wilhelm-Gesellschaft. Together they tackled again the problem of two hydrogen atoms with parallel spins, using now second-order perturbation theory. This, as Wang had found, is significantly more difficult than the first-order theory since it requires a knowledge of the energies and wave functions of all the excited states of the two unperturbed atoms, and not only those of the ground state, as suffices for the first-order theory. They were able to carry through the calculations using methods that have since been greatly simplified. They verified Wang's conclusion that there is an attractive potential at large distances that varies as $-C_6 r^{-6}$, and found a value of C_6 of 6.47 a.u., a result similar to, but significantly smaller than Wang's estimate of 8.6 a.u. Lennard-Jones immediately confirmed this result by a simpler perturbation calculation [13], while Hassé [14] and Slater and Kirkwood [15, 16] used the other main branch of approximated quantum mechanics, variational theory, to find a value of 6.4976 a.u. It is of the essence of this second method that one chooses a wave function for the interacting pair of atoms or molecules, of whatever form seems to be appropriate, with a set of initially undetermined parameters. These are then varied so as to minimise the energy, since we know that there is a rigorous theorem that says that the minimum so found is never lower than the true energy. In this case the variational method was slightly better than the second-order perturbation theory. Pauling and Beach found the definitive result for this artificially simple system a few years later [7]; C_6 is 6.499 03 a.u.

The origin of the attraction is purely quantal – it arises from the application of the rules of quantum mechanics established in the 1920s – and so a verbal description of it is even more imperfect than one for a classical electrostatic force. For hydrogen atoms it can be ascribed to the motion of the two electrons around their two nuclei. At any instant each atom has a dipole moment, although the time average of the moment is zero. The instantaneous dipole on one atom produces a field at the second atom proportional to r^{-3}, where r is the separation of the nuclei. This field modifies the dipole moment of the second atom by an amount proportional to this field. The energy of the whole system is reduced by an amount proportional to the product of this change of moment and the energy of interaction of this change with the first or inducing moment, an energy which is also proportional to r^{-3}. The reduction of the energy of the two atoms is therefore proportional to r^{-6}. The fact that the mutual action of the two oscillating dipoles is always a reduction of energy implies that there is a coupling of the phases of their motions, and so might be thought to lead to the same difficulty as was clear with classical induction effects, namely that what is effective in an isolated pair becomes neutralised in a symmetrical cluster of atoms. To some extent this is true but it is not sufficient to prevent a substantial extent of 'additivity' of pair potentials in condensed systems. A group of three molecules at the corners of an equilateral triangle at their equilibrium separations has typically an energy that is 95% of the sum of the three pair-energies. If the three molecules are in a straight line then there is a small enhancement of the coupling and the attractive energy is a little stronger than the sum of the three pair-energies. We return to this point later.

The simplest theoretical description of this attractive force was put forward by London [17] within a few months of his paper with Eisenschitz. It is based on a model of an atom or molecule that is usually associated with Paul Drude, although his picture was pre-quantal and, indeed, pre-electronic [18]. The spherical molecule is supposed to comprise a massive charged nucleus about which there oscillates a body of smaller mass m and charge q, equal and opposite to that on the nucleus. If the force constant of the oscillatory motion is c then the frequency of the simple harmonic oscillation is ν_0, where

$$2\pi \nu_0 \equiv \omega_0 = (c/m)^{1/2}, \tag{5.2}$$

where ω_0 is the often more convenient angular frequency. An electric field ξ displaces the charge q through a distance s, proportional to ξ, thus creating a dipole μ, where

$$\xi q = cs \quad \text{and} \quad \mu = qs = \xi q^2/c, \tag{5.3}$$

so that the polarisability of the molecule, α, which is the ratio of the scalar quantities μ/ξ, is

$$\alpha = q^2/c = q^2/m\omega_0^2. \tag{5.4}$$

Consider now two such molecules, a and b, whose centres are separated by r and where, at a given time, the displacements of the two equal charges $q_a = q_b = q$ from their centres are r_a and r_b. When the separation of the two molecules is large then Schrödinger's equation for the wave function ψ is

$$(h^2/8\pi^2 m)(\nabla_a^2 + \nabla_b^2)\psi + (E - \tfrac{1}{2}cr_a^2 - \tfrac{1}{2}cr_b^2)\psi = 0, \tag{5.5}$$

where E is the energy and ∇^2 are the operators

$$\nabla_a^2 = \partial^2/\partial x_a^2 + \partial^2/\partial y_a^2 + \partial^2/\partial z_a^2, \tag{5.6}$$

and x_a, y_a, and z_a are the cartesian components of r_a. This wave equation is separable into two independent equations for identical three-dimensional harmonic oscillators. The ground state of the system has therefore the energy of six oscillators each of energy $h\omega_0/4\pi$; that is, $E = 3h\omega_0/2\pi$. This result holds when the separation of the two molecules, r, is infinite. When r is finite then we must insert the energy of interaction of the two instantaneous dipoles into the wave equation; it is

$$(q^2/4\pi\varepsilon_0 r^3)(x_a x_b + y_a y_b - 2z_a z_b),$$

where the z-axis is chosen to lie along the line joining the centres. The new wave equation is obtained by adding this term into the second, or energy term in eqn 5.5. A change to normal coordinates transforms this into another equation for six one-dimensional oscillators, but now not all of the same frequency. Let

$$\mathbf{R} = (\mathbf{r}_a + \mathbf{r}_b)/\sqrt{2}, \qquad \mathbf{S} = (\mathbf{r}_a - \mathbf{r}_b)/\sqrt{2}, \tag{5.7}$$

when the equation becomes

$$(h^2/8\pi^2 m)(\nabla_a^2 + \nabla_b^2)\psi + \Big(E - \tfrac{1}{2}c_x^+ R_x^2 - \tfrac{1}{2}c_y^+ R_y^2 - \tfrac{1}{2}c_z^+ R_z^2$$
$$- \tfrac{1}{2}c_x^- S_x^2 - \tfrac{1}{2}c_y^- S_y^2 - \tfrac{1}{2}c_z^- S_z^2\Big)\psi = 0. \tag{5.8}$$

The six frequencies are therefore

$$\omega_x^\pm = \big(c_x^\pm/m\big)^{1/2}, \quad \omega_y^\pm = \big(c_y^\pm/m\big)^{1/2}, \quad \omega_z^\pm = \big(c_z^\pm/m\big)^{1/2}, \tag{5.9}$$

or

$$\big(\omega_x^\pm\big)^2 = \big(\omega_y^\pm\big)^2 = (c/m)(1 \pm q^2/4\pi\varepsilon_0 r^3),$$
$$\big(\omega_z^\pm\big)^2 = (c/m)(1 \pm q^2/2\pi\varepsilon_0 r^3), \tag{5.10}$$

and the energy is

$$E = (h/4\pi)(\omega_x^+ + \omega_y^+ + \omega_z^+ + \omega_x^- + \omega_y^- + \omega_z^-). \tag{5.11}$$

Inserting eqn 5.10 into eqn 5.11, and expanding the square roots, since r is large, gives the energy of the ground state of the system as

$$E = (3h\omega_0/2\pi)[1 - (q^2/8\pi\varepsilon_0 r^3)^2]. \tag{5.12}$$

The second term is the energy of interaction of the two molecules which can be written more simply in terms of the unperturbed frequency and the polarisability of eqns 5.3 and 5.4;

$$u(r) = -3h\omega_0\alpha_V^2/8\pi r^6 = -3h\nu_0\alpha_V^2/4r^6, \tag{5.13}$$

where $\alpha_V = \alpha/4\pi\varepsilon_0$ is the polarisability expressed in the dimensions of volume. This is the simplest form of the interaction energy, obtained by London in 1930. The supposed frequency of oscillation of the Drude model, ω_0, is related to the dispersion of light in this model, that is to the change of the refractive index with the frequency of the light. This change is associated in real molecules with the outermost electrons since they are the most polarisable. London therefore christened this attractive term the 'dispersion energy', and the term is now used generally; an alternative is the 'London energy'. The factor $h\omega_0/2\pi$, or $h\nu_0$, can be replaced, to a rough approximation, by the ionisation energy, I, the energy needed to remove an electron from the molecule, since this is determined primarily by the tightness of the binding of the outer electrons. Hence, as London observed, the attractive energy can be calculated approximately from two observable physical properties, the polarisability and the ionisation energy. Slater and Kirkwood's variational treatment, when similarly approximated, leads to the slightly different result that the dispersion energy varies not as $I\alpha^2$ but as $(N\alpha^3)^{\frac{1}{2}}$, where N is the number of electrons in the outer shell of the atom. This Drude model is only a simple but convenient representation of the quantum mechanics behind the dispersion forces. The actual calculations for light atoms such as hydrogen and helium were, from the first days, more fundamentally based on a proper quantum mechanical basis.

The oscillating electrons in a molecule generate not only instantaneous dipoles but also quadrupoles and higher multipoles. It is to be expected, therefore, that the London dispersion energy is only the first term in a series expansion for the attractive energy;

$$u(r) = -C_6 r^{-6} - C_8 r^{-8} - C_{10} r^{-10} - \text{etc.} \tag{5.14}$$

This extension was first considered by Henry Margenau [19] who found that the inclusion of the higher terms lowered the minimum of the He–He potential by a factor of about 3/2 [20]. A large correction was also found also for H–H by Pauling and Beach [7], but the change was believed to be much smaller for heavier atoms and molecules, such as in the Ar–Ar potential [21]. Quantitative work was difficult and for practical purposes it was assumed that a single term in r^{-6} was an adequate representation of the potential, at least at separations equal to or greater than that of the minimum in the total potential. It was a reasonable assumption at the time, but one that was later found to be flawed.

In the early 1930s quantal calculations of the dispersion forces could not go beyond approximations such as those of London or Slater and Kirkwood. The repulsive forces needed to balance these at short distances and give the molecules 'size' were even more of a problem. Heitler and London had shown that the origin of these lay in the Pauli exclusion principle that prevented the electron clouds from overlapping when there were no unpaired electron spins to lead to chemical bonding, but quantitative calculations were difficult except for hydrogen atoms which had only one electron on each atom. The simplest case that could be studied experimentally was helium, with two spin-paired electrons on each atom. An early triumph of the new theory was the good agreement between the purely quantal calculations of the attractive and repulsive parts of its potential and the parameters of a Lennard-Jones (12, 6) potential determined from the physical properties of the gas. The quantal calculation of Slater and Kirkwood [16] gave a potential

$$u(r) \cdot 10^{17}/\text{J} = 7.7 \exp(-2.43r/a_0) - 0.68(r/a_0)^{-6}, \qquad (5.15)$$

where a_0 is again the Bohr radius of the hydrogen atom. This potential is essentially that of Slater in 1928 but with an attractive parameter of 0.68 rather than 0.67. Kirkwood and his former research supervisor at the Massachusetts Institute of Technology, F.G. Keyes, calculated the second virial coefficient for this potential and showed that there was reasonable agreement ($\sim$5%) with experiment [22]. Meanwhile Lennard-Jones, in work that he reported in a lecture to the Physical Society in May 1931, had compared this potential with the (12, 6) potential that he had already fitted to the second virial coefficient [23]. Similar comparisons were made by R.A. Buckingham in 1936 and 1938 [24]. Table 5.1 shows a comparison of some of the pre-War calculations.

Here d is the 'collision diameter', or the separation at which the attractive and repulsive potentials are in balance, that is $u(d) = 0$, r_m is the separation at the minimum energy where the attractive and repulsives forces are in balance, $u'(r_m) = 0$, ε is the depth of the energy minimum, conveniently expressed in kelvin by dividing it by Boltzmann's constant, k, and C_6 is the coefficient of r^{-6} expressed in atomic units (Fig. 5.1). The quantal calculations in Table 5.1 are those by Slater and Kirkwood, eqn 5.15, and of C_6 (only) by Baber and Hassé [25]. (The accepted value of this coefficient is now 1.4615 ± 0.0004 a.u. [26]. It is smaller than that calculated for two hydrogen atoms, for although helium has two electrons to hydrogen's one, they are more tightly bound.) The 'experimental' values of the parameters were obtained by fitting the (12, 6) potential to the second virial coefficient [24] and

Table 5.1

Source	$d/\text{Å}$	$r_m/\text{Å}$	$(\varepsilon/k)/\text{K}$	$C_6/\text{a.u.}$
1931 quantal calculation, eqn 5.15	2.62	2.95	9.10	1.56
1937 quantal calc., Baber and Hassé	–	–	–	1.43
1931 exp. second virial coeff., via (12, 6)	2.60	2.92	7.33	1.30
1938 exp. Joule–Thomson coeff., via (12, 6)	2.57	2.88	9.56	1.59

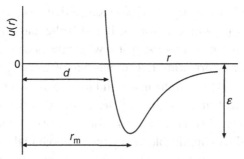

Fig. 5.1 The conventions used to describe the parts of a spherical intermolecular potential, $u(r)$, which is a function only of the one variable, the separation, r. The potential is zero at the collision diameter, d, and has its minimum value of $-\varepsilon$ at a separation r_m. It is at this separation that the intermolecular force is zero.

the Joule–Thomson coefficient at low pressures [27] with, in the second case, a correction for the quantal departures from the classical values that arise from the light mass of the helium atom [28]. As we saw earlier (Section 4.2), the information provided by the Joule–Thomson coefficient is formally the same as that provided by the second virial coefficient, since they are directly related by the laws of thermodynamics.

The agreement shown for the parameters of the He–He potential obtained in different ways is surprisingly good in view of the approximations made in the quantal calculations, the neglect of quadrupole and higher multipole terms, and the restriction imposed on the interpretation of the experimental results by the use of a (12, 6) potential. It is seen that the parameters ε and C_6 obtained in the last line of Table 5.1, in which quantal corrections have been applied in the interpretation of the physical property, are closer to those calculated theoretically in lines 1 and 2 than are the uncorrected classically obtained parameters in line 3. Thus for the interaction that gives what we can write as He_2, the simplest 'van der Waals molecule', there was at last a convincing link between calculations that started only from the assumption that a helium atom has two electrons and a relatively massive nucleus (and the laws of quantum mechanics) and a macroscopic physical property that can be measured in the laboratory. Newton had declared that it was the business of experimental philosophy to discover the "agents in Nature" that made matter stick together, and 'in principle', as Boltzmann and Dirac might have said, that aim was achieved in the early 1930s. In practice much remained to be done. Even for helium the agreement between theory and experiment was imperfect, although good enough to show that the interpretation was on the right lines. No other molecule is as simple as the helium atom; molecular hydrogen might be thought to be similar since it too has only two electrons, but it is not spherical. The second virial coefficient and its equivalent, the Joule–Thomson coefficient at zero pressure, are the simplest macroscopic physical properties and the only ones that could be calculated in this way in the 1930s, since statistical mechanics provides, as Maxwell and Boltzmann had shown, an exact route from $u(r)$ to this physical property, for which the necessary integral had been calculated. The other information-rich properties of gases, the viscosity and the coefficients of self- and thermal-diffusion, did not receive the same attention as the second virial coefficient in the 1930s. We have seen (Section 4.5) that Hassé and Cook had, in 1929, calculated the viscosity for an (8, 4) potential and had pointed out that their method was applicable to any $(n, \frac{1}{2}n)$ potential, but after 1927, when quantum mechanics led to $\frac{1}{2}n = 6$, their hint was not followed up; it is hard to see why. Instead, H.S.W. Massey and C.B. Mohr, then both 1851 Exhibitioners at Trinity College, Cambridge, went straight to a quantal calculation of the angles of deflection of two colliding helium atoms between which there acted the Slater–Kirkwood potential of eqn 5.15 [29]. Their calculated values of the viscosity were too high by 7% at room temperature and too high by 20% below 20 K. A standard textbook of the time said that such agreement would not normally be considered very good but noted the approximations in the theory and again stressed that all that had been assumed was that the helium "nucleus is much heavier than an electron and carries a charge numerically twice as great" [30]. These results could not be extended to other molecules.

There was, therefore, a big programme ahead before what had been achieved in principle could be shown convincingly in practice. First, molecules more complicated than helium must be tackled and, if possible, with greater accuracy. Second, the whole range of physical properties discussed in previous chapters must be brought within the scope of kinetic and statistical mechanical calculations. All this was what Kuhn has called 'normal science'; the problems were difficult but the principles were now known. Progress was, however, neither as rapid nor as steady as we, looking back from seventy years later, might have expected. The waywardness that marks the progress of science was again apparent. Distractions of different fields, fashionable and attractive ideas about the structure of liquids that were later shown to be wrong, and the small number of leading players, all contributed to the hesitancy of the advance.

The extension of theory from helium and hydrogen to more complicated atoms and molecules, and the struggle to extend statistical mechanics to more important physical properties than the second virial coefficient, can both be demonstrated by taking argon as an example, as was suggested by Nernst as early as 1913 [31]. It has an atom with enough electrons to challenge the quantum mechanics community, but one that is spherical and heavy, so that those working in classical statistical mechanics could not ask for an easier system. Moreover it is readily available from the distillation of liquid air, so measurements of every physical property of interest were made in the early years of the 20th century. Restricting our choice to argon leads naturally to what came to be the two matters of prime importance in the years after the end of the Second World War, the accurate determination of the intermolecular pair potential and the development of a satisfactory theory of the liquid state. The contribution to the intermolecular forces of the classical electrostatic effects so extensively discussed by Keesom, Debye and others early in the century is therefore ignored for the moment, not because the substances in which such forces act are uninteresting – one of them is water – but because the essence of the problems is best exemplified by the properties of the inert gases and of argon in particular. This was the way that the matter was seen at the time, and is the way that is most natural for a retrospective discussion.

5.2 Argon

The physical properties of argon were thought to be well known by the 1930s. The structure, lattice spacing, and energy of evaporation of the crystal had apparently been established by Simon and von Simson [32] and F. Born [33], although, as we now know, not with quite sufficient accuracy for acceptable deductions to be made about the intermolecular forces. The second virial coefficient had been measured several times; the most widely quoted results were those of Holborn and Otto in Berlin which extended from $-100\,°C$ to $+400\,°C$ [34]. The viscosity of this gas (and of many others) had been measured up to 1000 K by Max Trautz and his associates at Heidelberg [35]. These were thought to be the most reliable and most extensive then available, but were later found to have misleading errors. The vapour pressures of the liquid and the solid had been established in Leiden by 1914 [36], and the x-ray diffraction pattern of the liquid was studied by Keesom and De Smedt in 1922 [37]. The interpretation of this pattern as a pair distribution function, $g(r)$, for the atoms in the liquid, followed in 1927 when Zernike and Prins showed how to use a Fourier transform to obtain this function from the x-ray pattern [38]. The function $g(r)$ measures the normalised probability of finding a molecule with its centre at a distance r from any chosen molecule. It is now the most commonly used measure of the structure of a liquid but, as we shall see in Section 5.5, it was some years after 1927 before its use became widespread. In fact little use could then be made of any of the structural or thermophysical properties of the liquid state because of the primitive state of that branch of statistical mechanics. Only for gases and solids were there thought to be safely navigable paths from experiment through theory to intermolecular information.

Theoretical results for argon were more sparse. The strength of the dispersion force could be estimated from the atomic polarisability via Slater and Kirkwood's expression, from the ionisation potential which is approxiately equal to $h\omega_0/2\pi$ in London's expression, eqn 5.13, or from the dispersion coefficients themselves, which was London's preferred route. There was no way of testing the accuracy of these approximations. The dipole–quadrupole dispersion force, that is the coefficient C_8 of eqn 5.14, could be estimated similarly but with even less confidence. It was often convenient to express the importance of this term in the attractive potential by calculating a modified dipolar dispersion term, C_6^*, defined by

$$C_6^* = C_6 + C_8 r_{\mathrm{m}}^{-2}, \qquad (5.16)$$

where r_{m} is the separation of the molecules at the minimum of the potential. The coefficients are defined to be positive so the amount by which C_6^* exceeds C_6 is a measure of the dipole–quadrupole term. There was no way of calculating the repulsive potential for a system with as many electrons as a pair of argon atoms, so this part of the potential was estimated by comparing the predictions of model potentials containing several adjustable parameters with the equilibrium physical properties of the gas and solid.

Let us consider first the attractive potential where the consensus (Table 5.2) was that C_6 was about 60 a.u. and C_6^* about 70 a.u. The only dissent from these and similar results was a value of C_6 nearly twice as large as these figures found by Alexander Müller at the Royal Institution from a route due originally to Kirkwood, via the diamagnetic susceptibility of the argon atom, but he himself said that the value was clearly too high [41].

It was recognised by this time that an inexact knowledge of the second virial coefficient over a finite range of temperature does not determine a unique form of potential. The usual procedure was to require the chosen potential to yield also the correct lattice spacing and energy of evaporation of the crystal, extrapolated to zero temperature. It was tacitly assumed that the crystal energy could be found by adding the interactions of all pairs of atoms, with no multi-body effects. It was known that the observed energy at zero temperature would be numerically smaller than this sum because of the zero-point energy of oscillation of the atoms about their lattice

Table 5.2 [a]

	Source	C_6/a.u.	C_6^*/a.u.
1937	London, from dispersion coefficients [39]	58.0	–
	Buckingham, from polarizabilities [40]	66.3	76.4
1939	Margenau, from dispersion coefficients [21]	58.0	66.5
	Margenau, from Slater–Kirkwood approx. [21]	72.6	–

[a] The values of C_6^* have been calculated with $r_{\mathrm{m}} = 3.824\,\text{Å}$.

sites, a quantal effect that could be adequately accounted for in terms of the Debye frequency of the lattice vibrations. It was known also that the lattice spacing was not exactly at the minimum of the pair potential because of the attractions of the non-nearest neighbours (which reduces the lattice spacing), and the anharmonic nature of the zero-point oscillations (which increases the spacing); the second effect is the greater [42].

Two kinds of empirical functions were used to represent the whole intermolecular potential function, attractive and repulsive. The first was the Lennard-Jones $(n, 6)$ potential, in which n was often given the convenient and apparently acceptable value of 12. The second was a more realistic function much used by Buckingham and generally associated with his name and that of John Corner [43]:

$$u(r) = A \exp(-r/\rho) - C_6 r^{-6} - C_8 r^{-8}. \tag{5.17}$$

If the term in r^{-8} is omitted, as in the Slater and Kirkwood equation for helium, eqn 5.15, then this is usually called the (exp, 6) potential. The work of Heitler and London, and others, had suggested that the repulsive or overlap branch of the potential could be represented by a polynomial in r multiplied by a rapidly decreasing exponential factor. In practice, the polynomial was replaced by a single constant, A. This potential, eqn 5.17, like the $(n, 6)$ potential, has three adjustable parameters if the ratio C_8/C_6 is fixed, but the repulsive branch rises less steeply than in a $(12, 6)$ potential if ρ is given the often-used value of $(r_m/14)$. Some of the results of fitting these potentials to the experimental properties of gaseous and solid argon are given in Table 5.3. Herzfeld and Goeppert Mayer used two (exp, 6) potentials in which two different values were chosen for the parameter ρ in eqn 5.17. They took these from work on the properties of the salt KCl, since the ions K^+ and Cl^- are iso-electronic with Ar and so might be supposed to show similar repulsion between their overlapping electron clouds [49]. Kane's two sets of figures follow from the same two choices of ρ. Lennard-Jones in 1937 (and Corner in 1939 [42]) used a $(12, 6)$ potential. The others used (exp, 6) or (exp, 6, 8) potentials. Some of the

Table 5.3

	Source	$d/Å$	$r_m/Å$	$(\varepsilon/k)/K$	$C_6/$a.u.	$C_6^*/$a.u.
1934	Herzfeld and Goeppert Mayer [44]	3.48	3.83	120	82	–
		3.43	3.94	103	116	–
1937	Lennard-Jones [45]	3.41	3.83	120	108	–
1938	Buckingham [46]	3.40	3.82	135	107	–
1939	Kane [47]	3.48	3.83	134	91	–
		3.43	3.94	115	131	–
1948	Corner [48]	3.43	3.87	125	95	114

figures are not in the original papers but have been calculated from the parameters quoted there.

The most notable feature of Table 5.3 is the consistency of the results, obtained from three different forms of potential, over a 14-year period. By 1950 it had become generally accepted that the Ar–Ar potential had a depth of about 120 K at a separation of 3.82–3.86 Å. A second feature of the results in the table is that the values of C_6 are substantially larger than the theoretical values calculated from the dispersion coefficients. The former are in the range 80–130 a.u. and the latter about 60–70 a.u. This discrepancy was often ignored but when it was noted it was ascribed either to the approximations needed to obtain the theoretical results, or to faults in the forms of the fitted potentials, such as the inadequacy of the repulsive part of a (12, 6) potential, or to the neglect of the C_8 term. The first argument could not easily be faulted since, as with many quantal calculations, the approximations needed could not be independently assessed. Neither part of the second argument holds water, however, since the discrepancy is present also with exponential repulsion and with the inclusion of the C_8 term. Two further possible origins of the discrepancy received less attention. One was that the experimental properties of the gas and the solid were not known as accurately as was believed, and a second was that the energy of the crystal could not be calculated by adding the pair interaction energies but that there were significant contributions from three-body and maybe higher terms. Both effects were later found to be significant.

The results in Table 5.3 are not a complete account of all attempts to find the pair potential for argon but they are typical of work up to 1954, a year that saw the publication of a massive treatise: *Molecular theory of gases and liquids*, by J.O. Hirschfelder, C.F. Curtiss and R.B. Bird of the University of Wisconsin [50]. This book of 1219 pages marked the end of an era. It set out all that had been achieved in the 1920s and 1930s and brought it up to date with the substantial amount of new work that been done in the nine years since the end of the War, much of it at Wisconsin. It had as great an influence in the 1950s and 1960s as R.H. Fowler's books had had in the 1930s and 1940s. It probably did more than any other single text to establish a belief in the correctness of the parameters shown above for argon, and to reinforce the view that the properties of simple substances could, for all practical purposes, be calculated from a model that used the (12, 6) or the (exp, 6) potential. The former is the easier to use and became the model of choice for most research. Hirschfelder and his colleagues noted the discrepancy between the values of the coefficient C_6 calculated quantally, and those determined empirically, for all simple substances except hydrogen and helium, for which the aggreement was reasonable. They wrote:

The significance of this deviation is not understood. It may be that the short-range forces fall off faster than the $1/r^{12}$ term in the Lennard-Jones (6-12) potential would indicate, so that the attractive forces need not be so large in order to give the same total potential. [51]

We have seen, however, that this explanation was not supported by experience with the (exp, 6) potential.

One obvious property was missing from the study of argon in the 1930s, the viscosity of the gas. The natural step of extending Hassé and Cook's calculation for the (8, 4) potential to the (12, 6) was not taken. These years were marked by many fruitful applications of the new quantum mechanics to a great range of molecular problems; classical statistical mechanics and kinetic theory were relatively neglected except for a few workers in the U.S.A. and a small body of enthusiasts at Cambridge. This gap in our theoretical armoury became obvious after the War and in three laboratories there were independent calculations of the transport integrals for the (12, 6) potential in the years 1948–1949 [52]. These workers had, in fact, been preceded by a Japanese team in Tokyo in 1943 but that calculation was unknown to them until their work was finished [53]. These theoretical results were soon compared with the experimental work of Trautz [35] and with some more recent measurements of the viscosity of argon at low temperatures [54]. The conclusion was that the viscosity could be fitted to (12, 6) parameters similar to those that fitted the second virial coefficient [55]. A few years later, E.A. Mason, also then at Wisconsin, calculated the transport integrals for the (exp, 6) potential [56] and he and W.E. Rice used them, the second virial coefficient, and the properties of the crystal to obtain (Table 5.4) a new set of parameters [57]. The results are essentially the same as those obtained in the 1930s and 1940s, before it was possible to use the viscosity of the gas as a source of information.

Another satisfying confirmation of these parameters came from the newly introduced technique of the computer simulation of molecular systems [58]. Such simulations were first made during the second World War for studying the problem of the rate of diffusion of neutrons in a nuclear reactor and, from 1947 onwards, were applied to the problem of the equation of state and structure of simple fluids. The method is straightforward in principle; a model intermolecular potential is

Table 5.4

	Source	$d/\text{Å}$	$r_\text{m}/\text{Å}$	$(\varepsilon/k)/\text{K}$	$C_6/\text{a.u.}$
1954	Hirschfelder, Curtiss and Bird (12, 6)	3.418	3.837	124	114
1954	Mason and Rice (exp, 6)	3.437	3.866	123.2	104

chosen, an assembly of such molecules is 'created' in the computer memory, and
the physical state of the system is found either by solving Newton's equations of
motion to see how the system evolves with time, or by using a weighted sampling
method (the Monte Carlo method) that generates molecular configurations with
the same frequency of occurrence as is found in such a model fluid at equilibrium.
Such simulations quickly became an invaluable tool in the development and testing
of theories of the liquid state, the state of matter for which statistical mechanical
theories had made little advance since the time of van der Waals. The simulations
generated pseudo-experimental values for the macroscopic physical properties such
as density, vapour pressure, energy and heat capacity for systems of prescribed intermolecular potentials. Hitherto the testing of any theory of liquids or dense gases
had been a hazardous business because of the uncertainty in our knowledge of the
intermolecular forces. Any failure could either be one in the statistical mechanical theory or one of an inappropriate choice of intermolecular potential, or, of
course, of both. The method of computer simulation eliminated the second source
of uncertainty.

An early and influential application of this method was a Monte Carlo simulation
of a (12, 6) fluid undertaken by W.W. Wood and F.R. Parker at Los Alamos, who
calculated the pressure as a function of gas density for a reduced temperature
of $kT/\varepsilon = 2.74$. The first results were obtained in October 1954 [58] but their
paper did not appear until September 1957 [59]. They chose this temperature since
if ε/k is 120 K it corresponds to a laboratory temperature for argon of 55 °C,
and at that temperature there were measurements of the density to high pressures.
P.W. Bridgman at Harvard had measured the density up to 15 000 atm in 1935 [60]
and A. Michels at Amsterdam, with what appeared to be greater accuracy, to 2000
atm in 1949 [61]. The simulated results fitted the isotherm of Michels and his
colleagues but fell below that of Bridgman, by up to 30% in the pressure at the
highest density. This result was held to confirm the higher accuracy of the Dutch
results and to validate the choice of the (12, 6) potential.

The principle of corresponding states provided further evidence that a (12, 6)
potential might be adequate. When we left the discussion of this principle it was
an empirical correlation put forward by van der Waals behind which Kamerlingh
Onnes had discerned a principle of mechanical similitude in the intermolecular
forces. Within either Boltzmann's or Gibbs's formulation of classical statistical
mechanics this perception could readily have been made more precise by a simple
manipulation of the known form of canonical partition function at any time onwards
from the earliest years of the 20th century. Such a step was not taken, however, until
1938 and 1939 when first J. de Boer and A. Michels [62] and then K.S. Pitzer [63]
showed independently that the molecular condition for the principle to hold is that
the (assumed spherical) intermolecular potential of all substances can be written in

a common form;

$$u(r) = \varepsilon f(r/d), \tag{5.18}$$

where ε is an energy and d a length, both characteristic of any substance. They may conveniently be chosen to be the depth of the minimum of the potential and the collision diameter; $u(r_m) = -\varepsilon$ and $u(d) = 0$ (Fig. 5.1). The principle holds for any group of substances if the function $f(r/d)$ is the same for all of them. It had been observed that argon, krypton and xenon conform closely to the principle in all three phases of matter, and that neon shows small departures at low temperatures and helium large ones, as would be expected for systems for which quantal effects cannot be neglected [64]. If the potential is of the (n, m) form then the principle requires that n and m be the same for all conforming substances. The attractive index, m, was known to be 6 for all substances, so the conformation of argon, krypton and xenon argued for a common value of n, and 12 seemed to be the best choice. The argument is only indicative; there is no requirement for the function $f(r/d)$ to be of the (n, m) form – many other functions could be devised – but at least the evidence from the principle of corresponding states was consistent with the choice of a $(12, 6)$ potential for the inert gases.

A second quantal calculation led to another discrepancy which became apparent after the War, but to which most in the field turned a blind eye. London had established the crucial distinction between the attractive exchange force and the much weaker attractive dispersion force. The first 'saturates', that is, once it has formed a chemical bond between a pair of atoms it cannot use the same electrons to form further bonds. The second does not saturate, that is, an atom that is attracting a second one is not precluded from acting as strongly with a third, or a fourth, etc. This distinction was accepted throughout the 1930s, but during the War two attempts were made to test the validity of the second proposition, that is, what we now call the principle of pair-wise additivity. B.M. Axilrod and his then research supervisor, Edward Teller, in Washington, took London's perturbation theory to third order and calculated the energy of a group of three atoms [65]. The same calculation was made independently in Japan by Yosio Muto [66]. Both parties found that this energy departed from the sum of the two-body (or second order) terms by a three-body dipole–dipole–dipole energy:

$$u_3(r_{12}, r_{13}, r_{23}) = \left(9 I \alpha_V^3 / 16 r_{12}^3 r_{13}^3 r_{23}^3\right)(1 + 3\cos\theta_1 \cos\theta_2 \cos\theta_3), \tag{5.19}$$

where I is the ionisation energy, α_V is the polarisability volume, and θ_i is the angle of the triangle formed by the three atoms, at the corner of atom i. The corresponding expression for each of the three dipole–dipole energies is, from eqn 5.13,

$$u_2(r_{12}) = -3 I \alpha_V^2 / 4 r_{12}^6. \tag{5.20}$$

For three atoms at the corners of an equilateral triangle we have for the ratio of the three-body term to the sum of the three two-body terms:

$$u_3 / \left(- \sum u_2 \right) = 11\alpha_V / 32 r^3, \tag{5.21}$$

where r is the length of the side of the triangle. For three atoms in a straight line with the two nearest neighbours at a common separation r, we have

$$u_3 / \left(- \sum u_2 \right) = -4\alpha_V / 43 r^3. \tag{5.22}$$

For argon $\alpha_V / r_{\mathrm{m}}^3$ is 0.031, so the three-body term is positive and 1% of the sum of the two-body terms for the equilateral triangle, and negative and -0.3%, for three atoms in a line. At first sight these figures look reassuring; the effect of the three-body term is going to be negligible. In the crystal, however, the atoms are closely packed and the total effect is more serious. Axilrod estimated that the overall effect is then positive and that the magnitude of the crystal energy is diminished by about 2% in neon, 5% in argon, and 9% in xenon. His principal concern, however, was not the magnitude of these changes in the crystal energy, but whether this three-body effect could explain a minor anomaly of the crystal structures of the inert-gas solids. There are two close-packed lattices for spherical particles, the face-centred cubic (or fcc) lattice and the hexagonal close-packed (or hcp) lattice. Helium crystallises in the hcp structure but the others in the fcc structure. A simple summation of the pair energies shows that for static atoms the fcc is the less stable; its energy is higher by 0.01%. This small but irritating anomaly is not removed by calculations that allow for the vibrational energy of the atoms about their lattice sites. Axilrod had thought that the triple-dipole energy might remove the anomaly, but found that it did not. There is still no simple and convincing explanation, but there are many small higher-order terms in both the attractive and repulsive energies that have not been discussed here. One suggestion has been that the strength of the dispersion forces is changed by the presence of p-orbitals in neon and the heavier atoms, and that this change stabilises the fcc lattice [67], but the point is not settled and many dismiss the anomaly as too small to be worth worrying about. It may, however, have been the distraction of hunting down this minor problem that led to insufficient attention being paid to the quantitative effect on the calculated lattice energy of argon (5% as estimated by Axilrod, and now believed to be about 7%) and the consequences of this change for the many determinations of the intermolecular potential that relied on the crystal energy as an important input into the calculations.

   There were therefore at least two problems for the (12, 6) and (exp, 6) potentials lurking in the wings in the early 1950s: the large discrepancy between the quantal and the 'experimental' values of the dispersion coefficient C_6, and the need to include the triple-dipole term, and perhaps other minor terms, in the calculation of the crystal energy. The first serious doubt was raised by E.A. Guggenheim of Reading University at the Jubilee Meeting of the Faraday Society in London in April 1953 [68]. His criticism was based on a belief that the (12, 6) potential gave the wrong curvature of the potential at its minimum. He later found that he appeared to be wrong on this point, but his forceful criticisms opened up the subject for discussion. Seven years later he fulfilled his promise of 1953 to make a more detailed study of the problem and now his criticisms were more cogent [69]. He and M.L. McGlashan accepted the quantal value of C_6 and so were led to a deeper minimum in the potential than the generally accepted value of ε/k of 120 K; they found 138 K at a separation of 3.81–3.82 Å. This distance was close to that of the (12, 6) and (exp, 6) potentials. An over-simplified treatment of the viscosity of the gas at high temperatures (the known measurements of which were, in fact, in error) led them, however, to conclude that that the diameter d, at which the potential is zero, was 3.1–3.2 Å, a value that was much lower than anything previously proposed, and which is now known to be wrong. Their whole analysis rested heavily on the properties of the crystal but they made no use of, or even mention of, the three-body term of Axilrod and Teller.

It is difficult to give a comprehensive account of the often conflicting experimental evidence and fluctuating theoretical views on the argon potential from 1953 until about 1972; only representative papers can be cited. These came from a small number of centres in the United States and in the United Kingdom, with some important contributions from Australia and Japan. Continental Europe stood aside. By 1972 the problem of the argon potential was substantially solved although minor improvements followed for another few years, when the consensus was reviewed in a substantial monograph of 1981, *Intermolecular forces: their origin and determination*, by G.C. Maitland, M. Rigby, E.B. Smith and W.A. Wakeman [70]. Smith was in Oxford and the other three authors in London; Maitland and Rigby had been research students with Smith.

Confidence in the (12, 6) and (exp, 6) potentials was slowly undermined by new and apparently more accurate measurements of some of the physical properties, and doubts about some of the older measurements. Mason and Rice had found in 1954 that the viscosity of the gas at high temperatures calculated from the (exp, 6) potential lay above the experimental values [57]. This was probably the first tentative indication that the experimental values might be in error. Such a discrepancy implied a steeper repulsive potential than the one they had chosen, but such a change conflicted with Mason's own measurements, when working with I. Amdur at the Massachusetts Institute of Technology, of the scattering of high-energy beams of argon atoms off other argon atoms [71]. These required a repulsive wall of the potential at short separations that was softer than any hitherto proposed; it varied approximately as $r^{-8.3}$. Mason and Rice noted also that at low temperatures the calculated viscosity fell below the observed values, but said that "we can think of no explanation for this". Some years later it was shown that the limiting behaviour of the viscosity at low temperatures, which is related directly to the coefficient C_6, is consistent with the quantal calculations but not with the larger values required by the (12, 6) and (exp, 6) potentials [72]. A similar problem arose with the second virial coefficient. Michels and his colleagues in Amsterdam measured this down to 118 K and found that their results were lower than those calculated from the (12, 6) potential that they had used to fit successfully their results at ambient and higher temperatures [73]. The discrepancies became worse when measurements down to 80 K became available [74].

The first attempts to solve these problems came from an unexpected direction, namely from attempts to devise potentials for polyatomic molecules. In molecules such as CH_4, CF_4 and SF_6 the polarisable electrons are disposed symmetrically about the central atom and at some distance from it. It was a simple and obvious step to asume that such molecules could be described by a shell from which a potential of (12, 6) or similar form 'emanated'. Several such shell models were devised [75], the most detailed of which was that of Taro Kihara in which the force was assumed to arise from the points on the two shells that had the smallest separation. This potential became widely known through his review of 1953 [76]. It was not his intention to apply this model with a spherical shell to the inert gases. It was an italicised conclusion of that review that the potential for argon had a "*wider bowl and harder repulsive wall*" than that of the conventional (12, 6) potential, whereas it is characteristic of shell models that they have deeper and narrower bowls when these are described in terms of the centre–centre separation of the molecules. Some years later, however, A.L. Myers and J.M. Prausnitz at Berkeley [77] found that the low-temperature measurements of the second virial coefficient that Michels had found to be incompatible with the conventional (12, 6) potential could be fitted with a Kihara shell model;

$$u(r) = \varepsilon[(\rho_m/\rho)^{12} - 2(\rho_m/\rho)^6], \qquad \rho = r - 2a, \qquad (5.23)$$

where the shell radius $a = 0.175$ Å. The minimum of the potential they found to be at a separation of $r_m = \rho_m + 2a = 3.678$ Å and at a depth of $\varepsilon/k = 146.1$ K. They were not the first to suggest a depth about 20% greater than the conventional 120 K; as we have seen Guggenheim and McGlashan had suggested 138 K two years earlier, and in 1961 D.D. Konowalow and J.O. Hirschfelder had proposed 145 K [78], but neither of these potentials was in the main line of development. Guggenheim and McGlashan had tried to determine the form of the potential only near its minimum, and Konowalow and Hirschfelder had used a Morse potential – a double exponential form that lacked any r^{-6} term and so was suitable for a chemical bond but not for the potential of the van der Waals forces. What was becoming clear,

Table 5.5

Source	a/Å	d/Å	r_m/Å	(ε/k)/K	C_6/a.u.
Barker *et al.* [79]	0.168	3.363	3.734	142.9	63
Sherwood and Prausnitz [80]	0.184	3.314	3.675	147.2	56

however, was that algebraically simple forms of potential were unlikely to suffice. More than two adjustable parameters were needed for an accurate potential that fitted all the experimental evidence.

The second virial coefficient at low temperatures showed clearly that a depth of not less than 140 K is needed, but a full test cannot be made from one physical property alone. When the transport integrals were calculated it was evident that the potential of Guggenheim and McGlashan did not fit the viscosity of the gas, but that the Kihara (12, 6) potential, eqn 5.23, although not perfect, was an improvement on the Lennard-Jones (12, 6) potential [79]. Two sets of figures for Kihara potentials from 1964 are given in Table 5.5.

The values derived for C_6, the coefficient of the dipole dispersion force, are close to those of the quantal calculations of the 1930s listed in Table 5.2, but this apparent agreement has no significance since the Kihara potential has a spurious r^{-7} term. These potentials could not themselves account for the properties of the crystal. The greater depth of the Kihara potential led to an overestimate of the magnitude of the crystal energy of about 15%. This change was of the right sign to be accounted for by the triple-dipole term but was two to three times the expected magnitude for this correction.

More subtle tests of the Lennard-Jones and Kihara potentials arose from the interrelation of three properties of the liquid state that could be used for this purpose even in the absence of a fully-developed theory of the liquids. The three properties are, first, $u(r)$, the pair potential, second, its logarithmic derivative, the pair virial function, $v(r)$, and, third, the logarithmic derivative of the virial function, $w(r)$, which has no name:

$$v(r) = r[du(r)/dr], \qquad w(r) = r[dv(r)/dr]. \qquad (5.24)$$

The corresponding instantaneous values of the sums of these functions in a macroscopic portion of matter are U^*, V^*, and W^*, where

$$U^* = \sum\sum u(r_{ij}), \quad V^* = -(1/3)\sum\sum v(r_{ij}),$$

$$W^* = (1/9)\sum\sum w(r_{ij}), \qquad (5.25)$$

where the double sums are taken over all pairs of molecules. If we ignore any multi-body potentials then the mean or thermodynamic values of U^* and V^* are well known;

$$\langle U^* \rangle = U, \qquad \langle V^* \rangle = pV - NkT, \tag{5.26}$$

where U is the internal or configurational energy of the system, and p is the pressure of N molecules in a volume V at a temperature T. The mean value of W^* is not so easily accessible, but if the potential $u(r)$ is of the Lennard-Jones (n, m) form then

$$\langle W^* \rangle = -(nm/9)U + [(n+m)/3](pV - NkT). \tag{5.27}$$

This result is exact in a classical system of (n, m) particles [81]. A similar, but not quite so rigorously derived result holds for a Kihara (n, m) potential:

$$(1 - \gamma^2)\langle W^* \rangle = -(nm/9)U + [(n+m+\gamma)/3](1-\gamma)(pV - NkT), \tag{5.28}$$

where $\gamma = a/d$ [82]. A purely thermodynamic discriminant, based on the mathematical necessity for the average value of certain mean-square fluctuations to be positive, puts a lower bound on $\langle W^* \rangle$, and so on the value of n, if m is put equal to 6. The minimum value of $\langle W^* \rangle$ that is acceptable for liquid argon at its triple point is 4.49×10^4 J mol^{-1}, while a Lennard-Jones $(12, 6)$ potential yields the unacceptable value of 4.33×10^4 J mol^{-1} [83]. Kihara's potential, with $\gamma = 0.1$, gives a value of 5.40×10^4 J mol^{-1} which satisfies the thermodynamic discriminant. There is, however, an experimental route to $\langle W^* \rangle$ that requires only that U^* is composed of pair potentials. This route requires the knowledge of a quantal effect, the differences of the ratios of the abundance of the isotopes of argon of different mass in the liquid and in its co-existent vapour [84]. Its use needs only a value for the collision diameter, d, which is fortunately the least uncertain of the molecular parameters. This route yields $\langle W^* \rangle = 4.53 \times 10^4$ J mol^{-1}. This satisfies the thermodynamic discriminant, that is, it is greater than 4.49×10^4 J mol^{-1}, but it differs significantly from that calculated from Kihara's potential. The conclusion from the two tests, the simple one of the energy of the crystal and the less direct one of the thermodynamic discriminant for the liquid, is that neither the Lennard-Jones nor the Kihara $(12, 6)$ potential satisfies the properties of the condensed phases. The obvious culprit is again the neglect of the multi-body potentials, and, in particular, the three-body triple-dipole potential.

The most direct experimental route to the three-body potential is a measurement of the third virial coefficient of a gas. If we write the equation of state in the virial form,

$$pV/NkT = 1 + B(N/V) + C(N/V)^2 + D(N/V)^3 + \cdots, \tag{5.29}$$

then the second coefficient, B, is determined by the force between a pair of molecules; the higher coefficients, C, D, etc., are similarly, and exactly, related to the forces within clusters of three, four, etc., molecules. So if we seek to understand the three-body force we should measure the third coefficient, C, as a function of temperature. Unfortunately this is difficult to do. The pressure of a gas at low densities can be measured accurately and leads to a value of B that, with care, is good to $\pm 1\%$. At higher pressures it is not easy to determine C since the contribution of the terms in D, E, etc., is difficult to 'remove', and since any error in fixing B leads to a larger error in C. Nevertheless reasonable reliable values ($\pm 10\%$) were available for argon over a wide range of temperature, principally from the work of Michels and his colleagues. These were larger, by 50 to 100%, than those calculated from the conventional (12, 6) and (exp, 6) potentials, but could be accounted for by quantal estimates of the triple-dipole potential [80, 85]. Such results confirmed what was becoming clear from the study in parallel of the crystal and the dilute gas, that a pair potential that fitted the gas could not account unaided for the properties of the crystal. Throughout the 1960s many made the provisional compromise of using a (12, 6) potential as an 'effective' pair potential that gave a reasonable account of the properties of all three phases of matter without having to invoke the awkward three-body term [86]. This attitude was reinforced when the technique of computer simulation became sufficiently routine to generate a body of pseudo-experimental properties of the condensed phases [87]. These results for a (12, 6) potential were increasingly used to test statistical theories of liquids without worry about the unresolved difficulties of the three-body potential.

Any improvement in our knowledge of the true two-body potential must therefore come from the precise study of two-body properties, that is the second virial coefficient and the viscosity of the gas at low pressures. The assistance that it was hoped to find from the properties of the solid had proved to be misleading. Other two-body properties such as the thermal conductivity and the coefficients of self- and thermal-diffusion of the gas are, in principle, also available and were occasionally used but they could not be measured with the same accuracy as the primary pair. By the middle of the 1960s it was agreed that the (12, 6) potential was inadequate but there was no agreement over what should take its place. The Kihara (12, 6) potential was an improvement but did not account completely for the viscosity at high and low temperatures, and its form, with a spurious term in r^{-7} on expansion, was theoretically unappealing.

Better quantal calculations soon gave more confidence in the reliability of the size of the coefficient of the dispersion force. In 1964 A.E. Kingston found a value of C_6 of 65.4 a.u. and wrote that the "absolute error [is] certainly less than 10% and may be considerably smaller" [88]. This and similar calculations were confirmed the next year by measurements of the scattering cross-section of an argon atom when it meets another at a low speed. The cross-section is then determined only by the long-ranged part of the potential; one of the form $-C_m r^{-m}$ gives a cross-section proportional to $C_m^{2/(m-1)}$. In this way E.W. Rothe and R.H. Neynaber in California found, after an initial false start, a value of C_6 of 72 a.u. [89]. The accuracy was probably not high since the result depends on the experimentally measured area to the power of $2\frac{1}{2}$, so errors are magnified, but the figure was consistent with the best quantal calculations. Attempts were made to reconcile this value of C_6 with the observed values of the second virial coefficient and the viscosity. Some progress was made with potentials less simple than the Lennard-Jones and the Kihara but no consensus was reached [90]. This can be seen from the papers at a Discussion of the Faraday Society at Bristol in September 1965 [91]. There R.J. Munn of that University made, in discussion, the suggestion that one problem might be simply that the experimental results for the viscosity at high temperatures were wrong [92].

A breakthrough came in 1968 when J.A. Barker and A. Pompe in Melbourne decided that this solution of the problem was the only way forward [93]. It was a bold step to take since there were two independent sets of measurements that agreed well. Trautz was the accepted authority in the field and his measurements up to 1000 K [35] led smoothly into those that Virgile Vasilesco made in Paris during the War, and which extended to 1868 K [94]. Little was known of this (Romanian?) physicist but his experiments seemed to have been well performed and gained acceptance because of their agreement with those of Trautz. The only disagreement came from results obtained in 1963 by Joseph Kestin at Brown University which were up to 2% higher than Trautz's, but which extended only to 550 K [95]. Barker and Pompe were encouraged in their decision by early knowledge of experimental work from Los Alamos that suggested that the accepted values of the viscosity of helium were too low, and before their paper was published they were able to add a 'Note in proof' to say that they now had had confirmation that new and higher values for argon were about to be published from Los Alamos [96]. These followed the next year [97]. R.A. Dawe and E.B. Smith in Oxford soon confirmed this revision of the accepted values with measurements up to 1600 K [98]. The errors of the older work were found to be large – up to 8% at 1900 K. Barker and Pompe combined the quantal calculation of C_6, the observed second virial coefficient, the viscosity to 600 K, and information from beam scattering at high energies, which probes the repulsive wall of the potential, to produce an algebraically complicated potential, but one that fitted all the established 'two-body' results. It had a collision diameter, d, of 3.756 Å, and a depth, ε/k, of 147.7 K. They calculated successfully the properties of the crystal by adding the triple-dipole term. Other three-body terms had been suggested, such as a three-body repulsive or overlap term [99], but they found no evidence that they were needed and later work has confirmed this simplification. It may well be that each of the other three-body terms is not negligible but that there is a mutual cancellation. The situation has not been explored systematically; scientists are as happy as anyone else to let sleeping dogs lie. After a little further refinement Barker and his colleagues decided that 147.7 K was too big and reduced the depth to 142.1 K, with the distances a little larger in compensation, $d = 3.361$ Å and $r_m = 3.761$ Å [100]. These figures were now based in part on the properties of the liquid, as modelled by a computer simulation, with allowance for the triple-dipole energy.

At this point another physical technique entered the picture. The bulk properties of matter are determined by the intermolecular forces, but the links are far from simple. The newly introduced technique – spectroscopy – probes the interactions more directly. It was known from mass-spectroscopic studies that the van der Waals forces lead to a small part of gaseous argon being composed of dimers, Ar_2, not of single Ar atoms [101]. Such dimers have a vibrational energy which is quantised, and so there are discrete bound states, each with a different amount of vibrational energy. Is it possible to observe transitions between such states and so obtain directly information about the pair potential? Such measurements had been made for nearly fifty years on chemically bound diatomic molecules, and had produced a mass of precise information. Unfortunately the Ar_2 dimer has no dipole moment and so transitions between different vibrational levels neither emit nor absorb infra-red radiation. Moreover the concentration of the dimer is low, less than 1% at 120 K and atmospheric pressure. Y. Tanaka and K. Yoshino at the U.S. Air Force Laboratory in Massachusetts overcame both difficulties; the first by observing the ultra-violet excitation of Ar_2 to a high electronic state, when the precise energy of the transition depends on the ground vibrational state that the excitation starts from, and the second by using long path-lengths in the gas by means of multiple reflections between parallel mirrors [102]. In theory a knowledge of the vibrational energy levels of the dimer tells one how wide is the 'bowl' of the potential as a function of the height above the minimum. Tanaka and Yoshino did not try to extract the information in this form but fitted a Morse curve, a sum of two exponentials, to their results. This is a curve that is appropriate for a chemically bound pair of atoms but not for what is now usually called a van der Waals molecule. They obtained a depth of the minimum, ε/k, of 132 K. Spectroscopists sometimes think of themselves as an élite and are apt to overlook old-fashioned measurements of gas imperfection or viscosity made by the 'rude mechanicals'. Their potential was totally at variance with the known values of the second virial coefficient. Maitland and Smith realised the value of the information in the results of Tanaka and Yoshino and made a proper 'inversion' of these to get the bowl as a function of energy, constraining their fitting, however, to satisfy the traditional information from the virial coefficient and the viscosity [103]. The potential that they obtained had $d = 3.555$ Å, $r_m = 3.75$ Å, and a depth of 142.1 K. It was virtually the same as that of Barker, Fisher and Watts [100]. A few years later E.A. Colbourn and A.E. Douglas in Ottawa obtained a better spectrum in which the rotational lines of the vibronic transition were resolved [102]. An inversion could now be carried out to extract even more information. This they did, reporting a well-depth with a claimed precision of one part in 10^4 but, being spectroscopists, they again did not try a check by computing the bulk properties of the gas. Their potential, $d = 3.347$ Å, $r_m = 3.75$ Å, and $\varepsilon/k = 143.2$ K, was no advance on that of Barker or Smith and their associates. There have been a few improvements since then but the problem of the argon potential was essentially solved by 1971.

One further method of attack came just in time to help with the refinements. We have seen repeatedly how the determination of intermolecular potentials from bulk physical properties has been hampered by the fact that the only feasible routes were from the potentials to the properties. It was therefore always necessary to guess at model forms of potential, calculate the properties, and see if these agreed with what had been measured. In a Popper-like way this technique could show that a model was wrong, but it could never give assurance that it was correct, however good the apparent fit to the experiments. For the spectroscopic measurements there was an established inverse route, from the properties to the potential, or at least to some features of the potential. It had been known at least since 1950 that there is also, in principle, an inverse route from the second virial coefficient to the pair potential [104]; this seems to have been first noticed publicly by J.B. Keller and B. Zumino in 1959 [105]. The coefficient can be written, from eqn 4.39,

$$B(T) = -(2\pi N/3)e^{\varepsilon/kT} \int_0^\infty (r_+^3 - r_-^3)e^{-x}\mathrm{d}x, \qquad (5.30)$$

where $x = [u(r) + \varepsilon]/kT$, and r_+ and r_- are the outer and inner separations in the potential bowl for all negative values of $u(r)$. In the repulsive region of the potential r_+ is taken to be zero. This expression has the form of a Laplace transform of $(r_+^3 - r_-^3)$ and, since Laplace transforms can be inverted, there seems to be here a way of obtaining directly $(r_+^3 - r_-^3)$ as a function of x and so of the energy u. This route was first followed in practice for the simple case of helium for which the negative region of $u(r)$ is so small that it was possible to 'correct' for its presence and so obtain directly the repulsive separation as a function of energy [105]. Unfortunately, for argon, and for other substances for which the attractive part of the energy is at least as important as the repulsive, the direct inversion of the Laplace transform proved to be unstable; it would require a precision of one part in 10^4 in the virial coefficient for the method to succeed [106]. All was not lost, however, since it proved possible to find empirically ways of suppressing the instability and obtaining useful results [106]. It has also been possible to devise an iterative scheme for inverting the viscosity and other transport properties [107]. The potentials so obtained confirmed those arrived at by the older and less direct methods in 1971. These inversions have also proved useful for other less exhaustively studied systems [108].

Little use was made in these determinations of quantal calculations of the repulsive branch of the potential which arises from the overlap of the electronic orbits

Table 5.6

	$d/\text{Å}$	$r_\text{m}/\text{Å}$	$(\varepsilon/k)/\text{K}$	$C_6/\text{a.u.}$
1953	3.41	3.82	120	110
1977	3.36 ± 0.05	3.76 ± 0.02	143 ± 1	65

around each of the atoms. Such calculations are difficult because of the correlation of the motions of the electrons arising from their Coulombic repulsions. There is no difficulty of principle but the computational problems are formidable. By the 1970s the best calculations were approaching the same order of accuracy as the determinations from spectroscopy, beam scattering and from the physical properties of the dilute gas, but they did not displace these properties as determinants of the potentials of choice [109].

Thus, after a long and tortuous process, the argon problem was solved by the early 1970s. It is interesting, Table 5.6, to compare the accepted values of 1953, that is those of the Lennard-Jones (12, 6) potential, with the consensus of 1970–1977. The new potential could account, almost always within experimental error, for such molecular properties as the spectrum of the dimer and the beam-scattering cross-sections, for the macroscopic two-body properties such as the second virial coefficient and transport properties (of which only the most important, the viscosity, has been discussed here), and for the structural and thermodynamic properties of the liquid and solid when augmented with the triple-dipole term. One nagging doubt remains. This three-body term deals well with the difference found between the observed crystal energy and third virial coefficient and the values calculated from the now well-established pair potential, but many apparently reliable quantal calculations and some spectroscopic evidence suggests that the three-body exchange energy is equally important and of the opposite sign. The agreement obtained with the triple-dipole term alone seems too good to gainsay, and is provisionally accepted, but the doubt remains [110].

Argon is not the most important molecule that we encounter, indeed it must be one of the least important for most physicists and chemists. It was something of an accident, born of convenience, simplicity, and habit, that made it the chosen test-bed for experiments and theories on intermolecular forces. For twenty years the 'argon problem' attracted much of the effort of a relatively small but dedicated group of physical chemists. Many of them made important contributions in other fields also, principally in statistical mechanics, but they returned time and time again to argon. The wider group of physicists and chemists were often not in sympathy with this obsession. One senses something almost of a mild exasperation in the opening and closing papers at the Faraday Discussion on intermolecular forces of 1965. These were given by H.C. Longuet-Higgins and C.A. Coulson respectively, both of whom worked primarily in quantum mechanics, and both of whom tried to raise the discussion to wider issues [111]. Nevertheless the solution of the problem of argon was a necessary step in the quantitative study of intermolecular forces, and those who worked on the problem were certainly not wasting their time on a triviality.

It is one of the comforting self-delusions to which some academic scientists are prone, to believe that once a problem is solved in principle it is straightforward to extend that principle to other applications, or, if not entirely straightforward, then that such extension is unrewarding work that can safely be left to others. It was natural to feel that with the satisfactory determination of the argon potential the field had lost its most exciting moment. Those who had laboured hard here did not put the same effort into other practically more important cases although the lessons that had been learnt from argon could be and usually were applied to the other inert gases. Beyond argon and the inert gases lie the diatomic molecules, hydrogen, nitrogen, oxygen, etc., and then the polyatomic molecules such as the hydrocarbons, the polar molecules such as hydrogen fluoride and hydrogen chloride, and, more important, ammonia and water. Beyond these lie the even more complex problems of polymers, micelles, colloids, and the interactions in biologically important systems. These fields are immense and much work is now being done, but progress towards their solution (in the argon sense) is slow and necessarily far from elegant. Here, however, we shall shelter behind the delusion that the accurate determination of the force between two argon atoms is the breakthrough 'in principle', and not pursue the complications of the real world. Indeed, the writing of the history of the interaction of more complicated molecules cannot yet be done, for the whole field is still one of intermittent action, tentative conclusions and innumerable loose ends. Only one example will be given, that of water whose importance justifies the possibly premature attempt. One of the byways of the interaction of more complex, and indeed of macroscopic entities, is, however, also worth exploring since it led to a resolution of the old problem of action-at-a-distance in this field. We return to that subject after the discussion of water.

5.3 Water

Water is unique in its importance and in its properties. No other substance has been the subject of so much study and speculation, nor has any been harder to understand at a molecular level. The contrast with argon could not be greater, for in studying argon we are studying matter and its cohesion at its simplest, the very essence of the problem before us; in studying water we are studying a substance so atypical that every inch of progress is peculiar to it and often has no relevance to any other substance. The force between a pair of argon atoms is a function of one variable, the separation of the nuclei; the force between a pair of water molecules is a function of their separation and of the five angles needed to describe their mutual orientation. In saying that five angles are needed we are presuming that we know that the molecule is H_2O and that it has a triangular shape. The constitution was well established by the start of the 20th century but the shape was not. Kossel was arguing for a linear structure in 1916 [112], but a symmetrical linear structure with the central oxygen atom equidistant from each hydrogen is not compatible with what was then known of the infra-red absorption spectrum of the vapour, which required that the molecule has three different moments of inertia [113], nor with the fact that the molecule has a strong dipole. The evidence for this dipole became available early in the century. In 1901 Bädeker measured the 'dielectric constant' (now called the relative permittivity) of the vapour as a function of temperature [114]. His range was small, from 140.0 to 148.6 °C, but it was sufficient to show a rapid change with temperature. He did not then know how to interpret this result and fitted his experimental points to a function of the form $(a + bT)$. Langevin and Debye had yet to show that the appropriate form was $(a + c/T)$ where, as we have seen, the parameter c is proportional to the square of the dipole moment, μ. This interpretation of his result was made by J.J. Thomson in 1914 and by Holst in 1917 [115], who derived from it values of the dipole moment of 2.1 and 2.3 D respectively [116]. Holst sought also to determine the moment by seeing what value was needed to fit the second virial coefficient if this was to be interpreted in terms of Keesom's model of a hard sphere with a dipole at its centre; this calculation gave him a moment of 2.62 D. A more reliable value became available two years later when Jona measured the dielectric constant from 117 to 178 °C and showed that this led to a value of μ of 1.87 D [117]. The value accepted today is 1.84 D. It was possible that the molecule could have been linear but unsymmetrical and so have had a non-zero dipole moment and only one moment of inertia, but this seemed unlikely, and Debye claimed in 1929 that such a structure would be unstable [118].

The x-ray diffraction pattern of the crystal shows only the position of the oxygen atoms. These are arranged in an open structure with each atom having four nearest neighbours. William Bragg [119] interpreted this structure in 1922 as one composed of negatively charged oxygen ions, with the hydrogen ions, or protons, at the midpoints of the lines joining them. No doubt he was attracted to this interpretation by his son's success in determining the structure of the crystal of common salt and showing that it was formed not of NaCl molecules but of Na^+ and Cl^- ions, a result that upset some of the more traditionally minded chemists. For water, however, Bragg's proposal was a step too far; the ice crystal is formed of discrete H_2O molecules but these are orientated so that the hydrogen atoms are along the lines joining the oxygen atoms, as he surmised.

This structure, with the OH bond of each molecule directed towards the O atom of a neighbouring molecule, was consistent with what the chemists had deduced from other evidence. In 1912 T.S. Moore [120] showed that the degree of ionisation of aqueous solutions of amines could be understood if there were a weak bond or attraction between the H atom of a water molecule and the N atom of, for example, trimethylamine. This link could be represented N $\cdots$ H$-$O, where the full line is the covalent bond in the water molecule (the second bond not being shown) and the dashed line is the weaker attraction between the H and N atoms. This link could be be understood if there were a positive charge on the H atom and negative one on the nitrogen atom. The next year P. Pfeiffer suggested a similar link within one molecule [*innere Komplexsalzbindung*], in this case between the O atom of a carbonyl group and a nearby HO group in the same molecule [121]. Similar ideas arose, apparently independently, a few years later at Berkeley, first in an unpublished undergradate thesis of M.L. Huggins and then in a paper by Latimer and Rodebush [122] that is often taken as the first authoritative account of what now came to be called the 'hydrogen bond' [123]. The strength of this 'bond', typically about 20 kJ mol^{-1}, is large compared with the thermal energy, kT, at room temperature, 2.5 kJ mol^{-1}, and with the minimum potential between two argon atoms, 1.2 kJ mol^{-1}, but much smaller than that of a chemical bond, for example, 460 kJ mol^{-1} for the mean energy of the OH bond in water. Its origin is therefore primarily a classical electrostatic attraction between the partial positive charge on the hydrogen atom, which is here a proton with two electrons to one side of it and only partly shielding it, and a partial negative charge on the O, N, or F atom to which the bond is directed. The large size of the hydrogen-bond energy, compared, say, with the Ar–Ar energy, means that useful quantal calculations and estimations of the electrostatic interactions can be carried out more easily for this complicated molecule and its dimer than for the apparently simpler inert gases. This advantage goes a little way in compensating for the greater number of variables needed to define a potential.

A landmark was reached in 1933 with a long paper from Bernal and Fowler [124] on the structure and physical properties of liquid water which was published in the first volume of what soon came to be accepted as the leading journal for work in this field, the American *Journal of Chemical Physics*. It was agreed that in ice the oxygen atoms are arranged in a tetrahedral structure, that the angle of the HOH bonds in the isolated molecule (104.5°) was close enough to the tetrahedral angle $(2\cos^{-1}(1/\sqrt{3}) = 109.5°)$ for the hydrogen atoms to lie along the O–O lines, but there was no direct evidence for the precise position of the hydrogen atoms. Bernal and Fowler rejected Bragg's ionised structure and argued that the infra-red spectrum of the solid was close enough to that of the single molecule for it to be more likely that the H_2O molecule retained its integrity in both ice and water (Fig. 5.2). They interpreted the x-ray diffraction pattern of the liquid in terms of the then novel

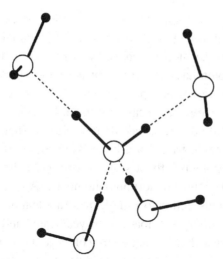

Fig. 5.2 A perspective sketch of five water molecules in ice. The oxygen atoms, shown by the large open circles, form a tetrahedral array in which each molecule has four nearest neighbours. Each of the hydrogen atoms, a small closed circle, is bonded to an oxygen atom, as is shown by a full line, and each of these bonds is directed towards another oxygen atom, so as to form a 'hydrogen bond' with it, as shown by a dashed line. The central water molecule is therefore linked to its four neighbours by two donor hydrogen bonds and two acceptor bonds. The arrangement shown in one of the many ways of assigning the hydrogen atoms to the O—O lines and in practice the molecules flip rapidly from one configuration to another in ice at the melting point, only becoming locked into one of the many alternative arrangements at low temperatures.

angle-averaged pair distribution function (see below, Section 5.5) and showed that its structure was predominantly of the quartz type, with a small fraction of the tridymite (or wurtzite) form near the freezing point, but that it changed into a more close-packed structure at higher temperatures. They were thus able to rationalise the occurrence of the density maximum at 4 °C and were able to give convincing accounts of the magnitude of the latent heat and a host of other properties, by supposing that the intermolecular potential was of a Lennard-Jones (12, 6) type with the addition of the electrostatic interaction of three discrete charges on each molecule, one positive one on each of the hydrogen atoms and a double negative charge on the far side of the oxygen atom. This was not quite consistent with the four charges arranged tetrahedrally that they used to justify the structure of ice and liquid water (Fig. 5.2). There are many different ways of orientating the water molecules in such a tetrahedral lattice, even with the restriction that there is only one hydrogen atom on each O—O line, and, unless this disorder is removed on cooling the crystal, there will be a residual entropy at 0 K. It was found that there was such an entropy, and Pauling showed in 1935 that its magnitude was accounted for by this disorder in the hydrogen bonds [125].

It is surprising that throughout the 1920s and 1930s there was no calculation of a second virial coefficient for a Lennard-Jones (n, m) potential with a point dipole at its centre. This natural advance was made by W.H. Stockmayer, then at M.I.T., in 1941 [126], and this potential is now known by his name. He chose a repulsive index, n, of 24 and fitted it to the then accepted values [127] of the second virial coefficient of water. Margenau had argued the case for including a quadrupole in the potential but had supported it only by crude calculations [128]. When Stockmayer's results became available Margenau tried again, but was constrained to use the angular form of the dipole–dipole potential for a single quadrupole–dipole interaction [129]. (The water molecule has three quadrupole moments.) A more correct angular integration, but with still a restriction to a single quadrupole moment that was supposed to have cylindrical symmetry, was made a few years later, and a 4-charge model was chosen to be consistent with the lattice energy of ice and the existence of its residual entropy [130]. All this work was undermined, however, when the quantal calculations of the electronic structure of the water molecule became sufficiently reliable for the resulting values of the three quadrupole moments to be trusted, and for the spectra to give a value for the average of the moments. Glaeser and Coulson [131] calculated the three moments about each of the axes of the molecule, and the mean of their values was soon confirmed from the spectra which yield the average $\langle r^2 \rangle$, where r is the distance of each electron from the centre of mass [132]. A more direct spectroscopic determination of the three moments followed a few years later [133]. These results were not consistent with what had been assumed in the calculations of the second virial coefficient but they confirmed, at least qualitatively, the 4-charge models.

From this time forward there were two different lines of advance. One group, who were interested primarily in the structure of liquid water, took advantage of the increasing power of computers to simulate its structure and calculate its properties. For this they needed a two-body intermolecular potential, but since an important part of this is the energy arising from the polarisation of one molecule by the electric field of its neighbour, and since this energy is far from pair-wise additive, the potentials that they devised to fit the structure were not true pair potentials but 'effective potentials' suitable for the problem in hand. There were a series of these, typically of the form of a Lennard-Jones (12, 6) potential centred on the oxygen atom with 3 or 4 charges appropriately distributed [134]. These were generally successful in reproducing many of the structural and thermodynamic properties of the liquid although usually not so successful with dielectric and transport properties. It was not surprising that an attempt to use one of these effective potentials to calculate a true pair property, the second virial coefficient of the gas, failed by a factor of two [135]. The aim of some of this work was to lead to molecular models of water that could be used in simulations of systems of biological interest [136], but the status of effective potentials is never wholly clear and these endeavours attracted fewer devotees after the 1980s.

The second line of advance was a spectroscopic attack on the water dimer, $(H_2O)_2$. We have seen that spectroscopy made a late but not negligible contribution to the problem of the Ar–Ar potential. With water, however, the position was different. Here the true pair potential, a function of six variables, can never be determined from the macroscopic properties alone. Fortunately both water and its dimer are polar and have information-rich microwave and infra-red spectra. For some years the spectra of what are usually called 'van der Waals molecules' have been studied in detail and have proved a powerful source of information on the potentials of some molecular pairs. Originally these pairs were naturally chosen for their ease of study and interpretation, and so told us a lot about interactions that were, however, of only specialised interest, such as Ar–CO [137]. The spectra are at their simplest if only one of the pair, CO in this example, has a dipole, and if the molecules are cooled to low temperatures so that they are in low vibrational and rotational states. This is brought about by expanding the mixed gases through a pinhole into a vacuum when a high-speed molecular beam is produced in which the random translational kinetic energy of the molecules and molecular clusters, which is a measure of their temperature, is converted into the ordered motion of the stream. Soon the ambitions of those working in this field went beyond the simplest cases and the water dimer was tackled. The first infra-red studies were inconclusive, but microwave spectroscopy, which measures transitions between rotational levels, showed more promise [138]. Later work involved highly resolved infra-red spectra and their detailed analysis. The culmination of this work was the determination of the pair potential of 'heavy water', D_2O, by R.J. Saykally in Berkeley and C. Leforestier of Montpellier, and their colleagues in 1999 [139]. Their potential was based on one originally derived from quantal calculations [140] and has no less than 72 parameters. It is a sign of the times that these were not given in the body of the paper but were listed on the Internet. (Heavy water has almost the same intermolecular potential as common water but a spectrum that is easier to interpret.) This impressive potential has the great virtue of yielding good values of the second virial coefficient, a delicate test that spectroscopists had often previously ignored. It is possible to prepare molecular beams with different ratios of single molecules, dimers, trimers, etc., by adjusting the pressure of the gas before expansion, and the size of the pinhole. In this way Saykally and his colleagues have obtained and analysed also the spectra of clusters containing three, four and five water molecules, but naturally the interpretation of these has not been carried out in the same detail as that of the dimer [141]. The power of these new spectroscopic techniques is only now being extended to other molecular systems and the exuberence of the field is shown in the increasing length of each of the three issues of *Chemical Reviews* that have been devoted to the subject of van der Waals clusters [137].

How far do these beautiful spectroscopic studies help us to understand the co-hesion of liquid water or of other liquids for which it is possible to determine the multi-dimensional potential surfaces of the dimer? In 1994 D.H. Levy addressed this question at the end of the Faraday Discussion on van der Waals molecules, and concluded that there was still a gap in our knowledge that we could not yet fill but that we were making progress [142]. The success of work on the water dimer confirms this but in 2001 there seems to be still some way to go.

5.4 Action at a distance

The natural philosophers of the late 17th and 18th centuries were much concerned with the metaphysical problem of action at a distance. They settled the matter by accepting that gravitational attraction was too successful a theory to be denied, but that there was no point in trying to understand what mechanism gave rise to it. Tacitly, and with less whole-hearted conviction, most came by the end of the 18th century to accept that cohesion is the result of attractive forces between some unknown basic particles out of which matter is formed. Laplace and his school became the most successful exponents of this idea. The counter-revolution started when it was found that electric and magnetic forces between moving charges or currents did not act along the lines joining the bodies in question. In Britain, Faraday's lines of force filled all space and were enshrined in mathematical form by Maxwell. William Thomson tried to replace the hard massy atoms by vortices in the aether. The current of ideas began to flow back again towards a Laplacian picture with the successes of the kinetic theory of gases from the middle of the 19th century onwards. By this time many scientists had lost interest in the metaphysical problem and were content to build theories as close as they could to the experimental facts. Maxwell was one who retained a concern with the question and was in a unique position to see the merits and defects of the kinetic model that relied on an apparent action at a distance between particles. In a Friday evening Discourse at the Royal Institution on 21 February 1873 he took the same pragmatic view that Newton had taken in his 'Query 31':

If we are ever to discover the laws of nature, we must do so by obtaining the most accurate acquaintance with the facts of nature, and not by dressing up in philosophical language the loose opinions of men who had no knowledge of the facts which throw most light on these laws. [143]

He outlined the arguments in favour of and against the idea of action at a distance, laying most emphasis on Faraday's view that even where there appears to be only empty space there can be lines of force with elastic properties. Had he been questioned closely it is almost certain that he would have prefered 'field' forces to simple 'action at a distance' but he is careful to balance the arguments and he ends cautiously: "Whether this resolution is of the nature of explication or complication, I must leave to the metaphysicians."

In the early 20th century there was little interest in the problem among those who were trying, without success, to determine the nature and form of the cohesive forces. They tacitly assumed that Coulombic interactions, like gravitational, acted at a distance, and that there was little to be gained by asking how they did it. When London found the quantal origin of the attractive forces then it was seen that they were electrical, and that they depended on the matching of the phases of the oscillating dipoles. It was assumed, therefore, although rarely explicitly stated, that they were propagated at the speed of light. The speed of light is 'large' and the separation of molecules in a solid or liquid is 'small', and so it was not thought necessary to raise the question of the time taken for the transmisssion of the interaction. The measures of largeness and smallness could easily have been quantified, and perhaps were, although never prominently. The relevant energy is approximately that of the ionisation energy, I, of the molecules involved, for example, 15.76 eV for argon. The distance at which one might have to ask about the time taken for the transmission of the interaction is therefore of the order of $hc/2\pi I$, where h is Planck's constant and c is the speed of light. This distance is 125 Å for argon and is so much larger than the effective range of the force, about 6 Å, that it is irrelevant.

Soon, however, there arose a situation in which the distance was relevant. During the 1930s and throughout the War there was a group in the Phillips Laboratories at Eindhoven who studied the problem of colloid stability. Colloidal particles are sometimes described as mesoscopic; they are small compared with the macroscopic lengths that characterise the surface behaviour of materials (for example, the capillary length of water at 3.8 mm) but large compared with the size of molecules. A typical colloidal particle might have a diameter of 1 μm, although the range of sizes and shapes is large. The forces between such particles in a liquid suspension are complicated since their surfaces are generally charged and these charges interact with each other and induce other electrostatic forces in the liquid. A major component of the forces between the particles is, however, the sum of the attractive dispersion forces between all the molecules in each. Once Wang and London had shown that the potential of the dispersion force fell off as the inverse sixth power of the separation of the molecules, with a coefficient that could be calculated, then it was a straightforward matter to find, by integration, the total dispersion force between two spherical colloidal particles. Prompted by London, such a calculation was made in 1932 by Kallmann and Willstaetter in Berlin [144], and also by Bradley in Leeds, who tried to measure directly the force of adhesion between two quartz spheres [145]. The best-known and most widely cited of such calculations was that made by H.C. Hamaker of the Phillips group and reported to the van der Waals centennial meeeting in Amsterdam in 1937; his name is now given to the constant or parameter that describes the integrated effect [146]. Bradley had considered attractive potentials proportional to r^{-m}, although he recognised that $m = 6$ was the appropriate value. Hamaker restricted himself to the sixth power. His colleagues continued their study of colloidal systems during the War, paying particular attention to the electrical forces and their modification in the presence of dissolved electrolytes. In the course of this work [147], J.Th.G. Overbeek came to the conclusion that the dispersion force between mesoscopic particles was much weaker than that calculated by integrating over all the inverse sixth-power potentials, as Hamaker had done. He thought that at large distances the dispersion force might be weakened because it was not an instantaneous action at a distance but must be transmitted at the speed of light. He put this point to his colleagues H.B.G. Casimir and D. Polder who confirmed that his hypothesis was correct [148].

It was not easy to understand this 'retardation' of the force since fourth-order perturbation theory is needed, in contrast to London's theory which requires only second order. Many routes to Casimir and Polder's result have now been found but none is simple. The physical origin can again be put into words in terms of Drude's model. The oscillating dipole in the first molecule interacts, in phase, with the oscillating dipole in the second, and it is this interaction that produces the r^{-6} potential at short separations. When the separation is large enough for the time taken for the signal to be transmitted from one molecule to the other to be an appreciable fraction of the reciprocal of the frequency of oscillation of either dipole then the oscillators can no longer remain in phase. The lag that ensues results in a weakening of the interaction and leads to a dispersion potential that falls as r^{-7}. The effect can be observed directly only if one can measure the force of attraction between mesoscopic or macroscopic bodies that contain a sufficiently large number of molecules for the force to be appreciable at large distances. A strictly quantitative study would then have to deal also with the fact that the sum over the two-body forces is an inadequate way of dealing with condensed matter. A treatment that encompassed this problem also was devised by E.M. Lifshitz in Moscow in 1954 [149]. He considered electrical fluctuations in bulk matter and did not break these down into their molecular components.

The experimental hunt for these retarded forces started soon after Casimir and Polder's paper of 1948. In the Institute of Physical Chemistry in Moscow, B.V. Deryagin and his student I.I. Abrikosova studied the force of attraction between a glass hemi-sphere and a flat plate, and found a force that fell off with l, the size of the gap, as l^{-3}, as required by Casimir and Polder's potential [150]. Other early experiments were attempts to study the adhesion of bodies 'in contact', but that is an ill-defined state and they were not very informative [151]. One cannot polish glass to produce a surface without irregularities of at least 100 Å, and so useful quantitative results could be obtained only for gaps of the order of 1000 Å or more. At this distance the force is weak but fully retarded and Abrikosova and Deryagin were soon claiming good agreement with theory [152]. Similar and contemporary experiments by Overbeek and his student at Utrecht, M.J. Sparnaay, led to appreciably stronger forces than would be expected even without retardation, which they did not mention in their first note [153]. Deryagin ascribed this failure to their inability to remove all electric charges from the surfaces and to a lack of sensitivity of their apparatus [154]. Independent measurements at Imperial College in London, with an apparatus similar to that of Overbeek, agreed broadly with Deryagin's results [155], which were also confirmed later by further measurements at Utrecht [156].

The real advance in technique came some years later when David Tabor in Cambridge replaced the glass surfaces with cleft sheets of mica bent into the shape of two crossed cylindrical surfaces. Split mica is smooth on an atomic scale over a length of the order of a few millimetres, and so the cylinders could be brought to within 15–20 Å. This reduction of working distance not only greatly increased the strength of the force to be measured but also allowed him and his students to explore the transition from the normal to the retarded force [157]. They were able to show that below about 100 Å the force is normal and that above about 200 Å it is fully retarded, a transition range that is consistent with the transmission of the interaction at the speed of light. This powerful technique was soon extended by spreading layers of other materials on the mica sheets, and by immersing the cylinders in water and in solutions. In this way much has been learnt by direct experiment of the cohesive forces in many systems of great physical, technological and biological interest [158].

With the work of Deryagin, Overbeek, Tabor and their associates, cohesive forces have been measured at what Laplace might just have recognised as 'sensible distances'. As so often in scientific arguments, both sides in the action-at-a-distance debate have been proved right. Descartes, Locke, Newton and Leibniz have all been vindicated in thinking that 'a body cannot act where it is not'; an electromagnetic mechanism has been found for the transmission of cohesive attraction from one body to another at the speed of light. Yet those innumerable scientists from Newton and Freind onwards who claimed that knowledge would be best advanced by ignoring such metaphysical niceties have also been amply justified. It is only a rare problem in physics, chemistry or biology for which the retardation of the dispersion forces must be taken into account. The position parallels that with the gravitational force where practical and theoretical astronomy flourished for centuries before any plausible mechanism for the transmission of this force could be devised [159].

5.5 Solids and liquids

We have seen that the investigation of intermolecular forces has been a two-way process. The experimental study of matter as gas, liquid and solid provides the evidence for the existence of the forces and, in principle, a means of measuring them but, conversely, this measurement can be carried out only if we have already a good theoretical picture of what properties of matter are implied by a given system of intermolecular forces. So far in this chapter we have looked only at the problem of the form and strength of the forces, using as evidence mainly the simply interpretable properties of the gas at low densities. We must now complete the picture by seeing how a knowledge of these forces was used in the 20th century to interpret the properties of solids and liquids.

During the 18th century, from Newton to Laplace, the study of the forces was primarily a study of their manifestation in the properties of liquids and, in particular, in those surface properties that result in capillarity. In the early and middle of the 19th century attention switched to the elastic properties of solids and to the propriety of interpreting these in terms of the attraction of Laplacian particles. Towards the end of the century gases and, to a lesser degree, liquids came to the fore, and in the early years of the 20th century it was realised that it was the properties of gases at low densities that provided the most direct and unambiguous link to the force between a pair of molecules. This realisation would doubtless have come sooner had the relevant properties of gases been easier to measure with a useful accuracy. Solids then played a minor role and one that was blighted by ignorance of the fact that classical mechanics, although adequate for most gases and liquids, is not so appropriate for solids. Liquids were generally ignored by the leaders of the field since they recognised the imperfections of theory in this area. Lesser lights, however, wrote innumerable papers on their physical properties in the early years of the 20th century and made many attempts to interpret these in terms of the properties of the molecules. The simple picture of van der Waals and his school had given a strong impetus to this part of the field. It had led to the best estimates yet of the range and the strength of the intermolecular forces and had established in the minds of most scientists that all three states of matter should, in principle, be explicable in terms of the same one set of molecules and the forces between them. But it had no rigorous foundation in the newly developing subject of the statistical mechanics of Boltzmann, Gibbs, Einstein and Ornstein, and so the simple picture could not be developed further.

With the establishment of the quantal theory of crystals in the 1920s and 1930s the way was apparently open again for the properties of non-metallic solids to contribute quantitatively to the study of intermolecular forces. (Metals raise other problems, outside the scope of this study.) The most useful properties of the inert gas crystals were, as in the classical picture, the lattice spacing and the crystal energy, which are related reasonably directly to the separation at the minimum of the pair potential and to its greatest depth. These properties are simplest to interpret if available for the crystal at zero temperature [160], and since they change little with temperature, such extrapolated values are easily found. As we have seen, these properties were used by Lennard-Jones in the 1920s and 1930s, and by Corner in the 1940s (among others) and became a part of the evidence that the (12, 6) and (exp, 6) potentials were apparently good representations of the inert gases in both gas and solid states. Later work showed the inadequacy of that conclusion [161].

The use of other mechanical and thermal properties is more difficult. Some obvious ones, like the strength of a solid, cannot be used since, even for a single crystal, the strain that occurs before breakage is too complicated to be interpreted directly in terms of the intermolecular forces [162]. Other properties such as the coefficient of thermal expansion and the heat capacity vanish at zero temperature and an interpretation of their values at non-zero temperatures needs a knowledge of the modes of vibration of the atoms in the crystal which, in turn, depend on the intermolecular forces. This interpretation is a non-trivial quantal problem to which the early and partial solutions of Einstein, Debye and of Born and von Kármán [163] were not a sufficient answer. It was inevitable that measurements of the heat capacity were used more to refine our knowledge of the frequency spectrum of the lattice vibrations than as a tool for studying the intermolecular forces, although some did attempt the second task [69, 164].

The elastic constants of a crystal are a more direct route to the intermolecular forces and, in particular, those at zero temperature are related to the curvature of the potential near its minimum. There are, however, two experimental problems here. The first is that the two most useful tools for measuring these constants for a material as difficult to work with as solid argon are the speed of sound and the inelastic scattering of neutrons. Both measure the adiabatic coefficient not the more useful isothermal coefficient. (The same distinction is found in liquids and gases and led to Laplace's correction of Newton's calculation of the speed of sound in air.) The second experimental difficulty is that the elastic constants change rapidly with temperature and so it is hard to extrapolate them to zero temperature. The compressibility of solid argon at its triple point of 84 K is nearly three times as large as the extrapolated value at zero temperature. Both difficulties can be overcome if measurements can be made at sufficiently low temperatures, generally 10–20 K, since the extrapolation becomes easier, and the difference between the adiabatic and isothermal coefficients vanishes at zero temperature. Barker and others used such results as were to hand but the really useful measurements were not made until the question of the argon potential had been virtually settled. In 1974 a team at the Brookhaven National Laboratory measured the elastic constants of argon at

10 K by using neutron scattering [164]. Argon has a cubic crystal and so has three independent elastic constants, c_{11}, c_{12} and c_{44}. The reciprocal of the isothermal coefficient of compressibility (or bulk modulus), κ_T^{-1}, is a weighted mean of the first two;

$$\kappa_T^{-1} = -V(\partial p/\partial V)_T = (c_{11} + 2c_{12})/3. \qquad (5.31)$$

The Brookhaven results were

	c_{11}	c_{12}	c_{44}	$\frac{1}{2}(c_{11} - c_{12})$	
^{36}Ar (10 K)	42.4 ± 0.5	23.9 ± 0.5	22.5 ± 0.1	9.3	kbar

These figures imply a value of κ_T^{-1} of 30.1 kbar which is a little larger than a contemporary directly measured value of 28.6 kbar at 4 K [165].

The question that naturally arises is what do these figures tell us about the hotly debated problems of the 19th century of the stability, isotropy, the Cauchy relations and the Poisson ratio of the crystal (see Section 3.6). The first is no problem; stability requires only that $c_{11} > c_{12} > 0$, and these inequalities are amply satisfied. A cubic crystal has a certain isotropy in the sense that a spherically symmetrical or hydrostatic stress induces a spherically symmetrical strain, but at a more subtle level it may be anisotropic. The elastic constants that govern the two possible shear modes of deformation are c_{44} and $\frac{1}{2}(c_{11} - c_{12})$ and it is seen that these are not equal. The Cauchy relation for a cubic crystal is $c_{12} = c_{44}$, and this is close to being satisfied. Poisson's ratio for the polycrystalline solid, extrapolated to zero temperature, had been measured in 1967 and was found to be 0.253 ± 0.006 [166], that is, it has the value of $\frac{1}{4}$ deduced for an isotropic material. The ratio for xenon is similar, and those for neon and krypton about 0.27. A neo-Laplacian could not ask for more! A Poisson's ratio of $\frac{1}{4}$ is consistent only with $c_{11} = 3c_{12} = 3c_{44}$, and the Brookhaven results for a single crystal do not satisfy the first of these equations. Thus the polycrystalline material seems to have a gross isotropy that is not present in the individual crystal. If we return to the theoretical criteria that Born and his predecessors established as the conditions to be satisfied for Cauchy's relation to hold then we see that argon would conform to them only if we were justified in using classical mechanics and if we could neglect the three-body term in the intermolecular energy. In practice we cannot do this. It seems as if the effect of the three-body term on the elastic constants is similar to its effect on the crystal energy, about 7% in the difference between c_{12} and c_{44}, but the difference here seems less important since we are not aiming at so high an accuracy.

The properties of the inert-gas solids made, in the end, a useful contribution to the determination of the two- and three-body potentials, but with liquids the position was reversed; they were borrowers from, not contributors to, the stock of knowledge of the potentials. The phrase 'theory of liquids' is used to describe the calculation of structure and macroscopic properties of simple liquids from a knowledge of their intermolecular potentials. Its history from the early years of the 20th century until about 1970 has been a curious one [167].

A portion of liquid at equilibrium and well removed from its surface and its bounding solid walls is both isotropic (that is, the same in all directions) and homogeneous (the same at all points) on a macroscopic scale, that is on a scale of, say, 1000 Å or more. On a microscopic scale of 1–20 Å it is neither isotropic nor homogeneous at any instant of time, but again has both properties if an average is taken over an interval of greater than about 1 ns. We must ask, therefore, in what sense a liquid can be said to have a structure, and how can that structure be observed. The answer, briefly mentioned at the opening of Section 5.2, is found by considering any one molecule and asking how, on average, the other molecules are distributed around it. If the molecules are spherical, as in argon and as will be assumed here, then this distribution is again isotropic; it has spherical symmetry. It is not, however, microscopically homogeneous. The average local density is a function of the distance from the first or test molecule. If we take an element of volume $d\boldsymbol{r}$, at a distance $r = |\boldsymbol{r}|$ from the test molecule that is large compared with the range of the intermolecular force, then the chance of finding another molecule with its centre in $d\boldsymbol{r}$ is $(N/V)d\boldsymbol{r}$, where there are N molecules in a total volume V. The ratio (N/V) is the number density and is denoted n. If the distance r is within the range of the intermolecular force then the chance may be greater or less than this random value. The ratio of this chance or probability to the random value is called the radial or pair distribution function and is denoted $g(r)$. We can infer at once some of the characteristics of this function. If r is small compared with the size of the molecule then $g(r)$ is zero; we cannot have two molecules with their centres in the same or nearly the same place. If r is close to the distance, r_{m}, at which the pair potential $u(r)$ has its minimum then $g(r)$ is larger than unity, both because the attractive potential makes it more likely that two molecules will be close together (the same effect that makes the second virial coefficient negative at most accessible temperatures) and because the packing of spherical molecules in a liquid, at a density not much above that of a close-packed solid, requires that each molecule is surrounded by a 'shell' of up to 12 nearest neighbours. This packing effect is equally strong in a dense fluid of hard spheres without attractive forces when, as we shall see, it can be interpreted as the consequence of an indirect 'potential of average force'. Just beyond this shell $g(r)$ dips below its random value of unity, and may then show weaker oscillations until it finally reaches the random value of unity, as r becomes infinite (Fig. 5.3).

The pair distribution function, at a given pressure and temperature, is a function of only one variable, the separation, r, of two points in the liquid one of which contains the centre of a molecule. It is the simplest measure of the structure of a liquid; it generally tells us all we need to know, and it is experimentally accessible.

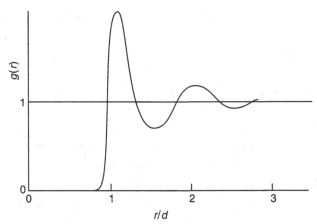

Fig. 5.3 A typical pair distribution function in a liquid, $g(r)$, as a function of the separation, shown here in units of the collision diameter, d.

It is, however, not the only measure. We can ask for the probability of finding three molecules with their centres in $d\mathbf{r}_1$, $d\mathbf{r}_2$ and $d\mathbf{r}_3$ and how this probability is related to its random or long-range value, $n^3 d\mathbf{r}_1 d\mathbf{r}_2 d\mathbf{r}_3$. We need to know this probability if there are three-body forces in the liquid, but this a refinement that we can ignore for the moment. For most of the 20th century the phrase 'theory of liquids' was understood to mean a satisfactory route from the intermolecular pair potential to the structural and macroscopic properties and, in particular, to $g(r)$.

Van der Waals's picture of a liquid was a body with no structure; the molecules are distributed at random with only the restriction that two of them could not be at the same place at the same time since they had 'size'. This restriction was embodied in the co-volume, b, and the lack of structure in what we now recognise as a mean-field approximation, namely that the pair distribution function has its random value, $g(r) = 1$. We can see how this assumption leads to his equation, as follows. The cohesive or internal energy of a system of molecules between which there is a pair potential, $u(r)$, can be written

$$U = \tfrac{1}{2}(N/V)^2 \int\!\!\int u(r_{12})g(r_{12})\,d\mathbf{r}_1 d\mathbf{r}_2. \qquad (5.32)$$

Within the integral we have $g(r_{12})$, the probability of finding a pair of molecules in $d\mathbf{r}_1$ and $d\mathbf{r}_2$, and the energy $u(r_{12})$ that such a pair contributes to the system. The integrations are taken over the volume of the liquid, and the factor of $\frac{1}{2}$ prevents the double counting of the energy of each pair. The differential elements can be written $d\mathbf{r}_1 d(\mathbf{r}_2 - \mathbf{r}_1)$, where the second element of volume is now in a coordinate system in which molecule 1 is at the origin. We take this integration first and let molecule 2 move through all space around molecule 1, then we take the first integration and let molecule 1 move through the whole volume. Since both u and g depend only the scalar distance r_{12} the second element of volume can be written in spherical coordinates as $4\pi r_{12}^2 dr_{12}$, and since $u(r_{12})$ goes rapidly to zero as r_{12} increases we can now again invert the order of the integrations and take first that over dr_1. Hence

$$U = \tfrac{1}{2}(N^2/V) \int_0^\infty u(r_{12})g(r_{12})4\pi r_{12}^2 dr_{12}. \tag{5.33}$$

The upper limit can be taken to be infinite since $u(r)$ is sufficiently short-ranged. We do not know exactly how $g(r)$ depends on the separation, r, nor how it changes with density and temperature, and so cannot proceed further with the integration without some additional information or approximation. Van der Waals's assumption is that $g(r)$ is unity for all distances beyond a collision diameter, d, and zero at shorter distances. We have therefore,

$$U = -a/V, \tag{5.34}$$

where a is a positive constant,

$$a = -2\pi N^2 \int_d^\infty u(r)r^2 dr. \tag{5.35}$$

By purely thermodynamic reasoning we have

$$(\partial U/\partial V)_T = T^2(\partial/\partial T)_V(p/T) = a/V^2, \tag{5.36}$$

and by integrating the second equation,

$$(p + a/V^2) = T \cdot f(V), \tag{5.37}$$

where $f(V)$ is the constant of integration with respect to temperature, which van der Waals took to have its limiting form at low densities of $R/(V - b)$.

Van der Waals did not, of course, introduce $g(r)$ into his derivation; the usefulness of this function was not apparent until after Ornstein's work in 1908–1917 [168]. Ornstein, however, seems never to have written down eqns 5.32 and 5.33; his interests moved rapidly to the interpretation of density fluctuations in liquids in terms of the pair distribution. In a liquid at equilibrium the molecules are moving rapidly and so, on a small scale of length, there are rapid changes in the local density and other properties. The pair distribution function, $g(r)$, is an average over times that are long on a molecular scale. The study of these fluctuations became an active branch of physics in the first decade of the 20th century, after Gibbs and later Einstein had shown how to handle them within the new branch of science, statistical mechanics. In an open system, Gibbs's 'grand ensemble', a portion of fluid of volume V is described by the two intensive properties, the chemical potential, μ, and the temperature, T. The number of molecules in the system, N, can fluctuate, although the changes are not significant if V is of macroscopic size. In 1907 the Polish physicist Marian Smoluchowski showed that the fluctuations are proportional to the compressibility [169];

$$\langle (N - \langle N \rangle)^2 \rangle / \langle N \rangle^2 = -(kT/V^2)(\partial V/\partial p)_T, \tag{5.38}$$

where $\langle N \rangle$ is the average number, and the left-hand side of this equation is the conventional measure of how far the instantaneous number in the system, N, departs from this average value. In a perfect gas the right-hand side is $\langle N \rangle^{-1}$, which even in a portion of gas at atmospheric pressure as small as 1 mm^3 is only 4×10^{-17}. In a liquid the compressibility is smaller and the mean fluctuation is only about 2×10^{-21} for 1 mm^3. Fluctuations in number in a fixed volume imply fluctuations in density and so in the refractive index, which, in turn, leads to the scattering of light. However even in a volume of liquid with the linear dimensions of the wavelength of light there is an increase in the mean fluctuation from that for 1 mm^3 only by a factor of about 10^{10}, which is not enough to produce an easily observable effect. This accords with experience; liquids refract light but scarcely scatter it. If, however, we heat a liquid towards its gas–liquid critical point then its compressibility rises dramatically and, indeed, becomes infinite at the point itself. A critical fluid can scatter light so strongly that it appears totally opaque, as had been observed since the experiments in the early 19th century. It was an attempt to understand this phenomenon more deeply that led Ornstein and his younger colleague, Frits Zernike [170], to make the next advance. They were dissatisfied with Smoluchowski's use of eqn 5.38 near a critical point since its derivation assumes that fluctuations in neighbouring sub-volumes are independent. This is not so; a molecule that leaves one sub-volume enters a neighbouring one and this complication cannot be ignored when the fluctuations are large. They were, however, able to relate the fluctuations to the departure of the distribution function, $g(r)$, from its random value of unity [171]. This departure is now called the total correlation function and denoted $h(r)$;

$$h(r) \equiv g(r) - 1, \tag{5.39}$$

$$\langle (N - \langle N \rangle)^2 \rangle / \langle N \rangle^2 = \langle N \rangle^{-1} + V^{-1} \int h(r)\,\mathrm{d}r. \tag{5.40}$$

The first term on the right-hand side of eqn 5.40 is the perfect-gas term. In a liquid it is largely cancelled by the second term. Thus in a one-dimensional van der Waals fluid [172] we have in a mean-field approximation,

$$h(r) = -1, r < d, \quad \text{and} \quad h(r) = 0, r > d, \tag{5.41}$$

so that the right-hand side of eqn 5.40 is $\langle N \rangle^{-1}(1 - b/V)$. The volume of a van der Waals liquid at zero temperature is b and its compressibility is zero, so that the fluctuations vanish. Conversely, at the critical point the second term on the right-hand side is positive and infinite in size. Since $h(r)$ itself cannot be infinite, indeed it is always of the order of unity, this condition requires that its range becomes so large that the integral diverges. It is when $h(r)$ has a range of 4000 Å or more that light becomes strongly scattered.

Ornstein and Zernike were not satisfied with a correlation function that had this divergence and sought to break it down into simpler components. To this end they introduced another correlation function which we now call the direct correlation function and denote $c(r)$. As they put it succinctly in the summary at the end of their first paper:

Two functions are introduced, one relating to the direct interaction of the molecules [i.e. $c(r)$], the other to the mutual influence of two elements of volume [i.e. $h(r)$]. An integral equation gives the relation between the two functions. [171]

This equation, which we now call the Ornstein–Zernike equation and which defines $c(r)$, is

$$h(r_{12}) = c(r_{12}) + n \int c(r_{13}) \, h(r_{23}) \, \mathrm{d}r_3, \tag{5.42}$$

where n is again the number density, (N/V). The equation cannot be solved directly to give h in terms of c, or vice versa, since both functions appear within the integral. This integral is a 'convolution' of h and c and so the equation can be solved, as they showed, by taking the Fourier transform of each side. The 'meaning' of the equation becomes a little clearer if we substitute repeatedly for h within the integral. We get then

$$h(r_{12}) = c(r_{12}) + n \int c(r_{13}) \left[c(r_{32}) + n \int c(r_{24}) \, h(r_{34}) \, \mathrm{d}r_4 \right] \mathrm{d}r_3$$

$$= c(r_{12}) + n \int c(r_{13}) \, c(r_{32}) \, \mathrm{d}r_3$$

$$+ n^2 \int\!\!\int c(r_{13}) \, c(r_{34}) \, c(r_{42}) \, \mathrm{d}r_3 \mathrm{d}r_4 + \cdots, \tag{5.43}$$

that is, h can be decomposed into a direct correlation between positions 1 and 2, $c(r_{12})$, and a series of indirect correlations of chains of c, through position 3, through positions 3 and 4, through positions 3, 4 and 5, etc. The value of the direct correlation function in the eyes of Ornstein and Zernike is that it has generally only the range of the pair potential, $u(r)$. They believed that this limitation on the range held good even at the critical point where $h(r)$ is divergent. In this they were not quite correct for we now know that $c(r)$ is also divergent at the critical point, although only very weakly. Their assumption is again a manifestation of a mean-field approximation.

Their paper, published in Dutch and English in the Netherlands during the first World War, attracted little notice. They themselves said in 1918 that their work was "clearly not well known" and they published a summary of it in a leading German journal [173]. This repeats explicitly the fact that $c(r)$ has the virtue of a range no longer than that of $u(r)$, but this paper also seems to have had little effect on those working in statistical mechanics.

In a simple liquid at low temperatures the main features of $g(r)$ or $h(r)$ lie in the range of 1–10 Å; $h(r)$ is close to zero beyond about 20 Å. To study these short-range functions experimentally we need to probe the system with radiation of similar wavelength and study the scattered radiation. We need, therefore, to use x-rays whose wavelengths are typically 2 Å or less. In 1916 Debye and Scherrer studied the scattering pattern from liquid benzene, but this has a complicated molecule and the pattern arises not only from scattering from pairs of atoms in different molecules but also from pairs of carbon atoms in the same molecule [174]. Potentially more useful was the diffraction pattern of liquid argon obtained by Keesom and De Smedt in 1922–1923 [37]. Little quantitative could be done with this until Zernike and Prins [38] showed that $h(r)$ was a Fourier transform of the x-ray scattering pattern. Zernike did not use this result to obtain any explicit values of $h(r)$; that came a few years later when Debye and Menke exploited it to obtain this function for mercury, another monatomic liquid [175].

The seven-year spacing of these papers, 1916 to 1923 to 1930, is itself evidence that liquids were no longer at the centre of physicists' attention, at least outside this group of Dutch scientists. Critical points were also not an active area of research in the 1920s and Ornstein and Zernike's work was ignored. Fowler's great monograph on *Statistical mechanics* of 1929 has a chapter on 'Fluctuations' but he makes no mention of their work [176]; it is similarly missing from the later version of this book with Guggenheim in 1939 [24], and from the texts of Tolman in 1938 [177] and of Mayer and Mayer in 1940 [178], who have a chapter on the critical region. Gases and solids were more fruitful fields of research in the 1920s and early 1930s. When liquids were discussed they were regarded as disordered versions of the better understood crystals. Thus even when the pair distribution function was determined from x-ray scattering patterns it was assimilated into the dominant physics of the solid state by attempts to interpret it as an average over random orientations of an array of micro-crystals [179].

Those interested in determining the structures of liquids were a different group from the small group working on the statistical mechanics of gases. The main task of this second group in the 1920s and early 1930s was putting Kamerlingh

Onnes's virial expansion on a proper theoretical footing; first, so that it could be used to obtain information about the intermolecular forces and, second, in the unrealised hope that something useful could be made of the higher coefficients. The second was a difficult task at which even Fowler confessed to have failed [180]. H.D. Ursell [181] first found out in 1927 how to express the higher coefficients in terms of products of Boltzmann factors of the form $\exp[-u(r)/kT]$. Mayer and his colleagues amplified this work ten years later [182], and it was through Mayer's efforts that the virial expansion of the pressure and of the pair distribution function became widely known. The expansion of the latter in powers of the density was also found independently by J. Yvon in 1937 [183] and by J. de Boer in 1940 [184], but their work was not so accessible.

Thus in the 1930s and in the years immediately after the second World War there were two different approaches to the liquid state. The first tried to build on the resemblance of liquids to solids. Its experimental basis lay in the x-ray studies of the Dutch–German school and in particular in attempts to interpret their results as evidence for liquids as disordered solids. The statistical mechanics of this group in the late 1930s and after the War was based mainly in Cambridge and at Princeton. This was the dominant approach. There was, however, a less well-organised group who were trying to build on the successes of the statistical mechanics of gases and extend these to liquids via the virial expansion. There were a few others at work, not so skilled in statistical mechanics, but with an instinctive feeling that the analogy with solids was a misleading one. However the line of thought that had started with van der Waals, and which had generated the pregnant papers of Ornstein and Zernike, was almost ignored. Both the liquids-as-solids and the liquids-as-gases schools had, at the time, good reasons for their approaches and it is only with hindsight that we can see that they had strayed from what was to prove the successful path. The solid school held the field for nearly thirty years and their work was to become one of the great dead-ends of modern physics.

The solid-like or lattice theories, as they came to be known, started with chemists' attempts to understand the change in thermodynamic properties on mixing two liquids. This was both an academic subject of some popularity and a matter of practical importance in the operation of distillation columns. In 1932 Guggenheim put forward a model of a liquid mixture in which the molecules were confined to the neighbourhoods of an array of fixed sites of an unspecified geometry [185]. The need for a more explicit description of the supposed structure came a few years later when he went beyond a mean-field treatment with what he called a 'quasi-chemical' approximation [186]. This work marked the opening of a long series of papers, initially from the Cambridge school, on the combinatorial problem of assigning molecules of different energies and sizes to one or more sites of a lattice of given geometry [187]. The combinatorial problems were fascinating in their own right and, in Onsager's hands, played a crucial role in the theory of the critical point of a two-dimensional magnet, but they were not to prove a useful route to the understanding of the thermodynamics of liquid mixtures.

The parallel work on lattice theories of pure liquids started in 1937 with Lennard-Jones and Devonshire in Britain [188] and Eyring and Hirschfelder in America [189]. The field grew rapidly after the War with increasingly sophisticated models, in the later versions of which the lattices served mainly as mathematical devices to assist in trying to evaluate the statistical mechanical partition function. A review of this work just before the War was given by Fowler and Guggenheim who wrote:

We are therefore driven to the conclusion that a liquid is much more like a crystal than like a gas, and the structure which we shall accept as the most plausible for a liquid is conveniently referred to as quasi-crystalline.... the number of nearest neighbours has a fairly well-defined average value, and, although there are fluctuations about this average, these fluctuations are not serious, and the geometrical relationship of each molecule to its immediate neighbours is on the average very similar to that in a crystal. [190]

A book written in comparative isolation during the War by Ya.I. Frenkel was published in 1946. The Preface opens with similar words:

The recent development of the theory of the liquid state, which distinguishes this theory from the older views based on the analogy between the liquid and the gaseous state, is characterised by the reapproximation of the liquid state – at temperatures not too far removed from the crystallization point – to the solid (crystalline) state. . . . The kinetic theory of liquids must accordingly be developed as a generalisation and extension of the kinetic theory of solid bodies. [191]

By 1954 the amount of work in this field justified a review of fifty pages in the treatise of Hirschfelder, Curtiss and Bird [192], and in 1963 it received its final summary in Barker's monograph, *Lattice theories of the liquid state* [193]. By then it was clear that lattice theories were not the way forward, although, as always, the deficiences were not fully realised until better theories were developed. The obvious success of solid-state physics was, as we have seen, one of the starting points for the attempt to extend lattice theories to liquids, but there seems also to have been an obstinate refusal to learn from earlier work. In 1936 the Faraday Society held a meeting in Edinburgh on *Structure and molecular forces in (a) pure liquids and (b) solutions* [194], and the next year saw the Dutch celebration in Amsterdam of the centenary of the birth of van der Waals [195]. Reading the more theoretical papers presented at these meetings gives one an impression of a certain arrogance; it seems as if their authors believed that physics had started again in 1925 with the new quantum mechanics and that one could safely ignore anything done before then. Only two of the papers at Amsterdam were on the liquid–vapour transition and one of these was Lennard-Jones's opening acccount of a lattice theory which was certainly not in the van der Waals–Ornstein tradition.

Theories are not abandoned because they fail but because they are superseded by better ones. There was a slim trail of papers from the middle 1930s that did not follow the dominant lattice models but tried to calculate the pair distribution function, relate it to experiment, and use it to calculate the thermodynamic properties. The energy, for example, is given by the transparently obvious eqn 5.33, and the pressure by the parallel equation that is an expression of the virial theorem:

$$p = NkT/V - \tfrac{1}{6}(N/V)^2 \int_0^\infty r[du(r)/dr]g(r)4\pi r^2 dr. \qquad (5.44)$$

(This is usually called the virial equation for the pressure, but is not to be confused with Kamerlingh Onnes's virial *expansion* for the pressure which is the expansion of p in terms of the gas density, eqn 5.29.) Ornstein and Zernike had used $g(r)$ in statistical mechanical theory but it was only with its experimental determination in the late 1920s that it made its hesitant way into the main stream of the statistical literature. Only the low-density limit of eqn 5.44 is to be found in Fowler's book of 1929 [196], that is, the limit in which $g(r)$ is replaced by $\exp[-u(r)/kT]$. The general form was given by Yvon in 1935 [197]. Equation 5.33 seems to have been written down first by Hildebrand in 1933 [198], who used it some years later to find the intermolecular potential of mercury from an experimental determination of $g(r)$ [199]; it too was given by Yvon. Hildebrand was one of those who had grown up in the van der Waals and van Laar tradition, and who had an instinctive distrust of 'solid' theories of liquids. But he was not a skilled specialist in statistical mechanics and so his insight was not as fertile as it might have been.

Equations 5.33 and 5.44 show how $g(r)$ should be used, but do not tell us how it should be determined theoretically. In Gibbs's canonical ensemble the probability of all N molecules being simultaneously in volume elements $dr_1 dr_2 dr_3 \ldots dr_N$ is proportional to the Boltzmann factor $\exp[-U^*(r^N)/kT]$, where $U^*(r^N)$ is the configurational energy of the system when the molecules are so situated. By integrating this relation over all positions $dr_3 \ldots dr_N$ we obtain the probability that there are molecules in positions dr_1 and dr_2; that is, we obtain $g(r_{12})$. The equation is

$$g(r_{12}) = \frac{V^2 \int \ldots \int \exp[-U^*(r^N)/kT]dr_3 \ldots dr_N}{\int \ldots \int \exp[-U^*(r^N/kT]dr_1 \ldots dr_N}. \qquad (5.45)$$

This equation appears in a less transparent notation in Fowler's 1929 treatise, where $-kT\ln g(r)$ is called the potential of average force in the system [200]. This potential reduces to $u(r)$ in the dilute gas and is now used more often for complex systems than for simple monatomic liquids. The more modern form, that is, eqn 5.45, appeared in two papers of 1935 that we can now see as the foundation of an alternative approach to the theory of liquids that eschews the assumption of a lattice structure. One, by Yvon [197], appeared in an obscure French series of occasional publications and was overlooked for many years, the other by Kirkwood appeared in what was rapidly becoming the leading journal in this field [201]. Equation 5.45, although exact, is not immediately useful since neither integral can be evaluated as it stands. Yvon and Kirkwood both found ways of simplifying the right-hand sides so that $g(r)$ is expressed by an integro-differential equation that involves only $g(r_{12})$ and the three-body distribution function $g(r_{12}, r_{13}, r_{23})$. Their equations were different but equivalent. Yvon's equation was obtained independently after the War by Bogoliubov in Moscow [202] and by Born and Green in Edinburgh [203]. To solve either of these equations for $g(r)$ needs an approximation for the three-body function, the simplest of which is Kirkwood's 'superposition approximation' which represents the three-body function as a product of two-body functions:

$$g^{(3)}(\boldsymbol{r}_1, \boldsymbol{r}_2, \boldsymbol{r}_3) = g^{(2)}(\boldsymbol{r}_1, \boldsymbol{r}_2)g^{(2)}(\boldsymbol{r}_1, \boldsymbol{r}_3)g^{(2)}(\boldsymbol{r}_2, \boldsymbol{r}_3). \qquad (5.46)$$

The theory of liquids was not in a happy state in the ten years after the second World War. The lattice theories over-emphasised the analogy with solids and were not producing quantitatively acceptable results. Their neglect of the 'continuity' of the gas and liquid states was their weakest point; in their simplest form (that of Lennard-Jones and Devonshire) they led, for example, to a zero value for the second virial coefficient of the gas. They were, however, theories that lent themselves to many ingenious schemes for their improvement [204] and so they attracted many devotees. The 'distribution function' approach of Kirkwood, Yvon, Bogoliubov, and Born and Green was based firmly on an attack from the gas side. It gave exact values for the second and third virial coefficients (with the use of eqn 5.46) but failed at higher densities. It was regarded as the more difficult theory, one that did not lead easily to numerical results, and one that was hard to improve by ad hoc adjustments. It was not, therefore, in a position to challenge the dominant lattice theories in the early 1950s. The position changed with the re-discovery of the work of Ornstein and Zernike and the realisation that the direct correlation function, $c(r)$, is a simpler entity than the total function, $h(r) \equiv g(r) - 1$, and one that lends itself more readily to plausible approximation. The direct correlation function had been ignored in the 1920s, 30s and 40s. It is mentioned but not used constructively in a paper on critical phenomena in 1949 [205] and appears as an aside in a book on *The theory of electrons* in 1951 [206], but the credit for its re-introduction into the main stream of statistical mechanics belongs to Stanley Rushbrooke and his student H.I. Scoins, in Newcastle [207]. Rushbrooke's first work on liquids had been in the lattice tradition of Cambridge and of his first research supervisor, Fowler, then came his 'prentice work on the pair distribution with Coulson [208], but in his paper with Scoins he opened up a new and productive channel.

The Ornstein–Zernike equation, eqn 5.42, defines $c(r)$ in terms of $h(r)$, but gives no hint as to how either function might be determined theoretically. Progress comes from the authors' belief that $c(r)$ is short-ranged, that is, of the range of $u(r)$. We can write

$$c(r) = [1 - e^{u(r)/kT}]g(r) + d(r), \qquad (5.47)$$

where $d(r)$ is a new function, defined by this equation, and so still to be determined. The form of the first term on the right-hand side is chosen because $g(r)\exp[u(r)/kT]$ is a function that is always a continuous and, indeed, smooth function of r even at those points where $u(r)$ and hence $g(r)$ have discontinuities, such as at the diameter of a hard sphere. The range of the first term is clearly that of $u(r)$ since it vanishes when $u(r) = 0$. In their pioneering paper, Rushbrooke and Scoins approximated $c(r)$ by $\{\exp[-u(r)/kT] - 1\}$, which has the same range; but this is too simple. A better way of achieving Ornstein and Zernike's aim is to put $d(r) = 0$ in eqn 5.47. This, in effect, was the what J.K. Percus and G.J. Yevick brought about in 1958 [209]. Their argument was based on quite different grounds but it soon came to be seen [210] that their result could be expressed most simply in terms of the Ornstein–Zernike equation with the approximation $d(r) = 0$. This connection was amplified in two long articles in 1964 in a collective work on *The equilibrium theory of classical fluids* [211]. A surprising feature of the Percus–Yevick (or PY) equation of state that follows from this approximation is that it can be expressed in simple closed forms for a fluid composed of hard spheres. There are two commonly used routes to the pressure from $c(r)$ or $g(r)$; the first is the virial route of eqn 5.44, and the second, due to Ornstein and Zernike, follows from Smoluchowski's fluctuation expression, eqn 5.38:

$$kT(\partial n/\partial p)_T = 1 + n \int h(r)\,dr. \qquad (5.48)$$

This is now usually called the compressibility equation. Since the Percus–Yevick approximation of putting $d(r) = 0$ is not exact, the pressure calculated from the virial expression, p_V, does not agree with that found from the compressibility equation, p_C. For hard spheres we have [212]:

$$(p/nkT)_V = (1 + 2\eta + 3\eta^2)(1 - \eta)^{-2},$$
$$(p/nkT)_C = (1 + \eta + \eta^2)(1 - \eta)^{-3}, \qquad (5.49)$$

where η is a reduced density which is the ratio of the actual volume of N spheres of diameter d to the volume V; $\eta = \pi N d^3/6V$. On expansion, these two expressions agree as far as the third virial coefficient, but differ thereafter. When they are compared with the results of computer simulations, it is found

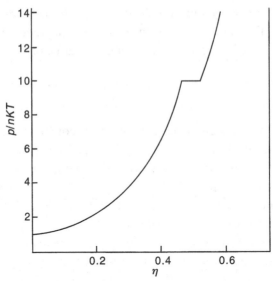

Fig. 5.4 The compression ratio, p/nkT, for an assembly of hard spheres, as a function of η, the reduced density. This density is defined so that η is unity at a density at which the volume of the system is equal to that of the spheres. In practice, such a density is unattainable and the maximum value of η is $(\pi\sqrt{2}/6) = 0.7405$, the density of a close-packed crystalline solid. The lower part of the curve represents the fluid state; crystallisation sets in at a reduced density of about 0.47 and is complete by 0.53. The upper curve represents the solid state and approaches an infinite value of the compression ratio as the density approaches the close-packed limit of 0.7405.

that the compressibility equation yields a pressure that is a little higher than the 'experimental' while the virial equation lies below it.

Interest in the hard-sphere model fluid had revived after the War because of the development of the technique of computer simulation which is at its simplest and most efficient for such a potential. There had been a few attempts to model mechanically the structure of such a fluid in the 1930s, either in two dimensions with round seeds or ball-bearings poured on to a flat plate [213], or in three dimensions with a suspension of coloured spheres of gelatine in water [214], but such experiments could tell us nothing of the thermodynamic properties of the system. Computer simulations not only yielded the structure, that is, $g(r)$, but also the pressure. It was found, moreover, that the fluid phase crystallised to a close-packed solid when the density η exceeded about 0.47 (Fig. 5.4). The notion that a system with a purely repulsive potential could crystallise was not new. Kirkwood had suggested it in 1940 from a study of his integral equation for $g(r)$ but the theory was not then good enough for the prediction to carry much weight [215]. A fluid of hard spheres shows no separation into gas and liquid phases, and so has no critical point; for that the attractive forces are needed also, as had been appreciated since the time of van der Waals. Indeed the critical temperature is itself a rough measure of the maximum energy of attraction, ε, of a pair of molecules; in general $\varepsilon \approx 0.9\ kT^c$. In a hard-sphere fluid the temperature is an irrelevant parameter that serves only to scale the pressure. The phase behaviour is governed by one parameter only, which can be taken to be either the density, η, or the ratio (p/T). A change of phase occurs when there can be a move, at a fixed temperature and pressure, to a state of equal Gibbs free energy, $G = U - TS + pV$, where U is the energy and S is the entropy. The energy of a system of hard spheres is purely kinetic, $3NkT/2$, and so is the same in any possible phase at a given temperature. The crystallisation of a hard-sphere fluid at a fixed pressure occurs therefore when the change ΔG, from liquid to solid, is zero, or when $\Delta S = (p/T)\Delta V$. Since ΔV is negative it follows that the entropy of the solid is less than that of the co-existing fluid. If, however, we were to compress the fluid to a metastable state in which its density was the same as that of the crystal then the irreversible change to the solid state would be accompanied by a fall in the Helmholtz free energy, $F = U - TS$, and, since ΔU is again zero, there is now an *increase* of entropy. Such a change is counter-intuitive for those brought up to think of the entropy as a measure of the disorder in the system, since the geometrical order of a crystal is certainly greater than that of the fluid of the same density from which it has been formed. The configurational order of statistical thermodynamics is, however, not a matter of simple geometry but takes account also of the freedom of motion, or 'free volume', of the particles in the system. At the density at which crystallisation sets in, $\eta \approx 0.47$, this freedom is greater if the particles are moving around the sites of an ordered lattice (for which the free volume goes to zero only when η reaches 0.74) than if they are moving in a dense amorphous or glassy state (for which the free volume goes to zero at $\eta \approx 0.64$) [216].

The assumption that $d(r)$ is zero in eqn 5.47, which underlies the PY equation of state, is not the only approximation that was tried, nor was it the first after the early choice of Rushbrooke and Scoins in 1953. Another choice followed in 1959, first from de Boer and his colleagues [217], but soon also from others in France, Japan, the U.S.A. and from Rushbrooke himself in Britain. This was

$$d(r) = y(r) - 1 - \ln y(r); \qquad y(r) = g(r)e^{u(r)/kT}. \qquad (5.50)$$

This became known as the 'hyper-netted chain' or HNC approximation, from the nature of the chains of linked molecules in the integrals used to express $g(r)$. Superficially it is more attractive than the PY approximation, rationalised in 1963 as $d(r) = 0$, since it includes more of these integrals and so makes an attempt to estimate the tail of $c(r)$ that extends beyond the range of the pair potential. For hard spheres, however, the HNC approximation is worse than the PY. The two values of the pressures calculated from eqns 5.44 and 5.48 are further apart and neither is close to the pressure found by computer simulation. For more realistic model fluids, such as a Lennard-Jones (12, 6) liquid at low temperatures, the HNC is better than the PY. Once it was found that approximations for the direct correlation function were a good route to reasonable forms of $g(r)$, and so to the physical properties, then the field was open to further and more realistic approaches, which generated an active line of research in the 1960s.

Assemblies of hard spheres are, however, model systems that apparently had little relation to real liquids. The results obtained by PY, HNC, and related theories for these systems were good enough to banish any lingering interest in lattice theories but did not, by themselves, constitute a theory of liquids. Direct solution of the equations for more realistic models is difficult and the results did not have the success of the hard-sphere models. A rather different way of using these results was needed.

We have seen that the essence of van der Waals's theory was the ascription to the system of a free volume in which the molecules moved at random subject only to the restriction imposed by their hard spherical cores, and that this movement took place in a uniform energy field, provided by the molecular attractions, and everywhere proportional to the overall density of the system, N/V. That is, the structure of the system is imposed by the hard cores; the attractive energy holds the system together but does not disturb this structure. In one sense this was also the view of those generating the lattice theories, but where we can now see that they went wrong was in supposing that this structure resembled closely that of a solid. It was not always the view of those who first developed the distribution-function theories, for they often believed that the attractive forces were also powerful determinants of the liquid structure [218]. Soon, however, the PY and later approximations began to generate pair distribution functions for hard spheres in which one could have reasonable confidence since they agreed with those found by computer simulation. It was then noticed how similar were the results of both the simulations and the theories to the pair distribution functions found for real simple liquids, such as argon, as found by x-ray scattering experiments. The large first peak in $g(r)$ in real liquids was not as sharp as that in a hard-sphere fluid but its similar size showed that it owed as much to the simple geometrical consequence of the dense packing of the molecules around any chosen molecule as to the direct effect of the attractive forces. The view grew in the early and middle 1960s that the way forward was a perturbation theory, in the general spirit of van der Waals, but based not on the total absence of structure beyond the collision diameter [i.e. $g(r) = 1$, for $r > d$] but on the realistic forms of $g(r)$ generated by computer simulation or by PY and other theories for the hard-sphere fluid [219].

Two steps are needed to turn a hard-sphere potential into a reasonably realistic one, such as a Lennard-Jones potential. First we must add the attractive part of the potential and, secondly, we must soften the repulsive core from that of a sphere [in effect, $(r/d)^{-\infty}$] to a more realistic form, say $(r/d)^{-n}$, where $n \approx 12$. Neither of these steps greatly perturbs the structure and it is this stability that makes perturbation theory appropriate. The first step was one that was well known in principle [220]. We can write the configurational part of the free energy, F_c, in Gibbs's canonical ensemble as

$$\exp(-F_c/kT) = (1/N!) \int \cdots \int \exp\left[-\sum\sum u(r_{ij})/kT\right] \mathrm{d}r^N, \quad (5.51)$$

where $u(r_{ij})$ is the potential energy of a pair of molecules, i and j, at a separation r_{ij}, and the double sum is over all pairs of molecules. The integrations are over all positions of all molecules within the volume V. The pressure and other thermodynamic properties follow at once from F_c, when this is known as a function of N, V, and T; for example, $p = -(\partial F_c/\partial V)_T$. We can now divide $u(r)$ into two parts, a positive or repulsive part, $u_+(r)$, and a negative or attractive part, $u_-(r)$. In a Lennard-Jones (n, m) potential these could be, for example, the terms in r^{-n} and r^{-m} respectively, but other divisions are possible. A better division in practice is to take u_+ as the whole of the potential for $r < d$, the collision diameter, and u_- to be the whole of the potential for $r > d$. With this second choice u_- is always bounded and so we can expand that part of the exponential in eqn 5.51 that contains u_- in powers of (u_-/kT);

$$(N!)\exp(-F_c/kT) = \int \cdots \int \prod\prod \exp[-u_+(r_{ij})/kT]\mathrm{d}r^N$$
$$- \int \cdots \int \prod\prod [u_-(r_{ij})/kT]\exp[-u_+(r_{ij})/kT]\mathrm{d}r^N$$
$$+ \cdots \text{terms in } T^{-2}, T^{-3}, \text{etc.,} \quad (5.52)$$

where the double products are again to be taken over all pairs of molecules. The first term is the exponential of the free energy of a system without attractive forces; the second is the average value of the attractive energies in a system whose structure is determined by the repulsive potentials only. Higher terms incorporate the small changes in this structure caused by the attractive forces. These are needed for an accurate representation of the properties of a liquid since $(-u_-/kT)$ can be as large as 2 near the freezing point.

A different method of perturbation is needed for the second step, that is, to assess the effect of going from a true hard-sphere potential to a more realistic repulsive potential such as r^{-n}. The first attempt was to expand the integrand in powers of n^{-1} since $n^{-1} = 0$ represents a hard sphere and $n^{-1} = 1/12$ is a small number [221]. This attempt met with only partial success; a more ingenious solution to the problem was needed by finding how to choose a temperature-dependent collision diameter and to combine this choice with a separation of u into u_+ and u_- that led to a rapid convergence of the expansion in eqn 5.51. This was first achieved by Barker and Henderson in 1967. Their results were given informally at the Faraday Discussion on *The structure and properties of liquids* held in April at Exeter. Henderson, who was at the meeting, read each morning a telegram from Barker in Melbourne in which the progress of the work was described. A short account of this appeared in the published proceedings [222] and a full account later in the year [223]. Other and even better ways of dividing u into u_+ and u_- followed soon afterwards [224], but Barker and Henderson's work was the decisive effort; for the first time one could go from a reasonably realistic model potential, in this case a (12, 6) potential, to a quantitatively acceptable determination of the structure of the liquid, as represented by $g(r)$, and of its thermodynamic properties. The 'experimental' values of these were provided by computer simulations since, by 1967, it had become clear that the (12, 6) potential is not an accurate representation of the interaction of real molecules, even those as simple as argon atoms. But what could be done for the (12, 6) potential could be done also for the more complicated potentials of the 1970s. Adding in the effects of the three-body potential is a little more difficult but, since it is much weaker than the two-body term, this is also a problem that can be handled by a perturbation treatment.

Thus by the early 1970s the core problems of 'cohesion' had been solved in principle. The attractive or dispersion forces could be calculated from a well-founded theory (quantum mechanics), the form and magnitude of the rest of the intermolecular potential could be found from the properties of the dilute gas, and this potential could be used in another well-founded theory (statistical mechanics) to calculate the properties of solids and, at last, of liquids also.

Only with the gas–liquid critical point was there still a problem. Here the perturbation methods break down since $g(r)$ has a range that becomes infinite at this point, in a complicated way. The solution of this difficulty required the importation into statistical mechanics of mathematical techniques hitherto quite foreign to the field. The details of the intermolecular forces become irrelevant; they determine the position of the critical point, that is, the values of p^c, V^c and T^c, but not how the physical properties behave as functions of $(p - p^c)$, $(V - V^c)$ and $(T - T^c)$; this behaviour is said to be 'universal'. This work also came to a satisfactory conclusion in the the early 1970s but the details need not be discussed here since the 'universality' means that the experimental characteristics of fluids near their critical points tells us nothing specific about the intermolecular forces [225]. It was in his treatment of the critical point that van der Waals's ideas have proved to be least correct. He insisted, rightly, that the force, or the potential $u(r)$, is of short range but did not know that such a force is incompatible with a simple analytic form of the equation of state of the kind that he put forward. Such equations become correct only if the attractive potential is everywhere weak but of infinite range, or if the potential is of short range but we live in a world of four or more dimensions.

The other important phase change, that from liquid to solid, still lacks a satisfactory interpretation in terms of the intermolecular forces. There are now good theories of both solid and liquid states, so that we can calculate the free energy of each state separately and then equate them to find the melting point where the two states are in equilibrium. But the theories of the two states are different and, indeed, incompatible, since one supposes a lattice structure that the other now eschews. The equating of the free energies, although effective in practice, is aesthetically displeasing. One would like to see a common treatment in which both states arise naturally from a particular assumed form of the intermolecular potential. Such a theory is under development as, for example, the so-called density-functional theory, which can be crudely thought of as an attempt to reverse the ideas of the lattice theories of liquids and instead treat the solid as a more structured form of the liquid. Some success has been achieved, but the matter is still 'unfinished business' [226].

Another problem that has been solved only partially is a theory of the structure and physical properties of the liquid–gas interface, which is the key to understanding the old problem of capillarity that played such an important rôle in the early years of the study of cohesion.

Laplace had identified correctly the link between the interparticle forces and the surface tension. His treatment was restricted by his static view of matter (his particles did not move), by what we can now recognise as a mean-field approximation (his liquid had no structure), and by his assumptions that the interface had negligible thickness and the gas density was zero (his density profile was a step-function). There were no direct attempts to remedy these defects in Laplace's treatment for over a century. Poisson had criticised the third assumption but his attempts to remedy it were not carried out effectively and led him to the mistaken conclusion that Laplace's assumption of a sharp interface led to a zero value of the surface tension. Maxwell discussed this point [227] but made no attempt to tackle the problem. In the 1930s there were some crude attempts to calculate the surface energy of a liquid, possibly made in the belief that this is easier to calculate than the surface tension, which is a surface free energy. This belief is not correct, but these papers [228], like many of those on the bulk properties of liquids in the same years, paid scant attention to what had been done previously. Laplace's second restriction was removed by Fowler in 1937 when he introduced the pair distribution for the uniform bulk liquid, $g(r)$. He obtained for the surface tension

$$\sigma = (n^2/32) \int r^2 u'(r) g(r) \, \mathrm{d}\boldsymbol{r}, \qquad (5.53)$$

where n is the number density, (N/V), and $u'(r)$ is the derivative of the potential, that is, the negative of the intermolecular force [229]. He left untouched the third restriction; his interface was still of zero thickness. We get Laplace's result again by putting $g(r) = 1$ in eqn 5.53, and integrating by parts,

$$\sigma = -(n^2/8) \int r u(r) \, d\mathbf{r}, \tag{5.54}$$

which correctly includes the factor of the square of the density, and where the integration must now be restricted to configurations in which the molecular cores do not overlap and in which $u(r)$ is therefore negative. The exact expression for the surface tension, to which these results are approximations, was found by Kirkwood and his then research student, Frank Buff, in 1949 [230]. They specified the structure of the fluid in the interface by a generalised two-body density $n^{(2)}(\mathbf{r}_1, \mathbf{r}_2)$ which reduces to $n^2 g(r_{12})$ in the bulk liquid or the bulk gas, where n is the liquid or gas density. Their expression for the surface tension is

$$\sigma = \pi \int_{-\infty}^{+\infty} dz_1 \int_0^\infty r_{12} u'(r_{12}) (r_{12}^2 - 3z_{12}^2) n^{(2)}(\mathbf{r}_1, \mathbf{r}_2) \, dr_{12}, \tag{5.55}$$

where r_{12} is the distance between $\mathbf{r}_1$ $(= x_1, y_1, z_1)$ and $\mathbf{r}_2$ $(= x_2, y_2, z_2)$, and $z_{12} = z_2 - z_1$. The whole contribution to the integral comes from the surface layer since, by symmetry, the mean value of $3z_{12}^2$ in a homogeneous liquid or gas is r_{12}^2. Fowler's result is recovered if one puts

$$n^{(2)}(\mathbf{r}_1, \mathbf{r}_2) = n(z_1) n(z_2) g(r_{12}), \tag{5.56}$$

where $n(z_i)$ is the density at height i and becomes zero if z_i lies in the gas phase. Eqn 5.55 is a formal solution of the problem, but not by itself a practically useful one until one knows something of the two-body density $n^{(2)}(\mathbf{r}_1, \mathbf{r}_2)$, that is, of the probability of finding molecules in these positions when $\mathbf{r}_1$ or $\mathbf{r}_2$ or both lie in the inhomogeneous surface layer between the liquid and the gas. Unlike $g(r)$ in the homogeneous liquid, this function cannot be determined directly from x-ray or neutron diffraction [231].

Quite a different route to the surface tension of an interface in which there is a continuous variation with height from the density of the liquid to that of the gas was found in the years 1888 to 1893, when Karl Fuchs, the Professor of Physics at Pressburg (now Bratislava in Slovakia), Lord Rayleigh, and van der Waals all realised that the energy of a molecule in such an interface would depend not only on the local density at that height but also on the densities of molecules in the layers above and below it, out to the range of the intermolecular force [232]. Since they knew that the thickness of the interface, away from the critical point, is of the same order as this range, they realised that the effect is a serious one; a molecule within the interface interacts with others below it in the dense liquid and with others above it in the gas. Van der Waals's treatment was the most thorough, being based on thermodynamic not mechanical arguments, that is, he explicitly recognised that the equilibrium in such a system is a dynamic one between moving molecules, not a static or mechanical one as the models of Fuchs and Rayleigh envisaged.

Laplace had obtained two integrals, the first of which, K, is a measure of the energy of a liquid, and the second of which, H, is a measure of its surface tension. In modern notation

$$K = -\tfrac{1}{2}n^2 \int u(r)\,\mathrm{d}r, \qquad H = -\tfrac{1}{4}n^2 \int ru(r)\,\mathrm{d}r. \qquad (5.57)$$

Thus K is the volume integral of $u(r)$ and H is the integral of its first moment, $ru(r)$. The treatment of Fuchs, Rayleigh and van der Waals in 1888 led to a different and apparently contradictory result. Since their profile of the fluid density was a continuous function they could expand the local energy density at height z, $\varphi(z)$, in terms of the derivatives of $n(z)$ with respect to z. By symmetry, the result contains only the even derivatives:

$$\varphi(z) = \tfrac{1}{2}n^2 \int u(r)\,\mathrm{d}r - \tfrac{1}{12}n(z)n''(z) \int r^2 u(r)\,\mathrm{d}r + O[n''''(z)]. \qquad (5.58)$$

The first term is again just Laplace's K, but his H is missing, and the next term is proportional to $r^2 u(r)$, or the second moment of the intermolecular potential. Since it is H that is the surface tension on Laplace's model it seems at first sight that, contrary to what Poisson surmised, it is the surface with a non-zero thickness that has zero surface tension. This however is not so; the two models cannot be compared so simply since a Taylor expansion of the kind of eqn 5.58 cannot be made if the density profile is a step-function. Van der Waals calculated the surface tension from the second term of eqn 5.58 and found it to be comparable with Laplace's H; as he put it, "these difficulties are imaginary" [233]. Rayleigh also noted the paradox and tried to resolve it [232], but a full explanation was not possible until there were exact expressions for the tension by both routes, the one that started with Laplace and the one that started with van der Waals. The first route was successfully followed by Kirkwood and Buff in 1949 and led to eqn 5.55, and the second route had already been reached by then, although few knew of it. Yvon had reported to a meeting in Brussels in January 1948 that the surface tension could be expressed as an integral that contained the product of the density gradients at two different heights in the interface [234], but he did not give a full derivation. The first derivation to be published was that of D.G. Triezenberg and Robert Zwanzig in 1972; this was followed at once by an alternative route to the same result by Ronald Lovett, Frank Buff and their colleagues [235]. This second exact expression for the surface tension is

$$\sigma = \tfrac{1}{4}kT \int_{-\infty}^{+\infty} n'(z_1)\,dz_1 \int \left(x_{12}^2 + y_{12}^2\right) n'(z_2) c(r_1, r_2)\,dr_2, \qquad (5.59)$$

where x_{12} and y_{12} are the transverse components of the vector $(r_2 - r_1)$, and where $c(r_1, r_2)$ is the direct correlation function between points r_1 and r_2. No more is known of this function than of the two-body density function in eqn 5.55, so the practical value of this expression is limited to approximations. The question naturally arose, however, of the equivalence of the two expressions, eqns 5.55 and 5.59, since by their derivations both claimed to be exact. They are the natural ends of the lines of argument that started with Laplace and with van der Waals. Many attempts were made to answer this question which was resolved only in 1979 when Peter Schofield at Harwell in Britain [236] showed that they were indeed equivalent, and so van der Waals was correct, if premature, in saying that the difficulty of reconciling his approach with that of Laplace was "imaginary".

There is a third way of formulating the surface tension and that is in terms of the stress or pressure at each point in the gas, liquid and interface. When the method is made precise it leads again to the 'virial' or Kirkwood–Buff expression, eqn 5.55, but for many years the method had an independent life of its own. Such a formulation is implicit in the very concept of surface tension and goes back to the work of Segner and Young, but it was only after the 'elasticians' of the 19th century had treated stress with proper mathematical rigour that this became a formal route to the surface tension. In a three-dimensional body the stress, or its negative, the pressure, can be expressed as a dyadic tensor with nine components. If the system is homogeneous, isotropic, and at equilibrium then the three diagonal terms p_{xx}, p_{yy}, and p_{zz} are all equal, and the off-diagonal terms, p_{xy}, p_{yz}, etc., are zero. That is, the pressure tensor can be written

$$\mathsf{P}(r) = p\mathbf{1}, \qquad (5.60)$$

where p is a constant (i.e. 'the pressure') and $\mathbf{1}$ is the unit tensor. If the system is at equilibrium but not homogeneous or isotropic, as is the case in a two-phase system of gas and liquid separated by an interface, then we know only that the gradient of the pressure tensor, itself a vector, is everywhere zero;

$$\nabla \cdot \mathsf{P}(r) = 0. \qquad (5.61)$$

For a planar interface between gas and liquid in the x–y plane this condition and the symmetry of the system require again that the off-diagonal terms are zero and that,

$$p_{xx}(z) = p_{yy}(z), \quad \text{and that } p_{zz}(z) = \text{constant.} \tag{5.62}$$

The last component, p_{zz}, is the pressure normal to the interface and is equal to the common value of the scalar pressure, p, in the bulk gas and liquid phases. It is usual to write $p_N(z)$ for this component and $p_T(z)$, for 'transverse', for p_{xx} and p_{yy}. The transverse components are again equal to p in the bulk phases but are large and negative, often around -100 bar, in the interface itself. The surface is now the integrated difference of the normal and transverse pressures (or stresses) across the thickness of the interface;

$$\sigma = \int [p_N - p_T(z)]dz. \tag{5.63}$$

Such an approach is implicit in the work of some of van der Waals's school, notably that of Hulshof, who derived this equation [237], but the formal use of the pressure tensor came later; it is to be found, for example, in Bakker's treatise of 1928 [238].

The tension $p_T(z)$ produces a moment about an arbitrarily chosen height, z, but there will be a certain height, z_s, called the 'surface of tension' about which this moment is zero. This is defined by a second integral across the interface,

$$\sigma z_s = \int z[p_N - p_T(z)]dz, \tag{5.64}$$

and may be regarded as the height at which the surface tension is presumed to act. We are now entering deep waters since these formal equations, 5.63 and 5.64, are useful only if we know how to calculate p_N and p_T from the intermolecular forces. The first presents little difficulty since it is equal to the pressure in the homogeneous gas and for that we have an adequate theory, for example the virial equation of state. The second, however, presents not only the problem of its calculation but even of its definition. Forces act on discrete molecules, but the concept of pressure or stress is one of continuum mechanics that calls for its definition at each point in space, whether there is a molecule there or not. In a homogeneous system this is no problem since every self-consistent way of summing and averaging the intermolecular forces gives the same answer, namely the 'virial' expression of eqn 5.44 for a system with forces acting centrally between spherical molecules. There is, however, no way of averaging the forces in an inhomogeneous system to give a uniquely-defined pressure tensor.

The first way the problem was tackled was to define the pressure across an element of area, dA, of given position and orientation, by erecting a cylinder on dA, perpendicular to its plane, and then calculating the interaction of the molecules

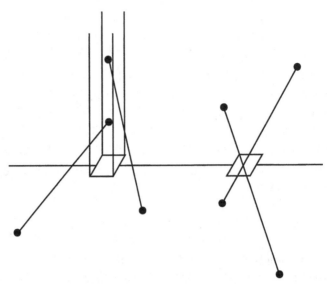

Fig. 5.5 Two ways of describing which pairs of molecules contribute to the stress (or pressure) across a small element of area in a surface. In the first case (*left*) it is the forces between the molecules in the thin column of material above and perpendicular to the element and all those in the bulk material below it (cf. Laplace's representation in Fig. 3.1). In the second case (*right*) it is the forces between all pairs of molecules, one above and one below the element, whose lines of centres pass through the element.

(or, more generally, of the matter) within this cylinder with all those in the half-space below $\mathbf{d}A$ (Fig. 5.5, *left*). This definition was adopted by Poisson [239], Cauchy [240], and Lamé and Clapeyron [241]. Its origin is not given but it may have derived from Laplace's treatment at the opening of his *Sur l'action capillaire* (see Section 3.2 and Fig. 3.1). A second way of calculating the pressure arose, according to Saint-Venant, from the parallel problem of the flow of heat across an element of area, as treated by Fourier [242]. Here one takes into account the forces between all pairs of molecules whose lines of centres pass through the element of area (Fig. 5.5, *right*). When he heard of this way of calculating the stress Cauchy wrote that it seemed to him to be "more exact" for a system of molecules interacting in pairs [243]. For the sake of definiteness, we may call the earlier pressure tensor the first, and the later the second. The first is, perhaps, the more natural if one is considering the stress arising from matter as an interacting continuum, and the second if one is considering it as composed of molecules interacting in pairs, but either may be used with both suppositions. It was the appearance of Cauchy's short paper that prompted Saint-Venant to give a brief history of the subject, saying that he had used the second definition since 1834, and that Duhamel had used it briefly in 1828 before reverting to the older one of Poisson and Cauchy [244]. In a homogeneous fluid they are equivalent, as Poisson proved in 1823 for the parallel problem of heat flow [245]. They differ if there is a density gradient, as in the interface between liquid and gas. The two expressions to which the definitions lead are, as follows [246]:

$$p_T(z) = kTn(z) - \frac{1}{4} \int u'(r_{12})\left[(x_{12}^2 + y_{12}^2)/r_{12}\right] n^{(2)}(r_{12}, z, z + z_{12}) d\mathbf{r}_{12}, \quad (5.65)$$

$$p_T(z) = kTn(z) - \frac{1}{4} \int u'(r_{12}) \left[(x_{12}^2 + y_{12}^2)/r_{12}\right]$$

$$\times \int_0^1 n^{(2)}(r_{12}, z - \alpha z_{12}, z + (1-\alpha)z_{12}) d\alpha d\mathbf{r}_{12}, \quad (5.66)$$

where $n^{(2)}(r_{12}, z', z'')$ is the probability of finding a pair of molecules at (x_1, y_1, z') and at (x_2, y_2, z'') and separated by the distance r_{12}. We can see at once that the first expression is formally simpler than the second. If z is situated in either of the homogeneous phases, gas or liquid, then $n^{(2)}$ becomes simply $n^2 g(r_{12})$ and both expressions reduce to eqn 5.44. Within the interface, however, eqns 5.65 and 5.66 lead to different results. If they are inserted into eqn 5.63 they lead to the same value of the surface tension, but in eqn 5.64 they give different values for the height of the surface of tension, z_s. The uncertainty in z_s is small, less than the range of the intermolecular force or the thickness of the interface, but the difference shows the arbitrariness of the choice of the definition of the pressure.

The same ignorance of the past that afflicted the statistical mechanics of liquids in the 1920s, 1930s and 1940s was now again apparent. The definitions of the elasticians of the 19th century were unknown to the physicists who, in the 1950s, turned again to the problems of capillarity. Kirkwood and Buff used the first form of $p_T(z)$ in their first paper of 1949 in which they obtained eqn 5.54, but a more 'statistical mechanical' derivation of this equation, free from any explicit introduction of the pressure tensor, soon followed [247]. McLellan used the same form of the tensor in 1953 [248]. In 1950 Irving and Kirkwood [249] introduced the second form. Some years later Harasima discussed both forms and, unknowingly echoing Cauchy, described the second as the "more reasonable" [250]. It is now conventional in this field to call the two forms of the tensor the Harasima pressure, $p_H(z)$, which is the first form, and the Irving–Kirkwood pressure, $p_{IK}(z)$, which is the second. It is a convenient convention even if it does not do justice to the history of the 1950s, and still less to that of the 19th century.

If there are two possible and apparently equally valid ways of defining the pressure, then does it follow that this concept is of little meaning in an inhomogeneous system? This seemed to be the case when, in 1982, P. Schofield and J.R. Henderson showed that there were arbitrarily many ways of defining the tensor, all of which led to the same value for the surface tension which is the only thermodynamic property of the interface that can be measured [251]. The root of the difficulty is that forces act on molecules and molecules occupy definable positions, at least in a classical mechanical system, whereas the tensor tries to define the pressure everywhere, whether there is a molecule there or not. Attempts are still being made to define the pressure in planar and curved interfaces in ways that overcome this difficulty, for example by arbitrarily requiring the components of the tensor to be derivatives of a vector field, as is necessary for the strain tensor (see Section 3.6), and other restrictions of this kind. These are still matters of unresolved discussion.

5.6 Conclusion

Is there a conclusion? In one sense there is not; no field of science can ever be said to be exhausted, and in the field of cohesion there are still many unsolved problems. We know the origins of the intermolecular forces, and in a few simple cases can calculate their magnitude from first principles. We can use this knowledge to calculate the properties of the monatomic gases at low and moderate densities, and the equilibrium properties of these gases at high densities and of liquid and solid substances composed of not-too-complicated molecules. Beyond these limits we are struggling. We cannot calculate with acceptable accuracy the viscosity, thermal conductivity and other transport properties of monatomic gases at high densities or of monatomic liquids. Even the transport properties of polyatomic gases at low densities are beyond us. Nevertheless the common perception is that the field is not at the moment one of the exciting areas of research. There are these fundamental limitations on our abilities to make accurate calculations, which no one yet knows how to overcome, and which few are willing to tackle. Much of the interest in the more active parts of the field is in the application of the theoretical knowledge that we now have to biological problems and to those of material science. Indeed much has already been done that has not been discussed here in such fields as the strength of metals, ceramics and composite materials and in understanding the phase behaviour of liquid crystals, colloids and other mesoscopic systems. The interpretation of such systems often requires an understanding of subtle indirect effects of the intermolecular forces. Here two examples may be cited from fields that are currently fashionable.

The first is what is called the hydrophobic effect, which describes the change in the structure of water on disolving in it molecules which, in whole or in part, have little affinity for forming hydrogen bonds. Such entities might be the lower hydrocarbon gases or molecules with a hydrocarbon chain attached to a strongly polar group. It is found that the structure of water around the non-polar groups is modified in ways that were difficult to predict and that one consequence of such modifications can be an apparent attractive force between the non-polar parts of different molecules. The results of the study of this effect has led to some advance in our understanding of the way that some systems of biological interest order their structures and, indeed, it is those interested in such problems as the folding of proteins who have driven much of the work in this field, although the first studies were on much simpler systems [252].

The second topic that involves indirect effects is what is now usually called the depletion force. We have seen (Section 5.5) that in a dense fluid the probability of finding two molecules at a separation of a little greater than their collision diameter is larger than random, and that this increase is found even in the absence of a direct attractive force between the pair. In 1948 de Boer pointed out that this effect occurs even for a system of two molecules in the presence of a third since at short distances each of the pair partially shields the other from collisions with the third, thus generating a value of the pair distribution function $g(r)$ larger than unity, or a negative or attractive value for the potential of average force, $-kT \ln g(r)$ [253]. The effect is stronger at high densities and stronger still in a dense assembly of large hard spheres in a 'sea' of smaller ones if the ratio of the diameters is about 10 to 1. It was first suggested by Biben and Hansen that the average force of attraction between the large spheres in such a system was strong enough to induce a separation into two fluid phases [254]. It now seems unlikely that this happens in an equilibrium state – the large spheres crystallise first as the density is raised – but it would probably occur in a metastable phase [255]. An example from the real world was put forward by Asakura and Oosawa in 1954, and independently by Vrij in 1976 [256]. Here the 'large spheres' were colloidal particles and the role of the small ones was taken by polymer molecules that could not insert themselves between the colloidal particles if these were close together. It is this lowering of the concentration of the particles of the smaller component in the space between the larger that gives rise to the attractive average force between the larger, and so to the name of 'depletion force'. Since the effective attraction has been produced without any direct attractive energy the effect is sometimes described as an entropic attraction. It is a modern version of Le Sage's theory of interparticle attraction (Section 2.4) with the polymer molecules playing the role of his 'ultramondane particles'.

Experimental advances are hard to predict since they often come from discoveries in fields remote from those under study. It is already clear, however, that the recent advances in molecular spectroscopy have opened the field of van der Waals or molecular clusters to a more detailed examination than seemed possible only a few years ago. It will, however, be our understanding of more complex systems that will benefit most from advances such as atomic force microscopy, scanning tunnelling microscopy, the ability to manipulate single atoms with intense laser beams – the so-called 'optical tweezers'– and other methods that may be devised for studying molecular systems directly in the laboratory.

It is hard to say how much we shall learn from computer modelling since the power of computers seems to grow without limit, but here I sense a feeling of satiation, at least for straightforward molecular systems. Much has been learnt, and simulation played a crucial role in solving many past problems, but today's work does not seem to have quite the same brightness and promise associated with the field twenty years ago. Again it is complex systems that are now attracting most attention, in which some of the 'fine-grained' molecular detail is suppressed and the model is chosen to do justice only to broad features of the system on a meso-molecular scale. There has, for example, been a recent announcement from the computer company IBM of a dedicated machine to predict the folding patterns of proteins from a knowledge of their sequence of amino-acids. When we remember that an accurate modelling of the water–water potential required 72 parameters then we can appreciate that the simulation of the interactions of chains of amino-acids in the presence of water can only be undertaken by essentially empirical methods. It will be interesting to see how far the modellers can go down such roads as protein folding.

Prophecy is impossible, however, and all that can be recorded is that the field of cohesion, which has had an episodic history of starting and then pausing again for the last three hundred years, has now reached, certainly not a conclusion, but a natural break in its development where the next advances will come in applications rather than in fundamental changes in our understanding. The most important attractive force, London's dispersion force, has been understood since 1930, and it is in this sense that this last chapter has been entitled 'Resolution'. The direct electrostatic forces that were so widely studied at the beginning of the 20th century also now present no fundamental problems. No doubt this is not the end of the story but new theories, and advances in understanding, supplement rather than supplant the old theories. Most of our day-to-day physical problems can still be resolved in terms of Newtonian mechanics and Maxwell's electromagnetic theory. These were subsumed into the quantum mechanics of the 20th century but they were not rendered false or obsolete. Quantum mechanics has changed fundamentally the way we think about things on a small scale but its limiting behaviour for atomically large masses and distances still allows us to retain many of our old ideas without leading us into error. We know now that the domain of validity of Newton's and Maxwell's work is limited but within their limits they retain their correctness and usefulness. The dispersion forces are outside the scope of the classical theories but they, in turn, can be adequately understood in terms of present-day quantal theory. When this eventually becomes absorbed into a 'theory of everything' [257], then we shall have a deeper understanding, but we shall surely still use the same conventional quantum mechanics and statistical mechanics for our calculations of intermolecular forces and the properties of gases, liquids and solids, in the same way that we continue to use Newtonian mechanics for the solution of the problems of the motion of planets and billiard balls.

Notes and references

1 L. Boltzmann, 'On certain questions of the theory of gases', *Nature* **51** (1895) 413–15; reprinted in his *Theoretical physics and philosophical problems*, ed. B. McGuinness, Dordrecht, 1974, pp. 201–9.
2 P.A.M. Dirac (1902–1984) O. Darrigol, DSB, v. 17, pp. 224–33; R.H. Dalitz and R. Peierls, *Biog. Mem. Roy. Soc.* **32** (1986) 139–85. P.A.M. Dirac, 'Quantum mechanics of many-electron systems', *Proc. Roy. Soc. A* **123** (1929) 714–33.
3 M. Born and R. Oppenheimer, 'Zur Quantentheorie der Molekeln', *Ann. Physik* **84** (1927) 457–84.
4 J.C. Slater (1900–1976) L. Hoddeson, DSB, v. 18, pp. 832–6; P.M. Morse, *Biog. Mem. U.S. Nat. Acad. Sci.* **53** (1982) 297–321. J.C. Slater, *Solid-state and molecular theory: a scientific biography*, New York, 1975; 'The normal state of helium', *Phys. Rev.* **32** (1928) 349–60.
5 Shou Chin Wang (b.1905). Wang was a Chinese student who took a Master's degree at Harvard in 1926 and then a Doctorate at Columbia. He made a few more contributions to molecular quantum mechanics but seems to have left the field in 1929; by 1934 he was back in China and I know nothing of his later career. J.C. Slater, ref. 4, 1975, pp. 151–5; S.G. Brush, *Statistical physics and the atomic theory of matter, from Boyle and Newton to Landau and Onsager*, Princeton, NJ, 1983, pp. 210, 355; *Harvard Alumni Directory*, 1934.
6 S.C. Wang, 'Die gegenseitige Einwirkung zweier Wasserstoffatome', *Phys. Zeit.* **28** (1927) 663–6.
7 The probable source of his error was found later by L. Pauling and J.Y. Beach, 'The van der Waals interaction of hydrogen atoms', *Phys. Rev.* **47** (1935) 686–92. L. C. Pauling (1901–1994) J.D. Dunitz, *Biog. Mem. Roy. Soc.* **42** (1996) 315–38 and *Biog. Mem. U.S. Nat. Acad. Sci.* **71** (1997) 221–61.
8 F. London (1900–1954) C.W.F. Everitt and W.M. Fairbank, DSB, v. 8, pp. 473–9; K. Gavroglu, *Fritz London, a scientific biography*, Cambridge, 1995.
9 As London later told A.B. Pippard, see Gavroglu, ref. 8, pp. 44, 51.
10 W. Heitler and F. London, 'Wechselwirkung neutraler Atome und homöopolare Bindung nach der Quantenmechanik', *Zeit. f. Phys.* **44** (1927) 455–72. G.N. Lewis first described covalent bonding in terms of shared pairs of electrons in 'The atom and the molecule', *Jour. Amer. Chem. Soc.* **38** (1916) 762–85. The best survey of valency under the old quantum theory is by N.V. Sidgwick, *The electronic theory of valency*, Oxford, 1927.
11 F. London, 'Die Bedeutung der Quantentheorie für die Chemie', *Naturwiss.* **17** (1929) 516–29.
12 R.K. Eisenschitz (1898–1968) Eisenschitz left Germany in 1933 and worked for thirteen years at the Royal Institution in London. In 1946 he moved to London University and finished his career as Professor of Theoretical Physics at Queen Mary College. His later work was mainly on problems of classical physics. *Who was who,*

1961–1970, London, 1972. R. Eisenschitz and F. London, 'Über das Verhältnis der van der Waalsschen Kräfte zu den homöopolaren Bindungskräften', *Zeit. f. Phys.* **60** (1930) 491–527. For a modern account of all forms of intermolecular forces, see A.J. Stone, *The theory of intermolecular forces*, Oxford, 1996.

13 J.E. Lennard-Jones, 'Perturbation problems in quantum mechanics', *Proc. Roy. Soc. A* **129** (1930) 598–615.

14 H.R. Hassé, 'The calculation of the van der Waal [sic] forces for hydrogen and helium at large inter-atomic distances', *Proc. Camb. Phil. Soc.* **27** (1931) 66–72.

15 J.G. Kirkwood (1907–1959) J. Ross, DSB, v. 7, p. 387; S.A. Rice and F.H. Stillinger, *Biog. Mem. U.S. Nat. Acad. Sci.* **77** (1999) 162–74.

16 J.C. Slater and J.G. Kirkwood, 'The van der Waals forces in gases', *Phys. Rev.* **37** (1931) 682–97.

17 F. London, 'Über einiger Eigenschaften und Anwendungen der Molekularkräfte', *Zeit. phys. Chem.* **B11** (1930) 222–51. For another review, see 'Zur Theorie und Systematik der Molekularkräfte', *Zeit. f. Phys.* **63** (1930) 245–79.

18 P. Drude, *The theory of optics*, New York, 1902, p. 382ff. The original German edition was published in 1900.

19 H. Margenau (1901–1997) Pogg., v. 6, pp. 1647–8; v. 7a, pp. 199–200. Henry Margenau was born in Germany and spent his career from 1939 at Yale.

20 H. Margenau, 'The role of quadrupole forces in van der Waals attractions', *Phys. Rev.* **38** (1931) 747–56. This work was undertaken after a suggestion to the author from Ya. Frenkel that quadrupolar forces might not be negligible.

21 H. Margenau, 'Van der Waals forces', *Rev. Mod. Phys.* **11** (1939) 1–35.

22 J.G. Kirkwood and F.G. Keyes, 'The equation of state of helium', *Phys. Rev.* **37** (1931) 832–40.

23 J.E. Lennard-Jones, 'Cohesion', *Proc. Phys. Soc.* **43** (1931) 461–82.

24 R.A. Buckingham (1911–1994) *Who was who, 1991–1995*, London, 1996. He became Professor of Computing Science at University College, London, in 1963. R.A. Buckingham, 'The classical equation of state of gaseous helium, neon and argon', *Proc. Roy. Soc. A* **168** (1938) 264–83. He had earlier reported different values of the parameters in the second edition of R.H. Fowler, *Statistical mechanics*, Cambridge, 1936, p. 306. The 1938 value for C_6 of helium was itself corrected by 4% for a "numerical slip" in R.H. Fowler and E.A. Guggenheim, *Statistical thermodynamics*, Cambridge, 1939, p. 285.

25 T.D.H. Baber and H.R. Hassé, 'A comparison of wave functions for the normal helium atom', *Proc. Camb. Phil. Soc.* **33** (1937) 253–9.

26 G. Starkschall and R.G. Gordon, 'Improved error bounds for the long-range forces between atoms', *Jour. Chem. Phys.* **54** (1971) 663–73.

27 J.O. Hirschfelder, R.B. Ewell and J.R. Roebuck, 'Determination of intermolecular forces from the Joule–Thomson coefficients', *Jour. Chem. Phys.* **6** (1938) 205–18. For J.O. Hirschfelder (1911–1990), see R. B. Bird, C.F. Curtiss and P.R. Certain, *Biog. Mem. U.S. Nat. Acad. Sci.* **66** (1995) 191–205. Hirschfelder soon became a prominent player in this field. After the War he directed the Naval Research Laboratory at the University of Wisconsin.

28 G.E. Uhlenbeck and E. Beth, 'The quantum theory of the non-ideal gas; I. Deviations from classical theory', *Physica* **3** (1936) 729–45; '...; II. Behaviour at low temperatures, *ibid.* **4** (1937) 915–24. The second paper was the first in a symposium held in Amsterdam to mark the centenary of van der Waals's birth.

29 H.S.W. Massey and C.B.O. Mohr, 'Free paths and transport phenomena in gases and the quantum theory of collisions. I. The rigid sphere model', *Proc. Roy. Soc. A* **141**

(1933) 434–53; '. . . . II. The determination of the laws of force between atoms and molecules', *ibid.* **144** (1934) 188–205.

30 E.H. Kennard, *Kinetic theory of gases*, New York, 1938, p. 160.

31 W. Nernst, 'Kinetische Theorie fester Körper', in M. Planck *et al.*, *Vorträge über die kinetische Theorie der Materie und der Elektrizität*, Leipzig, 1914, pp. 61–86, see p. 66.

32 F. Simon and C. von Simson, 'Die Kristallstruktur des Argons', *Zeit. f. Phys.* **25** (1924) 160–4.

33 F. Born, 'Über Dampfdruckmessungen an reinem Argon', *Ann. Physik* **69** (1922) 473–504.

34 L. Holborn and J. Otto, 'Über die Isothermen einiger Gase zwischen + 400° und −183°' [−100 °C for argon], *Zeit. f. Phys.* 33 (1925) 1–11. Earlier measurements by H. Kamerlingh Onnes and C.A. Crommelin, 'Isotherms of monatomic gases and their binary mixtures. VII. Isotherms of argon between +20 °C and −150 °C', *Proc. Sect. Sci. Konink. Akad. Weten. Amsterdam* **13** (1911) 614–25, extended to lower temperatures but were thought to be less accurate.

35 M.T. Trautz (1880–1960) Pogg., v. 4, p. 1521; v. 5, pp. 1267–8; v. 6, pp. 2683–4; v. 7a, pp. 705–6. M. Trautz and R. Zink, 'Die Reibung, Wärmeleitung und Diffusion in Gasmischungen; XII. Gasreibung bei höheren Temperaturen', *Ann. Physik* **7** (1930) 427–52.

36 C.A. Crommelin, 'Isothermals of monatomic substances and their binary mixtures. XV. The vapour pressure of solid and liquid argon, from the critical point down to −206°', *Proc. Sect. Sci. Konink. Akad. Weten. Amsterdam* **16** (1913) 477–85; '. . . . XVI. New determination . . . down to −205°', *ibid.* **17** (1914) 275–7. A useful *Bibliography of thermophysical properties of argon from 0 to 300 °K* was compiled by L.A. Hall, J.G. Hurst and A.L. Gosman, National Bureau of Standards, Tech. Note 217, Washington, DC, 1964, and was extended to a wider range of substances by V.A. Rabinovich, A.A. Vasserman, V.I. Nedostup and L.S.Veksler, *Thermophysical properties of neon, argon, krypton, and xenon*, Washington, DC, 1988, a translation of the Russian original of 1976 .

37 W.H. Keesom and J. De Smedt, 'On the diffraction of Röntgen-rays in liquids', *Proc. Sect. Sci. Konink. Akad. Weten. Amsterdam* **25** (1922–1923) 118–24; **26** (1923) 112–15.

38 F. Zernike and J.A. Prins, 'Die Beugung von Röntgenstrahlen in Flüssigkeiten als Effekt der Molekülanordnung', *Zeit. f. Phys.* **41** (1927) 184–94.

39 F. London, 'The general theory of molecular forces', *Trans. Faraday Soc.* **33** (1937) 8–26. This paper contains an English version of his calculation of the dispersion force from the Drude model.

40 R.A. Buckingham, 'The quantum theory of atomic polarization; I. Polarization in a uniform field', *Proc. Roy. Soc.* A **160** (1937) 94–113; '. . . ; II. The van der Waals energy of two atoms', *ibid.* 113–26.

41 A. Müller, Appendix to 'The van der Waals potential and lattice energy of a n-CH_2 chain molecule in a paraffin crystal', *Proc. Roy. Soc.* A **154** (1936) 624–39.

42 J. Corner, 'Zero-point energy and lattice distances', *Trans. Faraday Soc.* **35** (1939) 711–16. John Corner was a student of Fowler and Lennard-Jones at Cambridge who worked on ballistics during the War, see J. Corner, *Theory of the internal ballistics of guns*, New York, 1950.

43 R.A. Buckingham and J. Corner, 'Tables of second virial and low-pressure Joule–Thomson coefficients for intermolecular potentials with exponential repulsion', *Proc Roy. Soc.* A **189** (1947) 118–29.

44 K.F. Herzfeld and M. Goeppert Mayer, 'On the theory of fusion', *Phys. Rev.* **46** (1934) 995–1001.

45 J.E. Lennard-Jones, 'The equation of state of gases and critical phenomena', *Physica* **4** (1937) 941–56. The value of r_m in this paper is 3.819 Å, but 3.825 Å is consistent with the other parameters.

46 Buckingham, ref. 24 (1938), and the same figures in Fowler and Guggenheim, ref. 24, p. 293.

47 G. Kane, 'The equation of state of frozen neon, argon, krypton, and xenon', *Jour. Chem. Phys.* **7** (1939) 603–13.

48 J. Corner, 'Intermolecular potentials in neon and argon', *Trans. Faraday Soc.* **44** (1948) 914–27.

49 See M. Born and J.E. Mayer, 'Zur Gittertheorie der Ionenkristalle', *Zeit. f. Phys.* **75** (1932) 1–18, and W.E. Bleick and J.E. Mayer, 'The mutual repulsive potential of closed shells' [i.e. neon], *Jour. Chem. Phys.* **2** (1934) 252–9. Joseph Mayer (1904–1983) was the husband of the Nobel prize winner Maria Goeppert Mayer, ref. 44; see B.H. Zimm, *Biog. Mem. U.S. Nat. Acad. Sci.* **65** (1994) 211–20.

50 J.O. Hirschfelder, C.F. Curtiss and R.B. Bird, *Molecular theory of gases and liquids*, New York, 1954.

51 Hirschfelder, Curtiss and Bird, ref. 50, text and Table 13.3–1, p. 966.

52 J. de Boer and J. van Kranendonk, 'The viscosity and heat conductivity of gases with central intermolecular forces', *Physica* **14** (1948) 442–52; J.O. Hirschfelder, R.B. Bird and E.L. Spotz, 'The transport properties for non-polar gases', *Jour. Chem. Phys.* **16** (1948) 968–81; *ibid.* **17** (1949) 1343–4; J.S. Rowlinson, 'The transport properties of non-polar gases', *ibid.* **17** (1949) 101.

53 T. Kihara and M. Kotani, 'Determination of intermolecular forces from transport phenomena in gases. II', *Proc. Phys.-Math. Soc. Japan* **25** (1943) 602–14. There is an earlier paper, Part I, by Kotani, *ibid.* **24** (1942) 76–95, which is a calculation for the Sutherland or $(\infty, 6)$ potential, but without the assumption made previously that the attractive forces are weak. Taro Kihara (b.1917) became Professor of Physics at Tokyo in 1958.

54 H.L. Johnston and E.R. Grilly, 'Viscosities of carbon monoxide, helium, neon, and argon between 80° and 300 °K. Coefficients of viscosity', *Jour. Phys. Chem.* **46** (1942) 948–63.

55 Hirschfelder, Curtiss and Bird, ref. 50, pp. 561–2 and Appendix, Table 1-A, p. 1110.

56 E.A. Mason, 'Transport properties of gases obeying a modified Buckingham (exp-six) potential', *Jour. Chem. Phys.* **22** (1954) 169–86; W.E. Rice and J.O. Hirschfelder, 'Second virial coefficients of gases obeying a modified Buckingham (exp-six) potential', *ibid.* 187–92. The modification was the trivial one of removing a spurious maximum in $u(r)$ at very small values of r.

57 E.A. Mason and W.E. Rice, 'The intermolecular potentials for some simple nonpolar molecules', *Jour. Chem. Phys.* **22** (1954) 843–51.

58 For these simulations, see W.W. Wood, 'Early history of computer simulations in statistical mechanics' in *Molecular-dynamics simulation of statistical–mechanical systems*, Proceedings of the International School of Physics 'Enrico Fermi', Course 97, Amsterdam, 1986, pp. 3–14.

59 W.W. Wood and F.R. Parker, 'Monte Carlo equation of state of molecules interacting with the Lennard-Jones potential. I. A supercritical isotherm at about twice the critical temperature', *Jour. Chem. Phys.* **27** (1957) 720–33.

60 P.W. Bridgman (1882–1961) E.C. Kemble, F. Birch and G. Holton, DSB, v. 2, pp. 457–61; P.W. Bridgman, 'Melting curves and compressibilities of nitrogen and argon', *Proc. Amer. Acad. Arts Sci.* **70** (1935) 1–32.

61 A.M.J.F. Michels (1889–1969) Pogg., v. 6, p. 1726; v. 7b, pp. 3264–7. For an account of the life and work of Michels and of the laboratory that he developed, see J.M.H. Levelt Sengers and J.V. Sengers, 'Van der Waals Fund, Van der Waals Laboratory and Dutch high-pressure science', *Physica* A **156** (1989) 1–14, and J.M.H. Levelt Sengers, 'The laboratory founded by Van der Waals', *Int. Jour. Thermophysics* **22** (2001) 3–22. A. Michels, Hub. Wijker and Hk. Wijker, 'Isotherms of argon between 0 °C and 150 °C and pressures up to 2900 atmospheres', *Physica* **15** (1949) 627–33.

62 J. de Boer and A. Michels, 'Quantum-mechanical theory of the equation of state. Law of force of helium', *Physica* **5** (1938) 945–57. Jan de Boer (b.1911) studied at Amsterdam where he later became Professor of Theoretical Physics. For a review of his life's work at the meeting to mark his 70th birthday, see E.G.D. Cohen, 'Enige persoonlijke reminiscenties aan Jan de Boer', *Nederlands Tijdschrift voor Natuurkunde* **A47** (1981) 124–8.

63 K.S. Pitzer, 'Corresponding states for perfect liquids', *Jour. Chem. Phys.* **7** (1939) 583–90.

64 A. Byk, 'Das Theorem der übereinstimmenden Zustände und die Quantentheorie der Gase und Flüssigkeiten', *Ann. Physik* **66** (1921) 157–205; 'Zur Quantentheorie der Gase und Flüssigkeiten', *ibid.* **69** (1922) 161–201.

65 B.M. Axilrod and E. Teller, 'Interaction of the van der Waals type between three atoms', *Jour. Chem. Phys.* **11** (1943) 299–300; B.M. Axilrod, 'The triple-dipole interaction between atoms and cohesion in crystals of the rare gases', *ibid.* **17** (1949) 1349. Detailed calculations followed later, see B.M. Axilrod, 'Triple-dipole interaction. I. Theory', *ibid.* **19** (1951) 719–24; '. . . . II. Cohesion in crystals of the rare gases', *ibid.* 724–9.

66 Y. Muto, Letter to Axilrod in March 1948, see Axilrod, ref. 65 (1949). Muto's work was published in Japanese: Y. Muto, [The force between nonpolar molecules], *Nihon Sugaku Butsuri Gakkaishi* [*Jour. Phys.-Math. Soc. Japan*] **17** (1943) 629–31. The often-quoted reference to the European language journal, *Proc. Phys.-Math. Soc. Japan*, is incorrect. I thank Richard Sadus of Melbourne for a copy of Muto's paper and for the observation that there is an error of sign in his result, eqn 15.

67 K.F. Niebel and J.A. Venables, 'An explanation of the crystal structure of the rare gas solids', *Proc. Roy. Soc. A* **336** (1974) 365–77.

68 E.A. Guggenheim (1901–1970) F.C. Tompkins and C.F. Goodeve, *Biog. Mem. Roy. Soc.* **17** (1971) 303–26; E.A. Guggenheim, [no title] *Discuss. Faraday Soc.* **15** (1953) 108–10. The evidence in favour of the (12, 6) potential was reviewed by J.S. Rowlinson, [no title] *ibid.* 108–9.

69 E.A. Guggenheim and M.L. McGlashan, 'Interaction between argon atoms', *Proc. Roy. Soc. A* **255** (1960) 456–76. Guggenheim gave the substance of this paper in his Baker Lectures at Cornell in 1963 and repeated it in his *Applications of statistical mechanics*, Oxford, 1966. Max McGlashan (1924–1997), Guggenheim's only Ph.D. student, was later Professor of Chemistry at Exeter and at University College, London. What is essentially a revision of this calculation but with similar conclusions is in M.L. McGlashan, 'Effective pair interaction energy in crystalline argon', *Discuss. Faraday Soc.* **40** (1965) 59–68.

70 G.C. Maitland, M. Rigby, E.B. Smith and W.A. Wakeman, *Intermolecular forces: their origin and determination*, Oxford, 1981. There is a short history of recent work in Chapter 9 which is valuable since it was written by those in the thick of things. This account makes use of it. The same authors, but now Rigby, Smith, Wakeham and Maitland, later published a simpler version of this monograph as *The forces between molecules*, Oxford, 1986.

71 I. Amdur and E.A. Mason, 'Scattering of high-velocity neutral particles.
    III. Argon–argon', *Jour. Chem. Phys.* **22** (1954) 670–1.

72 R.J. Munn, 'On the calculation of the dispersion-forces coefficient directly from
    experimental transport data', *Jour. Chem. Phys.* **42** (1965) 3032–3; J.S. Rowlinson,
    'Determination of intermolecular forces from macroscopic properties', *Discuss.
    Faraday Soc.* **40** (1965) 19–26.

73 A. Michels, J.M. Levelt and W. de Graaff, 'Compressibility isotherms of argon
    at temperatures between −25°C and −155°C, and at densities up to 640 Amagat
    (pressures to 1050 atmospheres)', *Physica* **24** (1958) 659–71. After her marriage,
    Levelt published under the name of Levelt Sengers.

74 B.E.F. Fender and G.D. Halsey, 'Second virial coefficients of argon, krypton, and
    argon–krypton mixtures at low temperatures', *Jour. Chem. Phys.* **36** (1962) 1881–8;
    R.D. Weir, I.W. Jones, J.S. Rowlinson and G. Saville, 'Equation of state of gases at
    low temperatures. Part I. Second virial coefficient of argon and krypton', *Trans.
    Faraday Soc.* **63** (1967) 1320–9; M.A. Byrne, M.R. Jones and L.A.K. Staveley,
    'Second virial coefficients of argon, krypton and methane and their binary mixtures at
    low temperatures', *ibid.* **64** (1968) 1747–56. The change of the speed of sound with
    gas pressure can be measured with a higher accuracy than the change of density and
    yields the 'second acoustic virial coefficient' which can be expressed in terms of $B(T)$
    and its first two derivatives with respect to temperature. It has proved difficult to use it
    directly to determine intermolecular potentials but it serves as a valuable check; see,
    for example, M.B. Ewing, A.A. Owusu and J.P.M. Trusler, 'Second acoustic virial
    coefficients of argon between 100 and 304 K', *Physica A* **156** (1989) 899–908.

75 T. Kihara, 'The second virial coefficent of non-spherical molecules', *Jour. Phys. Soc.
    Japan* **6** (1951) 289–96; J.S. Rowlinson, 'Intermolecular forces in CF_4 and SF_6', *Jour.
    Chem. Phys.* **20** (1952) 337; S.D. Hamann and J.A. Lambert, 'The behaviour of fluids
    of quasi-spherical molecules, I. Gases at low densities', *Aust. Jour. Chem.* **7**
    (1954) 1–17; A.G. De Rocco and W.G. Hoover, 'Second virial coefficient for the
    spherical shell potential', *Jour. Chem. Phys.* **36** (1963) 916–26.

76 T. Kihara, 'Virial coefficients and models of molecules in gases', *Rev. Mod. Phys.* **25**
    (1953) 831–43. This review was written on a visit to Hirschfelder's laboratory at
    Wisconsin.

77 A.L. Myers and J.M. Prausnitz, 'Second virial coefficients and Kihara parameters for
    argon', *Physica* **28** (1962) 303–4.

78 D.D. Konowalow and J.O. Hirschfelder, 'Intermolecular potential functions for
    nonpolar molecules', *Phys. Fluids* **4** (1961) 629–36.

79 J.A. Barker, W. Fock and F. Smith, 'Calculation of gas transport properties and the
    interaction of argon atoms', *Phys. Fluids* **7** (1964) 897–903. For J.A. Barker
    (1925–1995) see J.S. Rowlinson, *Biog. Mem. Roy. Soc.* **42** (1996) 13–22. John
    Barker of Melbourne worked later in Canada and then in California, with
    I.B.M.

80 A.E. Sherwood and J.M. Prausnitz, 'Third virial coefficient for the Kihara, exp-6, and
    square-well potentials', *Jour. Chem. Phys.* **41** (1964) 413–28; 'Intermolecular
    potential functions and the second and third virial coefficients', *ibid.* 429–37.

81 W.B[yers]. Brown, 'The statistical thermodynamics of mixtures of Lennard-Jones
    molecules', *Phil. Trans. Roy. Soc. A* **250** (1957) 175–220, 221–46. Equation 5.27 is
    clearly related to the two equations of Simon and von Simson, eqns 4.58 and 4.59,
    but I do not think that the connection has been explored.

82 J.S. Rowlinson, 'A test of Kihara's intermolecular potential', *Molec. Phys.* **9** (1965)
    197–8.

83 W.B[yers]. Brown and J.S. Rowlinson, 'A thermodynamic discriminant for the Lennard-Jones potential', *Molec. Phys.* **3** (1960) 35–47.

84 J.S. Rowlinson, 'The use of the isotopic separation factor between liquid and vapour for the study of intermolecular potential and virial functions', *Molec. Phys.* **7** (1964) 477–80.

85 A.E. Sherwood, A.G. De Rocco and E.A. Mason, 'Nonadditivity of intermolecular forces: Effects on the third virial coefficient', *Jour. Chem. Phys.* **44** (1966) 2984–94.

86 See, for example, McGlashan, ref. 69, for the use of an 'effective' potential.

87 A. Rahman, 'Correlation in the motions of atoms in liquid argon', *Phys. Rev.* **136A** (1964) 405–11.

88 A.E. Kingston, 'Van der Waals forces for the inert gases', *Phys. Rev.* **135A** (1964) 1018–19. More recent calculations confirm this result. The consensus now is that $C_6 = 64$–65 a.u.; A.Kumar and W.J. Meath, 'Pseudo-spectral dipole oscillator strengths and dipole–dipole and triple-dipole dispersion energy coefficients for HF, HCl, HBr, He, Ne, Ar, Kr and Xe', *Molec. Phys.* **54** (1985) 823–33; M.P. Hodges and A.J. Stone, 'A new representation of the dispersion interaction', *ibid.* **98** (2000) 275–86.

89 E.W. Rothe and R.H. Neynaber, 'Atomic-beam measurements of van der Waals forces', *Jour. Chem. Phys.* **42** (1965) 3306–9. An earlier experiment had erroneously led to a value of C_6 that was at least as large as that from the conventional (12, 6) potential, see E.W. Rothe, L.L. Marino, R.H. Neynaber, P.K. Rol, and S.M. Trujillo, 'Scattering of thermal rare gas beams of argon. Influence of the long-range dispersion forces', *Phys. Rev.* **126** (1962) 598–602.

90 R.J. Munn, 'Interaction potential of the inert gases. I', *Jour. Chem. Phys.* **40** (1964) 1439–46; R.J. Munn and F.J. Smith, '. . . . II', *ibid.* **43** (1965) 3998–4002; E.A. Mason, R.J. Munn and F.J. Smith, 'Recent work on the determination of the intermolecular potential functions', *Discuss. Faraday Soc.* **40** (1965) 27–34; J.C. Rossi and F. Danon, 'Molecular interactions in the heavy rare gases', *ibid.* 97–109; J.H. Dymond, M. Rigby and E.B. Smith, 'Intermolecular potential-energy functions for simple molecules', *Jour. Chem. Phys.* **42** (1965) 2801–6; J.H. Dymond and B.J. Alder, 'Pair potential for argon', *ibid.* **51** (1969) 309–20.

91 See the papers in the *Faraday Discussion* in refs. 69, 72 and 90, and the discussion of them.

92 R.J. Munn, [no title], *Discuss. Faraday Soc.* **40** (1965) 130–2.

93 J.A. Barker and A. Pompe, 'Atomic interactions in argon', *Aust. Jour. Chem.* **21** (1968) 1683–94.

94 V. Vasilesco, 'Recherches expérimentales sur la viscosité des gaz aux températures élevées', *Annales Phys. Paris* **20** (1945) 137–76, 292–334. Vasilesco worked in the Laboratoire des Hautes Températures in the University of Paris.

95 J. Kestin and J.H. Whitelaw, 'A relative determination of the viscosity of several gases by the oscillating disk method', *Physica* **29** (1963) 335–56.

96 H.J.M. Hanley and G.E. Childs, 'Discrepancies between viscosity data for simple gases', *Science* **159** (1968) 1114–16.

97 F.A. Guevara, B.B. McInteer and W.E. Wageman, 'High-temperature viscosity ratios for hydrogen, helium, argon, and nitrogen', *Phys. Fluids* **12** (1969) 2493–505.

98 R.A. Dawe and E.B. Smith, 'Viscosity of argon at high temperatures', *Science* **163** (1969) 675–6; 'Viscosity of the inert gases at high temperatures', *Jour. Chem. Phys.* **52** (1970) 693–703. Dawe and Smith found that an unpublished Ph.D. thesis of N.L. Anfilogoff at Imperial College, London in 1932 had led to essentially the same results up to 1288 K as were now being obtained nearly forty years later. They speculated (Smith, private communication, 1998) that Anfilogoff's results had

remained unpublished because they disagreed with those just published by Trautz, the accepted authority in the field. The last word on the 'viscosity problem' was the paper of J.A. Barker, M.V. Bobetic and A. Pompe, 'An experimental test of the Boltzmann equation: argon', *Molec. Phys.* **20** (1971) 347–55.

99 L. Jansen and E. Lombardi, 'Three-atom and three-ion interactions and crystal stability', *Discuss. Faraday Soc.* **40** (1965) 78–96.

100 J.A. Barker, R.A. Fisher and R.O. Watts, 'Liquid argon: Monte Carlo and molecular dynamics calculations', *Molec. Phys.* **21** (1971) 657–73.

101 R.E. Leckenby and E.J. Robbins, 'The observation of double molecules in gases', *Proc. Roy. Soc. A* **291** (1966) 389–412. The calculation of that part of the second virial coefficient that is due to dimers was made by D.E. Stogryn and J.O. Hirschfelder, 'Contribution of bound, metastable, and free molecules to the second virial coefficient and some properties of double molecules', *Jour. Chem. Phys.* **31** (1959) 1531–45.

102 Y. Tanaka and K. Yoshino, 'Absorption spectrum of the argon molecule [i.e. Ar_2] in the vacuum–uv region', *Jour. Chem. Phys.* **53** (1970) 2012–30; E.A. Colbourn and A.E. Douglas, 'The spectrum and ground state potential curve for Ar_2', *ibid.* **65** (1976) 1741–5. Further confirmation was also provided by new scattering experiments, see J.M. Parson, P.E. Siska and Y.T. Lee, 'Intermolecular potentials from crossed-beam differential elastic scattering measurements. IV. Ar + Ar', *ibid.* **56** (1972) 1511–6. Smith reviewed the position for the van der Waals centennial meeting in 1973, see E.B. Smith, 'The intermolecular pair-potential energy functions of the inert gases', *Physica* **73** (1974) 211–25.

103 G.C. Maitland and E.B. Smith, 'The intermolecular pair potential for argon', *Molec. Phys.* **22** (1971) 861–8. An account of the Rydberg–Klein–Rees method of inversion that they used is in Maitland, Rigby, Smith and Wakeman, ref. 70, chap. 7.

104 J.G. Kirkwood, private communication, 1950.

105 J.B. Keller and B. Zumino, 'Determination of intermolecular potentials from thermodynamic data and the law of corresponding states', *Jour. Chem. Phys.* **30** (1959) 1351–3. The first application of this inversion was to helium, see D.A. Jonah and J.S. Rowlinson, [no title], *Discuss. Faraday Soc.* **40** (1965) 55–6; 'Direct determination of the repulsive potential between helium atoms', *Trans. Faraday Soc.* **62** (1966) 1067–71.

106 G.C. Maitland and E.B. Smith, 'The direct determination of potential energy functions from second virial coefficients', *Molec. Phys.* **24** (1972) 1185–201; H.E. Cox, F.W. Crawford, E.B. Smith and A.R. Tindell, 'A complete iterative inversion procedure for second virial coefficient data I. The method', *ibid.* **40** (1980) 705–12; E.B. Smith, A.R. Tindell, B.H. Wells and F.W. Crawford, '. . . II. Applications', *ibid.* **42** (1981) 937–42; E.B Smith, A.R. Tindell, B.H. Wells and D.J. Tildesley, 'On the inversion of second virial coefficient data derived from an undisclosed potential energy function', *ibid.* **40** (1980) 997–8.

107 D.W. Gough, G.C. Maitland and E.B. Smith, 'The direct determination of intermolecular potential energy functions from gas viscosity measurements', *Molec. Phys.* **24** (1972) 151–61.

108 Maitland, Rigby, Smith and Wakeham, ref. 70, pp. 136–43, 361–71, 491, and 602–4; G.C. Maitland, V. Vesovic and W.A. Wakeham, 'The inversion of thermophysical properties I. Spherical systems revisited'; '. . . II. Non-spherical systems explored', *Molec. Phys.* **54** (1985) 287–300, 301–19; J.P.M. Trusler, 'The inversion of second virial coefficients for polyatomic molecules', *ibid.* **57** (1986) 1075–81.

109 J.N. Murrell, 'Short and intermediate range forces', in *Rare gas solids*, ed. M.L. Klein and J.A. Venables, 2 vols., London, 1976, 1977, v. 1, chap. 3, pp. 176–211; Stone, ref. 12, chaps. 5, 6 and 11.

110 These problems are reviewed by M.J. Elrod and R.J. Saykally, 'Many-body effects in intermolecular forces', *Chem. Rev.* **94** (1994) 1975–97.

111 H.C. Longuet-Higgins, 'Intermolecular forces', *Discuss. Faraday Soc.* **40** (1965) 7–18; C.A. Coulson, 'Intermolecular forces – the known and the unknown', *ibid.* 285–90. For Coulson (1910–1974) see S.L. Altmann and E.J. Bowen, *Biog. Mem. Roy. Soc.* **20** (1974) 75–134, and for an account of Coulson's view of theoretical chemistry see A. Simoes and K. Gavroglu, 'Quantum chemistry *qua* applied mathematics. . . .', *Hist. Stud. Phys. Biol. Sci.* **29** (1999) 363–406. A.D. Buckingham similarly took a broader view of intermolecular forces, with particular emphasis on the electric and magnetic properties of molecules, see 'Permanent and induced molecular moments and long-range intermolecular forces', *Adv. Chem. Phys.* **12** (1967) 107–42 (this is chap. 2 of a volume of this series with the title *Intermolecular forces*, ed. J.O. Hirschfelder). A.D. Buckingham, 'Basic theory of intermolecular forces: applications to small molecules', pp. 1–67 of *Intermolecular interactions: from diatomics to biopolymers*, ed. B. Pullman, Chichester, 1978; A.D. Buckingham, P.W. Fowler and J.M. Hutson, 'Theoretical studies of van der Waals molecules and intermolecular forces', *Chem. Rev.* **88** (1988) 963–88.

112 W. Kossel, 'Über Molekülbildung als Frage des Atombaues', *Ann. Physik* **49** (1916) 229–362. This was the paper in which Kossel proposed that atoms in polar compounds gain or shed electrons so as to acquire an inert-gas structure.

113 A. Eucken, 'Rotationsbewegung und absolute Dimensionen der Moleküle', *Zeit. Elektrochem.* **26** (1920) 377–83.

114 K. Bädeker, 'Experimentaluntersuchung über die Dielektrizitätskonstante einiger Gase und Dämpfe in ihrer Abhängigkeit von der Temperatur', *Zeit. phys. Chem.* **36** (1901) 305–35.

115 J.J. Thomson, 'The forces between atoms and chemical affinity', *Phil. Mag.* **27** (1914) 757–89; G. Holst, 'On the equation of state of water and of ammonia', *Proc. Sec. Sci. Konink. Akak. Weten. Amsterdam* **19** (1917) 932–7.

116 The conventional unit for the strength of a dipole moment is the debye, symbol D, which is 10^{-18} e.s.u. cm, or 3.3356×10^{-30} C m.

117 M. Jona, 'Die Temperaturabhängigkeit der Dielektrizitätskonstante einiger Gase und Dämpfe', *Phys. Zeit.* **20** (1919) 14–21, from his Göttingen thesis of 1917.

118 P. Debye, *Polar molecules*, New York, 1929, pp. 63–8.

119 W.H. Bragg, 'The crystal structure of ice', *Proc. Phys. Soc.* **34** (1922) 98–102. For W.L. Bragg's determination of the structure of NaCl, see 'The structure of some crystals as indicated by their diffraction of x-rays', *Proc. Roy. Soc.* A **89** (1914) 248–77. He did not explicitly describe the units of his crystal as ions but used the conventional word 'atom'. The structure made sense, however, only if the units were Na^+ and Cl^- and this interpretation of bonding in such crystals was then becoming the norm; see, for example, Kossel, ref. 112, Thomson, ref. 115 and G.N. Lewis, ref. 10, and 'Valence and tautomerism', *Jour. Amer. Chem. Soc.* **35** (1913) 1448–55. P. Debye and P. Scherrer, 'Atombau', *Phys. Zeit.* **19** (1918) 474–83; English trans. in *The collected papers of Peter J.W. Debye*, New York, 1954, pp. 63–79. For an extreme response to such 'physical' intrusions into chemistry, see H.E. Armstrong, 'Poor common salt!', *Nature* **120** (1927) 478.

120 T.S. Moore and T.F. Winmill, 'The state of amines in aqueous solution', *Jour. Chem. Soc.* **101** (1912) 1635–76, see 1674–5. This section was written by Moore (1881–1966), then at Magdalen College, Oxford, and later at Royal Holloway College, London University; *Who was who, 1961–70*, London, 1972. There were others who had similar ideas at the same time, see L. Pauling, *The nature of the chemical bond*, Ithaca, New York, 1939, chap. 9, and G.C. Pimentel and A.L. McClellan, *The hydrogen bond*, San Francisco, 1960, pp. 3–4, but Pauling gives Moore the principal credit.

121 P. Pfeiffer, 'Zur Theorie der Farblacke, II', *Ann. Chem.* **398** (1913) 137–96, see 152.

122 W.M. Latimer and W.H. Rodebush, 'Polarity and ionization from the standpoint of the Lewis theory of valence', *Jour. Amer. Chem. Soc.* **42** (1920) 1419–33. When G.N. Lewis saw the last section of this paper in manuscript he advised that they delete the last part on associated liquids on the ground that there can be no 'hydrogen bond' since there are not enough electrons to form a secondary covalent link, see Pimentel and McClellan, ref. 120. Latimer and Rodebush acknowledge that the idea of this bond was also put forward by M.L. Huggins in his undergraduate thesis at Berkeley in 1919. Huggins worked on proteins in the 1930s and is remembered now for his mean-field expression for the entropy of polymer solutions – the Flory–Huggins equation.

123 Pauling, ref. 120, p. 281ff.

124 J.D. Bernal (1901–1971) C.P. Snow, DSB, v. 15, pp. 16–20; D.M.C. Hodgkin, *Biog. Mem. Roy. Soc.* **26** (1980) 17–84; J.D. Bernal and R.H. Fowler, 'A theory of water and ionic solution, with particular reference to hydrogen and hydroxyl ions', *Jour. Chem. Phys.* **1** (1933) 515–48.

125 L. Pauling, 'The structure and entropy of ice and of other crystals with some randomness of atomic arrangement', *Jour. Amer. Chem. Soc.* **57** (1935) 2680–4.

126 W.H. Stockmayer, 'Second virial coefficients of polar gases', *Jour. Chem. Phys.* **9** (1941) 398–402; J.S. Rowlinson, 'The second virial coefficients of polar gases', *Trans. Faraday Soc.* **45** (1949) 974–84.

127 F.G. Keyes, L.B. Smith and H.T. Gerry, 'The specific volume of steam in the saturated and superheated condition together with derived values of the enthalpy, entropy, heat capacity and Joule Thomson coefficients', *Proc. Amer. Acad. Arts Sci.* **70** (1934–1935) 319–64, see 327; S.C. Collins and F.G. Keyes, 'The heat capacity and pressure variation of the enthalpy for steam from 38 ° to 125 °C', *ibid.* **72** (1937–1938) 283–99. Later measurements showed that their values of the second virial coefficient were probably in error below 250 °C, see G.S. Kell, G.E. McLaurin and E. Whalley, '*PVT* properties of water. II. Virial coefficients in the range 150°–450 °C without independent measurement of vapor volumes', *Jour. Chem. Phys.* **48** (1968) 3805–13.

128 H. Margenau, 'The second virial coefficient for gases: a critical comparison between theoretical and experimental results', *Phys. Rev.* **36** (1930) 1782–90, and refs. 20 and 21.

129 H. Margenau and V.W. Myers, 'The forces between water molecules and the second virial coefficient for water', *Phys. Rev.* **66** (1944) 307–15.

130 J.S. Rowlinson, 'The lattice energy of ice and the second virial coefficient of water vapour', *Trans. Faraday Soc.* **47** (1951) 120–9.

131 R.M. Glaeser and C.A. Coulson, 'Multipole moments of the water molecule', *Trans. Faraday Soc.* **61** (1965) 389–91.

132 D. Eisenberg, J.M. Pochan and W.H. Flygare, 'Values of $\langle \Psi^\circ | \Sigma_i r_i^2 | \Psi^\circ \rangle$ for H_2O, NH_3, and CH_2O', *Jour. Chem. Phys.* **43** (1965) 4531–2; D. Eisenberg and W. Kauzmann, *The structure and properties of water*, Oxford, 1969, pp. 12–35.

133 J. Verhoeven and A. Dymanus, 'Magnetic properties and molecular quadrupole tensor of the water molecule by beam-maser Zeeman spectroscopy', *Jour. Chem. Phys.* **52** (1970) 3222–33.

134 F.H. Stillinger and A. Rahman, 'Improved simulation of liquid water by molecular dynamics', *Jour. Chem. Phys.* **60** (1974) 1545–57; 'Revised central force potentials for water', *ibid.* **68** (1978) 666–70; R.O. Watts, 'An accurate potential for deformable water molecules', *Chem. Phys.* **26** (1977) 367–77; J.R. Reimers, R.O. Watts and M.L. Klein, 'Intermolecular potential functions and the properties of water', *ibid.* **64** (1982) 95–114; H.J.C. Berendsen, J.P.M. Postma, W.F. van Gunsteren and J. Hermans, 'Intermolecular models for water in relation to protein hydration' in *Intermolecular forces*, ed. B. Pullman, Dordrecht, 1981, pp. 331–42; E. Clementi and P. Habitz, 'A new two-body water–water potential', *Jour. Phys. Chem.* **87** (1983) 2815–20; W.L. Jorgensen, J. Chandrasekhar, J.D. Madura, R.W. Impey and M.L. Klein, 'Comparison of simple potential functions for simulating liquid water', *Jour. Chem. Phys.* **79** (1983) 926–35; J.Brodholt, M. Sampoli and R. Vallauri, 'Parameterizing a polarizable intermolecular potential for water', *Molec. Phys.* **86** (1995) 149–58; I. Nezbeda and U. Weingerl, 'A molecular-based theory for the thermodynamic properties of water', *ibid.* **99** (2001) 1595–1606. A recent list and review of the some these potentials is in T.M. Nymand, P. Linse and P.-O. Åstrand, 'A comparison of effective and polarizable intermolecular potentials in simulations: liquid water as a test case', *ibid.* **99** (2001) 335–48.

135 Clementi and Habitz, ref. 134.

136 Berendsen *et al.*, ref. 134.

137 See the reviews 'Van der Waals molecules' in *Chem. Rev.* **88** (1988) 813–988; **94** (1994) 1721–2160, **100** (2000) 3861–4264, and the reports of the two meetings, 'Structure and dynamics of van der Waals complexes', *Faraday Discuss.* **97** (1994), and 'Small particles and inorganic clusters', *Zeit. f. Phys. D* **40** (1997). For a full list of papers on the much-studied 'molecule', Ar–CO, see I. Scheele, R. Lehnig and M. Havenith, 'Infrared spectroscopy of van der Waals modes in the intermolecular potential of Ar–CO,...', *Molec. Phys.* **99** (2001) 197–203, 205–9.

138 T.R. Dyke and J.S. Muenter, 'Microwave spectrum and structure of the hydrogen bonded water dimer', *Jour. Chem. Phys.* **60** (1974) 2929–30; T.R. Dyke, K.M. Mack and J.S. Muenter, 'The structure of water dimer from molecular beam resonance spectroscopy', *ibid.* **66** (1977) 498–510; J.A. Odutola and T.R. Dyke, 'Partially deuterated water dimers: Microwave spectra and structure', *ibid.* **72** (1980) 5062–70.

139 R.S. Fellers, C. Leforestier, L.B. Braly, M.G. Brown and R.J. Saykally, 'Spectroscopic determination of the water pair potential', *Science* **284** (1999) 945–8. The potential parameters were listed at www.cchem.berkeley.edu/~rjsgrp/

140 C. Millot and A.J. Stone, 'Towards an accurate intermolecular potential for water', *Molec. Phys.* **77** (1992) 439–62.

141 K. Liu, J.G. Loeser, M.J. Elrod, B.C. Host, J.A. Rzepiela, N. Pugliano and R.J. Saykally, 'Dynamics of structural rearrangements in the water trimer', *Jour. Amer. Chem. Soc.* **116** (1994) 3507–12; K. Liu, M.J. Elrod, J.G. Loeser, J.D. Cruzan, N. Pugliano, M.G. Brown, J.A. Rzepiela and R.J. Saykally, 'Far-I.R. vibration-rotation-tunelling spectroscopy of the water trimer', *Faraday Discuss.*, ref. 137, 35–41; J.D. Cruzan, L.B. Braly, K. Liu, M.G. Brown, J.G. Loeser and R.J. Saykally, 'Quantifying hydrogen bond cooperativity in water: VRT spectroscopy of the water tetramer', *Science* **271** (1996) 59–62; K. Liu, M.G. Brown, J.D. Cruzan and R.J. Saykally, 'Vibration-rotation tunneling spectra of the water pentamer:

structure and dynamics', *ibid.* 62–4. See also the review by U. Buck and F. Huisken, 'Infrared spectroscopy of size-selected water and methanol clusters', *Chem. Rev.* **100** (2000) 3863–90.

142 D.H. Levy, 'Concluding remarks', *Faraday Discuss.*, ref. 137, 453–6.

143 J.C. Maxwell, 'On action at a distance', a Friday evening Discourse, *Proc. Roy. Inst.* **7** (1873) 44–54. For other contemporary views, for and against, see W.R. Browne, 'On action at a distance', *Phil. Mag.* **10** (1880) 437–45, and O. Lodge, 'The ether and its functions', *Nature* **27** (1882–1883) 304–6, 328–30. For Faraday's less clear views of twenty years earlier, see 'On the conservation of force', also a Friday evening Discourse, *Phil. Mag.* **13** (1857) 225–39.

144 H. Kallmann and M. Willstaetter, 'Zur Theorie des Aufbaues kolloidaler Systeme', *Naturwiss.* **20** (1932) 952–3.

145 R.S. Bradley, 'The cohesive force between solid surfaces and the surface energy of solids', *Phil. Mag.* **13** (1932) 853–62.

146 H.C. Hamaker, 'The London–van der Waals attraction between spherical particles', *Physica* **4** (1937) 1058–72; 'London–v.d. Waals forces in colloidal systems', *Rec. Trav. Chim. Pays-Bas* **57** (1938) 61–72. J.M. Rubin had obtained the same results in 1933, see Hamaker, (1938) 65.

147 The matter is discussed briefly by E.J.W. Verwey in his paper, 'Theory of the stability of lyophobic colloids', *Jour. Phys. Coll. Chem.* **51** (1947) 631–6. See also, E.J.W. Verwey and J.Th.G. Overbeek, 'Long distance forces acting between colloidal particles', *Trans. Faraday Soc.* **42B** (1946) 117–23; *Theory of the stability of lyophobic colloids*, Amsterdam, 1948.

148 H.B.G. Casimir and D. Polder, 'Influence of retardation on the London–van der Waals forces', *Nature* **158** (1946) 787–8. They followed this brief note with the full paper, with the same title, in *Phys. Rev.* **73** (1948) 360–72. Their treatment of the problem is discussed, at different levels of difficulty, by H. Margenau and N.R. Kestner, *Theory of intermolecular forces*, Oxford, 1971, chap. 6; J. Mahanty and B.W. Ninham, *Dispersion forces*, London, 1976, chaps. 2 and 3; R.J. Hunter, *Foundations of colloid science*, Oxford, 1987, v. 1, chap. 4. A related effect, often called the Casimir force, is the long-range force between two electrically-conducting macroscopic objects, for example, two metal plates. This was hinted at in the 1946 note and first described by Casimir in 'On the attraction between two perfectly conducting plates', *Proc. Sec. Sci. Konink. Akad. Weten. Amsterdam* **51** (1948) 793–5. This force can be attractive or repulsive, depending on the shapes of the two metal objects. It was measured by S.K. Lamoreaux, 'Demonstration of the Casimir force in the 0.6 to 6 μm range', *Phys. Rev. Lett.* **78** (1997) 5–8, and the theory reviewed by D. Langbein in *Theory of van der Waals attraction*, Springer Tracts in Modern Physics, v. 72, Berlin, 1974, by E. Elizalde and A. Romeo, 'Essentials of the Casimir effect and its computation', *Amer. Jour. Phys.* **59** (1991) 711–19, and by V.M. Mostepanenko and N.N. Trunov, *The Casimir effect and its applications*, Oxford, 1997.

149 E.M. Lifshitz (1915–1985) Ya.B. Zel'dovich and M.I. Kaganov, *Biog. Mem. Roy. Soc.* **36** (1990) 337–57; E.M. Lifshitz [Theory of molecular attractive forces between condensed bodies], *Doklady Akad. Nauk SSSR* **97** (1954) 643–6; 'The theory of molecular attractive forces between solids', *Sov. Phys. JETP* **2** (1956) 73–83. The Russian original of this paper was submitted in September 1954 and published in *Zhur. Eksp. Teor. Fiz. SSSR* **29** (1955) 94–110. For a review, see I.E. Dzyaloshinskii, E.M. Lifshitz and L.P. Pitaevskii, 'The general theory of van der Waals forces', *Adv. Physics* **10** (1961) 165–209.

150 B.V. Deryagin and I.I. Abrikosova, [Direct measurement of the molecular attraction as a function of the distance between surfaces], *Zhur. Eksp. Teor. Fiz. SSSR* **21** (1951) 945–6.

151 P.G. Howe, D.P. Benton and I.E. Puddington, 'London–van der Waals attractive forces between glass surfaces', *Canad. Jour. Chem.* **33** (1955) 1375–83, and earlier work cited there.

152 I.I. Abrikosova and B.V. Deryagin, [On the law of intermolecular interaction at large distances], *Doklady Akad. Nauk SSSR* **90** (1953)1055–8. The same results were reported in § 3, pp. 33–7 of B.V. Derjaguin, A.S. Titijevskaia and I.I. Abricossova, 'Investigations of the forces of interaction of surfaces in different media and their application to the problem of colloidal stability', *Discuss. Faraday Soc.* **18** (1954) 24–41. New measurements, described later by Dzyaloshinskii, Lifshitz and Pitaevskii in their review, ref. 149, as the first accurate ones, were made for the force between a glass sphere and a glass plate by B.V. Deryagin and I.I. Abrikosova, 'Direct measurement of the molecular attraction of solid bodies. I. Statement of the problem and method of measuring forces by using negative feedback', *Sov. Phys. JETP* **3** (1957) 819–29; I.I. Abrikosova and B.V. Deryagin, '. . . II. Method for measuring the gap. Results of experiments', *ibid.* **4** (1958) 2–10.

153 J.Th.G. Overbeek and M.J. Sparnaay, 'Experimental determination of long-range attractive forces', *Proc. Sect. Sci. Konink. Akad. Weten. Amsterdam* **54** (1951) 386–7.

154 See the discussion between Deryagin and Overbeek at the Faraday Society meeting in Sheffield in September, 1954, reported on pp. 180–7 of ref. 152, 1954.

155 J.A. Kitchener and A.P. Prosser, 'Direct measurement of the long-range van der Waals forces', *Proc. Roy. Soc. A* **242** (1957) 403–9.

156 W. Black, J.G.V. de Jongh, J.Th.G. Overbeek and M.J. Sparnaay, 'Measurement of retarded van der Waals forces', *Trans. Faraday Soc.* **56** (1960) 1597–608.

157 D. Tabor and R.H.S. Winterton, 'Surface forces: Direct measurement of normal and retarded van der Waals forces', *Nature* **219** (1968) 1120–1; 'The direct measurement of normal and retarded van der Waals forces', *Proc. Roy. Soc. A* **312** (1969) 435–50; J.N. Israelachvili and D. Tabor, 'The measurement of van der Waals dispersion forces in the range 1.5 to 130 nm', *ibid.* **331** (1972–1973) 19–38; J.N. Israelachvili, 'The calculation of van der Waals dispersion forces between macroscopic bodies', *ibid.* 39–55. The smoothness of cleaved mica had previously been exploited in the same laboratory by J.S. Courtney-Pratt, 'Direct optical measurement of the length of organic molecules', *Nature* **165** (1950) 346–8; 'An optical method of measuring the thickness of adsorbed monolayers', *Proc. Roy. Soc. A* **212** (1952) 505–8.

158 See, for example, J.N. Israelachvili, 'Adhesion forces between surfaces in liquids and condensible vapours', *Surface Sci. Rep.* **14** (1992) 109–59; *Intermolecular and surface forces*, 2nd edn, London, 1992.

159 See ref. 348 in Section 2.5.

160 Nernst, in Planck *et al.*, ref. 31, p. 64.

161 See Section 5.2, and the reviews of J.A. Barker, 'Interatomic potentials for inert gases from experimental data', v. 1, chap. 4, pp. 212–64; P. Korpiun and E. Lüscher, 'Thermal and elastic properties at low pressure', v. 2, chap. 12, pp. 729–822, and B. Stoicheff, 'Brillouin spectroscopy and elastic constants', v. 2, chap. 16, pp. 979–1019, in *Rare gas solids*, ref. 109; R.A. Aziz, 'Interatomic potentials for rare gases: pure and mixed interactions, chap. 2, pp. 5–86 of *Inert gases. Potentials, dynamics and energy transfer in doped crystals*, ed. M.L. Klein, Berlin, 1984.

162 See, for example, the review of E. Orowan, 'Fracture and the strength of solids', *Rep. Prog. Phys.* **12** (1948–1949) 185–232.

163 A. Einstein, 'Die Plancksche Theorie der Strahlung und die Theorie der spezifischen
     Wärme', *Ann. Physik* **22** (1907) 180–90; 'Eine Beziehung zwischen dem elastischen
     Verhalten und der spezifischen Wärme bei festen Körpern mit einatomigem
     Molekül', *ibid.* **34** (1911) 170–4; reprinted in *The collected papers of Albert Einstein*,
     Princeton, NJ, v. 2, 1989, pp. 378–89; v. 3, 1993, pp. 408–14; *English translation*,
     v. 2, pp. 214–24; v. 3, pp. 332–5. P. Debye, 'Zur Theorie der spezifischen Wärmen',
     Ann. Physik. **39** (1912) 789–839; English trans. in his *Collected papers*, ref. 119,
     pp. 650–96; M. Born and Th.v. Kármán, 'Über Schwingungen im Raumgittern',
     Phys. Zeit. **13** (1912) 297–309; 'Zur Theorie der spezifischen Wärme', *ibid.* **14**
     (1912) 15–19; 'Über die Verteilung der Eigenschwingungen von Punktgittern', *ibid.*
     65–71. Born and von Kármán acknowledge Debye's priority for the theory of the
     specific heat, by "a few days".

164 Y. Fujii, N.A. Lurie, R. Pynn and G. Shirane, 'Inelastic neutron scattering from solid
     ^{36}Ar', *Phys. Rev.* **B10** (1974) 3647–59. Fujii was at Brookhaven on leave from
     Tokyo.

165 M.S. Anderson and C.A. Swenson, 'Experimental equations of state for the rare gas
     solids', *Jour. Phys. Chem. Solids* **36** (1975) 145–62.

166 A.O. Urvas, D.L. Losee and R.O. Simmons, 'The compressibility of krypton, argon,
     and other noble gas solids', *Jour. Phys. Chem. Solids* **28** (1967) 2269–81.

167 For a fuller account of some of the work in this Section, see J.S. Rowlinson, 'Van der
     Waals and the physics of liquids', pp. 1–119 of J.D. van der Waals, *On the continuity
     of the gaseous and liquid states*, ed. J.S. Rowlinson, Amsterdam, 1988. This is v. 14
     of the series, *Studies in statistical mechanics*.

168 See Section 4.4 and ref. 240 of Chapter 4.

169 M. Smoluchowski (1872–1917) A.A. Teske, DSB, v. 12, pp. 496–8; *Marian
     Smoluchowski: Leben und Werke*, Wroclaw, 1977. M. Smoluchowski, 'Théorie
     cinetique de l'opalescence des gaz à l'état critique et de certains phénomènes
     corrélatifs', *Bull. Int. Acad. Cracovie, Classe Sci. Math. Nat.* (1907) 1057–75, see
     eqn 7. Published in German as 'Molekular-kinetische Theorie der Opaleszanz von
     Gasen im kritischen Zustande, sowie einiger verwandter Erscheinungen', *Ann.
     Physik* **25** (1908) 205–26.

170 F. Zernike (1888–1966) J.A. Prins, DSB, v. 14, pp. 616–17; S. Tolansky, *Biog. Mem.
     Roy. Soc.* **13** (1967) 393–402.

171 L.S. Ornstein and F. Zernike, 'Accidental deviations of the density and opalescence at
     the critical point of a single substance', *Proc. Sect. Sci. Konink. Akad. Weten.
     Amsterdam* **17** (1914) 793–806; F. Zernike, 'The clustering-tendency of the
     molecules in the critical state and the extinction of light caused thereby', *ibid.* **18**
     (1916) 1520–7; L.S. Ornstein, 'The clustering tendency of the molecules at the
     critical point', *ibid.* **19** (1917) 1321–4. The first two papers are reprinted in *The
     equilibrium theory of classical fluids*, ed. H.L. Frisch and J.L. Lebowitz, New York,
     1964, pp. III 1–25. See also Zernike's Amsterdam thesis of 1915, published again as
     'Étude théoretique et expérimentale de l'opalescence critique', *Arch. Néerl.* **4** (1918)
     73–149. Ornstein and Zernike worked at Groningen.

172 This model is formed of hard rods of length *d* moving on a line, and between which
     there is an attractive pair potential of minute depth but infinite range, defined in such
     a way that the parameter *a* of eqn 5.34 is finite and non-zero. Van der Waals's
     equation is exact for this simple, if artificial model; M. Kac, G.E. Uhlenbeck and
     P.C. Hemmer, 'On the van der Waals theory of the vapor–liquid equilibrium.
     I. Discussion of a one-dimensional model', *Jour. Math. Phys.* **4** (1963) 216–28; '. . . .
     II. Discussion of the distribution functions', *ibid.* 229–47; ' . . . . III. Discussion of the

critical region', *ibid.* **5** (1964) 60–74; P.C. Hemmer, '.... IV. The pair correlation function and the equation of state for long-range forces', *ibid.* 75–84.

173 L.S. Ornstein and F. Zernike, 'Die linearen Dimensionen der Dichtsschwankungen', *Phys. Zeit.* **19** (1918) 134–7.

174 P. Debye and P. Scherrer, 'Interferenzen an regellos orientierten Teilchen im Röntgenlicht. I.', *Phys. Zeit.* **17** (1916) 277–83; English trans. in Debye's *Collected papers*, ref. 119, pp. 51–62.

175 P. Debye and H. Menke, 'Bestimmung der inneren Struktur von Flüssigkeiten mit Röntgenstrahlen, *Phys. Zeit.* **31** (1930) 797–8; English trans. in Debye's *Collected papers*, ref. 119, pp. 133–6; H. Menke, 'Röntgeninterferenzen an Flüssigkeiten', *Phys. Zeit.* **33** (1932) 593–604.

176 R.H. Fowler, *Statistical mechanics*, Cambridge, 1929, chap. 20, pp. 497–518.

177 R.C. Tolman, *The principles of statistical mechanics*, Oxford, 1938.

178 J.E. Mayer and M.G. Mayer, *Statistical mechanics*, New York, 1940.

179 J.A. Prins, 'Über die Beugung von Röntgenstrahlen in Flüssigkeiten und Lösungen', *Zeit. f. Phys.* **56** (1929) 617–48; O. Kratky, 'Die Struktur des flüssigen Quecksilbers', *Phys. Zeit.* **34** (1933) 482–7; J.A. Prins and H. Petersen, 'Theoretical diffraction patterns for simple types of molecular arrangement in liquids', *Physica* **3** (1936) 147–53.

180 Fowler, ref. 176, p. 169.

181 H.D. Ursell, 'The evaluation of Gibbs' phase-integral for imperfect gases', *Proc. Camb. Phil. Soc.* **23** (1925–1927) 685–97.

182 The first of these papers is J.E. Mayer, 'Statistical mechanics of condensing systems. I', *Jour. Chem. Phys.* **5** (1937) 67–73; see also, Mayer and Mayer, ref. 178.

183 J. Yvon (b.1903) Jacques Yvon was Professor of Physics at Strasbourg from 1938 to 1949, and later become the French Commissioner for Atomic Energy. J. Yvon, 'Théorie statistique des fluides et l'équation d'état', *Actual. Sci. Indust.* No. 203 (1935); 'Recherches sur la théorie cinétique des liquides', *ibid.* No. 542 (1937). These papers are reprinted in his *Oeuvre scientifique*, Paris, 1986, v. 1, pp. 35–83, 109–74, 175–252.

184 J. de Boer, *Contribution to the theory of compressed gases*, Thesis, Amsterdam, 1940. This thesis formed the basis of his later review, 'Molecular distribution and equation of state of gases', *Rep. Prog. Phys.* **12** (1948–1949) 305–74.

185 E.A. Guggenheim, 'On the statistical mechanics of dilute and of perfect solutions', *Proc. Roy. Soc. A* **135** (1932) 181–92.

186 E.A. Guggenheim, 'The statistical mechanics of regular solutions', *Proc. Roy. Soc. A* **148** (1935) 304–12; 'The statistical mechanics of co-operative assemblies', *ibid.* **169** (1938) 134–48. The same approximation, under a different name, was put forward also by H.A. Bethe, 'Statistical theory of superlattices', *ibid.* **150** (1935) 552–75.

187 The opening papers were R.H. Fowler and G.S. Rushbrooke, 'An attempt to extend the statistical theory of perfect solutions', *Trans. Faraday Soc.* **33** (1937) 1272–94, and G.S. Rushbrooke, 'A note on Guggenheim's theory of strictly regular binary liquid mixtures', *Proc. Roy. Soc. A* **166** (1938) 296–315. Some of the last attempts at this interpretation of the properties of liquid mixtures are to be found in E.A. Guggenheim, *Mixtures*, Oxford, 1952, chaps. 3 and 4; in Guggenheim, ref. 69, chaps. 6 and 7; and in I. Prigogine, *The molecular theory of solutions*, Amsterdam, 1957.

188 J.E. Lennard-Jones and A.F. Devonshire, 'Critical phenomena in gases, I' [and similar titles], *Proc. Roy. Soc. A* **163** (1937) 53–70; **165** (1938) 1–11; **169** (1938–1939) 317–38; **170** (1939) 464–84; A.F. Devonshire, '... V', *ibid.* **174** (1939–1940) 102–9.

189 H. Eyring (1901–1981) K.J. Laidler, DSB, v. 17, pp. 279–84. H. Eyring and
    J. Hirschfelder, 'The theory of the liquid state', *Jour. Phys. Chem.* **41** (1937) 249–57;
    F. Cernuschi and H. Eyring, 'An elementary theory of condensation', *Jour. Chem.
    Phys.* **7** (1939) 547–51.
190 Fowler and Guggenheim, ref. 24, p. 322.
191 J. Frenkel, *Kinetic theory of liquids*, Oxford, 1946. For Ya.I. Frenkel (1894–1952),
    see V.Ya. Frenkel, 'Yakov Ilich Frenkel: Sketches towards a civic portrait', *Hist. Stud.
    Phys. Biol. Sci.* **27** (1997) 197–236, an article which includes a short section
    describing the circumstances in which this book was written.
192 Hirschfelder, Curtiss and Bird, ref. 50, pp. 271–320.
193 J.A. Barker, *Lattice theories of the liquid state*, Oxford, 1963.
194 Published in *Trans. Faraday Soc.* **33** (1937) 1–282.
195 Published in *Physica* **4** (1937) 915–1180; **5** (1938) 39–45, 170, 718–24.
196 Fowler, ref. 176, p. 213.
197 Yvon, ref. 183, (1935).
198 J.H. Hildebrand (1881–1983) K.S. Pitzer, *Biog. Mem. U.S. Nat. Acad. Sci.* **62** (1993)
    225–57. Hildebrand was at the University of California at Berkeley from 1913 until
    his retirement in 1952, and beyond. J.H. Hildebrand and S.E. Wood, 'The derivation
    of equations for regular solutions', *Jour. Chem. Phys.* **1** (1933) 817–22.
199 J.H. Hildebrand, H.R.R. Wakeham and R.N. Boyd, 'The intermolecular potential of
    mercury', *Jour. Chem. Phys.* **7** (1939) 1094–6.
200 Fowler, ref. 176, pp. 180–2. For the 'potential of average force' see also L. Onsager,
    'Theories of concentrated electrolytes', *Chem. Rev.* **13** (1933) 73–89.
201 J.G. Kirkwood, 'Statistical mechanics of fluid mixtures', *Jour. Chem. Phys.* **3** (1935)
    300–13. This work was developed further; 'Molecular distribution in liquids', *ibid.* **7**
    (1939) 919–25; J.G. Kirkwood and E. Monroe, 'On the theory of fusion', *ibid.* **8**
    (1940) 845–6; 'Statistical mechanics of fusion', *ibid.* **9** (1941) 514–26; 'The radial
    distribution function in liquids', *ibid.* **10** (1942) 394–402. In the last paper Monroe
    has become E.M. Boggs, on her marriage.
202 N. Bogolubov, 'Expansions into a series of powers of a small parameter in the theory
    of statistical equlibrium', *Jour. Phys. USSR* **10** (1946) 257–64; 'Kinetic equations',
    ibid. 265–74. These articles are shortened versions of a longer monograph in Russian
    which appeared in an English translation as N.N. Bogoliubov, 'Problems of a
    dynamical theory in statistical physics', *Studies in statistical mechanics*, Amsterdam,
    1963, v. 1, pp. 1–118.
203 M. Born and H.S. Green, 'A general kinetic theory of liquids, I. The molecular
    distribution functions', *Proc. Roy. Soc. A* **188** (1946) 10–18; H.S. Green, '....
    II. Equilibrium properties', *ibid.* **189** (1947) 103–17; M.Born and H.S. Green, '....
    III. Dynamical properties', *ibid.* **190** (1947) 455–74; '.... IV. Quantum mechanics of
    fluids', *ibid.* **191** (1947) 168–81; 'The kinetic basis of thermodynamics', *ibid.* **192**
    (1947–1948) 166–80; H.S. Green, '....V. Liquid He II', *ibid.* **194** (1948) 244–58;
    A.E. Rodriguez, '....VI. The equation of state', *ibid.* **196** (1949) 73–92. The papers
    of Born and Green were reprinted with additional notes in their *A general kinetic
    theory of liquids*, Cambridge, 1949. A less technical account of some of this work
    was included in Born's Waynflete Lectures at Oxford, *Natural philosophy of cause
    and chance*, Oxford, 1949.
204 J.S. Rowlinson and C.F. Curtiss, 'Lattice theories of the liquid state', *Jour. Chem.
    Phys.* **19** (1951) 1519–29; J. de Boer, 'Cell-cluster theory for the liquid state. I',
    Physica **20** (1954) 655–64; and successive parts in collaboration with E.G.D. Cohen,
    Z.W. Salsburg and B.C. Rethmeier, '.... II', *ibid.* **21** (1955) 137–47;'.... III. The
    harmonic oscillator model', *ibid.* **23** (1957) 389–403; '.... IV. A fluid of hard

spheres', *ibid.* **23** (1957) 407–22; J.A. Barker, 'The cell theory of liquids', *Proc. Roy. Soc. A* **230** (1955) 390–8; ' . . . . II.', *ibid.* **237** (1956) 63–74; 'A new theory of fluids: the "Tunnel" Model', *Aust. Jour. Chem.* **13** (1960) 187–93; Barker, ref. 193.

205 M.J. Klein and L. Tisza, 'Theory of critical fluctuations', *Phys. Rev.* **76** (1949) 1861–8.

206 L. Rosenfeld, *Theory of electrons*, Amsterdam, 1951, chap. 5.

207 G.S. Rushbrooke and H.I. Scoins, 'On the theory of liquids', *Proc. Roy. Soc. A* **216** (1953) 203–18. For Rushbrooke (1915–1995), see C. Domb, *Biog. Mem. Roy. Soc.* **44** (1998) 365–84.

208 C.A. Coulson and G.S. Rushbrooke, 'On the interpretation of atomic distribution curves for liquids', *Phys. Rev.* **56** (1939) 1216–23.

209 J.K. Percus and G.J. Yevick, 'Analysis of classical statistical mechanics by means of collective coordinates', *Phys. Rev.* **110** (1958) 1–13.

210 G. Stell, 'The Percus–Yevick equation for the radial distribution function of a fluid', *Physica* **29** (1963) 517–34.

211 J.K. Percus, 'The pair distribution function in classical statistical mechanics', pp. II 33–170, and G. Stell, 'Cluster expansions for classical systems in equilibrium', pp. II 171–266, in the book edited by Frisch and Lebowitz, ref. 171.

212 M.S. Wertheim, 'Exact solution of the Percus–Yevick integral equation for hard spheres', *Phys. Rev. Lett.* **10** (1963) 321–3; E. Thiele, 'Equation of state for hard spheres', *Jour. Chem. Phys.* **39** (1963) 474–9.

213 Menke, ref. 175, (1932).

214 W.E. Morrell and J.H. Hildebrand, 'The distribution of molecules in a model liquid', *Jour. Chem. Phys.* **4** (1936) 224–7.

215 Kirkwood, ref. 201, (1940).

216 This limit was first established by J.D. Bernal in London and independently by G.D. Scott in Toronto by experiments on arrays of ball-bearings and by similar macroscopic studies. J.D. Bernal, 'A geometrical approach to the structure of liquids', a Friday evening Discourse at the Royal Institution on 31 October 1958, published in *Nature* **183** (1959) 141–7, and similar papers with his colleagues, *ibid.* **185** (1960) 68–70; **188** (1960) 910–11; **194** (1962) 957–8. See also the paper of J.D. Bernal, S.V. King and J.L. Finney, 'Random close-packed hard-sphere model. I . . . . . II.', *Discuss. Faraday Soc.* **43** (1967) 60–9 and the discussion that followed it, 75–85. For Scott's work, see G.D. Scott, 'Packing of equal spheres', *Nature* **188** (1960) 908–9, and similar papers by him and his colleagues, *ibid.* **194** (1962) 956–7; **201** (1964) 382–3.

217 J.M.J. van Leeuwen, J. Groeneveld and J. de Boer, 'New method for the calculation of the pair correlation function, I', *Physica* **25** (1959) 792–808.

218 See, for example, Kirkwood's first paper on this subject, ref. 201 (1935), or, for a later expression of the same view, E.B. Smith and B.J. Alder, 'Perturbation calculations in equilibrium statistical mechanics. I. Hard sphere basis potential', *Jour. Chem. Phys.* **30** (1959) 1190–9. Both soon modified their views, see J.G. Kirkwood and E. Monroe, ref. 201, and E.B. Smith, 'Equation of state of liquids at constant volume', *Jour. Chem. Phys.* **36** (1962) 1404–5.

219 Such views were discussed intently at the Gordon Conferences on the Physics and Chemistry of Liquids held in New Hampshire in 1963 and 1965; one discussion started in the bar in the evening and went on until breakfast.

220 R.W. Zwanzig, 'High-temperature equation of state by a perturbation method. I. Nonpolar gases', *Jour. Chem. Phys.* **22** (1954) 1420–6.

221 J.S. Rowlinson, 'The statistical mechanics of systems with steep intermolecular potentials', *Molec. Phys.* **8** (1964) 107–15; D. Henderson and S.G. Davison,

'Quantum corrections to the equation of state for a steep repulsive potential', *Proc. Nat. Acad. Sci. U.S.A.* **54** (1965) 21–3; D.A. McQuarrie and J.L. Katz, 'High-temperature equation of state', *Jour. Chem. Phys.* **44** (1966) 2393–7.

222 J.A. Barker and D. Henderson, [no title], *Discuss. Faraday Soc.* **43** (1967) 50–3.

223 J.A. Barker and D. Henderson, 'Perturbation theory and equation of state for fluids: The square-well potential', *Jour. Chem. Phys.* **47** (1967) 2856–61; '. . . . II. A successful theory of liquids', *ibid.* 4714–21.

224 The most widely used treatment is that of J.D. Weeks, D. Chandler and H.C. Andersen, 'Role of repulsive forces in determining the equilibrium structure of simple liquids', *Jour. Chem. Phys.* **54** (1971) 5237–47. For later developments, see C.G. Gray and K.E. Gubbins, *Theory of molecular fluids. Volume 1: Fundamentals*, Oxford, 1984, chap. 4, pp. 248–340, 'Perturbation theory'; vol. 2, in preparation.

225 C. Domb, *The critical point: a historical introduction to the modern theory of critical phenomena*, London, 1996.

226 An early attempt to marry the hard-sphere transition with a van der Waals-like mean-field approximation was made by H.C. Longuet-Higgins and B. Widom, 'A rigid sphere model for the melting of argon', *Molec. Phys.* **8** (1964) 549–56. For reviews, see M. Baus, 'The present status of the density-functional theory of the liquid–solid transition', *Jour. Phys. Condensed Matter* **2** (1990) 2111–26, P.A. Monson and D.A. Kofke, 'Solid–fluid equilibrium: Insights from simple molecular models', *Adv. Chem. Phys.* **115** (2000) 113–79, and H. Löwen, 'Melting, freezing and colloidal suspensions', *Phys. Reports* **237** (1994) 249–324. The last two reviews range more widely than density-functional theory.

227 J.C. Maxwell, art. 'Capillary action', *Encyclopaedia Britannica*, 9th edn, London, 1876. For Maxwell's own measurements, see I.B. Hopley, 'Clerk Maxwell's apparatus for the measurement of surface tension', *Ann. Sci.* **13** (1957) 180–7.

228 R.S. Bradley, 'The molecular theory of surface energy: the surface energy of the liquefied inert gases', *Phil. Mag.* **11** (1931) 846–8; H. Margenau, 'Surface energy of liquids', *Phys. Rev.* **38** (1931) 365–71; L.S. Kassel and M. Muskat, 'Surface energy and heat of vaporization of liquids', *ibid.* **40** (1932) 627–32; A. Harasima, 'Calculation of the surface energies of several liquids', *Proc. Phys.-Math. Soc. Japan* **22** (1940) 825–40.

229 R.H. Fowler, 'A tentative statistical theory of Macleod's equation for surface tension, and the parachor', *Proc. Roy. Soc. A* **159** (1937) 229–46; 'A calculation of the surface tension of a liquid–vapour interface in terms of van der Waals force constants', *Physica* **5** (1938) 39–45.

230 J.G. Kirkwood and F.P. Buff, 'The statistical mechanical theory of surface tension', *Jour. Chem. Phys.* **17** (1949) 338–43.

231 J. Penfold, 'The structure of the surface of pure liquids', *Rep. Prog. Phys.* **64** (2001) 777–814.

232 See the papers cited in refs. 241–3 of Chapter 4.

233 Van der Waals, ref. 243 of Chapter 4, English trans., p. 210.

234 J. Yvon, 'Le problème de la condensation de la tension et du point critique', *Colloque de thermodynamique*, Int. Union Pure and Applied Physics, Brussels,1948, pp. 9–15. Yvon does not explicitly invoke the direct correlation function by name, nor by formal definition, but he introduces an equivalent function, L_{12}, which is defined only by means of the first two terms of its density expansion without any indication of how the series should be continued. There were only 22 participants in the meeting and it is clear from the discussion, p. 16, that neither Born nor de Boer followed his derivation.

235 D.G. Triezenberg and R. Zwanzig, 'Fluctuation theory of surface tension', *Phys. Rev. Lett.* **28** (1972) 1183–5; R. Lovett, P.W. DeHaven, J.J. Vieceli Jr. and F.P. Buff, 'Generalized van der Waals theories for surface tension and interfacial width', *Jour. Chem. Phys.* **58** (1973) 1880–5. A formally similar but less useful equation was given earlier, without derivation, in F.P. Buff and R. Lovett, 'The surface tension of simple liquids', in *Simple dense fluids*, ed. H.L. Frisch and Z.W. Salsburg, New York, 1968, chap. 2, pp. 17–30.

236 P. Schofield, 'The statistical theory of surface tension', *Chem. Phys. Lett.* **62** (1979) 413–15.

237 H. Hulshof, 'The direct deduction of the capillary constant σ as a surface-tension', *Proc. Sect. Sci. Konink. Akad. Weten. Amsterdam* **2** (1900) 389–98; 'Ueber die Oberflächenspannung', *Ann. Physik* **4** (1901) 165–86.

238 G. Bakker, *Kapillarität und Oberflächenspannung*, v. 6 of the *Handbuch der Experimentalphysik*, ed. W. Wien, F. Harms and H. Lenz, Leipzig, 1928.

239 S.-D. Poisson, 'Mémoire sur l'équilibre et le mouvement des corps élastiques', *Mém. Acad. Roy. Sci.* **8** (1825) 357–570, 623–7, see 373; read in April and November 1828 and published in 1829.

240 A.-L. Cauchy, 'De la pression ou tension dans un système de points matérials', *Exercises de mathématiques*, 3rd year, Paris, 1828, pp. 213–36.

241 G. Lamé and E. Clapeyron, 'Mémoire sur l'équilibre intérieur des corps solides homogènes', *Mém. div. Savans Acad. Roy. Soc.* **4** (1833) 463–562, see 483; submitted in April 1828.

242 J. Fourier, *Théorie analytique de la chaleur*, Paris, 1822, § 96, pp. 89–91; *The analytical theory of heat*, trans. A. Ferguson, Cambridge, 1878, § 96, pp. 78–9.

243 A.-L. Cauchy, 'Notes relatives à la mécanique rationelle', *Compt. Rend. Acad. Sci.* **20** (1845) 1760–6, see 1765; see also his 'Observations sur la pression que support un élément de surface plane dans un corps solide ou fluide', *ibid.* **21** (1845) 125–33.

244 B. de Saint-Venant, 'Note sur la pression dans l'intérieur des corps ou à leurs surfaces de separation', *Compt. Rend. Acad. Sci.* **21** (1845) 24–6. See also the discussion by I. Todhunter and K. Pearson, *A history of the theory of elasticity*, Cambridge, 1886, v. 1, pp. 860–1, 863–4.

245 S.-D. Poisson, 'Sur la distribution de la chaleur dans les corps solides', *Jour. École Polytech.* 19me cahier, **12** (1823) 1–144, 249–403, see § 11, 272–3.

246 See e.g. J.S. Rowlinson and B. Widom, *Molecular theory of capillarity*, Oxford, 1982, pp. 85–93.

247 F.P. Buff, 'Some considerations of surface tension', *Zeit. Elektrochem.* **56** (1952) 311–13. This paper was read by Arnold Münster at a meeting of the Bunsen Gesellschaft in Berlin in January 1952. A.G. MacLellan [sic], 'A statistical–mechanical theory of surface tension', *Proc. Roy. Soc.* A **213** (1952) 274–84. McLellan was at Otago in New Zealand.

248 A.G. McLellan, 'The stress tensor, surface tension and viscosity', *Proc Roy. Soc.* A **217** (1953) 92–6.

249 J.H. Irving and J.G. Kirkwood, 'The statistical mechanical theory of transport processes. IV. The equations of hydrodynamics', *Jour. Chem. Phys.* **18** (1950) 817–29, see Appendix.

250 A. Harasima, 'Statistical mechanics of surface tension', *Jour. Phys. Soc. Japan* **8** (1953) 343–7; 'Molecular theory of surface tension', *Adv. Chem. Phys.* **1** (1958) 203–37. For the expression "more reasonable", see 223.

251 P. Schofield and J.R. Henderson, 'Statistical mechanics of inhomogeneous fluids', *Proc. Roy. Soc.* A **379** (1982) 231–46.

252 Probably the first conference of physical scientists on this subject was that held at
     Reading in December 1982: 'The hydrophobic interaction', *Faraday Symp. Chem.
     Soc.* **17** (1982). The development of the field is set out by F. Franks in his
     Introduction, 'Hydrophobic interactions – a historical perspective', pp. 7–10, which
     contains a list of the early key papers. An important later one is K. Lum, D. Chandler
     and J.D. Weeks, 'Hydrophobicity at small and large length scales', *Jour. Phys. Chem.*
     103 (1999) 4570–7. A recent simple account of the field is in P. Ball, *H_2O: a
     biography of water*, London, 1999, chap. 9, pp. 231–48.
253 De Boer, ref. 184, (1948–1949), pp. 359–60.
254 T. Biben and J.-P. Hansen, 'Osmotic depletion, non-additivity and phase separation',
     Physica A **235** (1997) 142–8.
255 M. Dijkstra, R. van Roij and R. Evans, 'Phase diagram of highly asymmetric binary
     hard-sphere mixtures', *Phys. Rev. E* **59** (1999) 5744–71.
256 S. Asakura and F. Oosawa, 'On the interaction between two bodies immersed in a
     solution of macromolecules', *Jour. Chem. Phys.* **22** (1954) 1255–6; A. Vrij,
     'Polymers at interfaces and the interactions in colloidal dispersions', *Pure Appl.
     Chem.* **48** (1976) 471–83, see § 4.
257 S. Weinberg, *Dreams of a final theory*, London, 1993. Weinberg observes that
     quantum mechanics is a 'rigid' theory, that is, it cannot be changed in an ad hoc way
     without the whole structure disintegrating. He suggests, therefore, that it would
     survive in its present form in any 'final' theory.

Name index

An entry of the form
Achard, F.C., 48, 52–3. **2**: *297*, 298
denotes that Achard is mentioned in the text on pages 48 and 52 to 53, and in references 297 and 298 of
chapter 2, where reference 297, in italics, contains some biographical information.

Subject index

Page numbers that fall in the 'Notes and References' section of each chapter are listed here only if there is matter there that cannot be inferred from the relevant text page.

Printed in the United States
By Bookmasters